Electronic Filter Analysis and Synthesis

For a complete listing of the *Artech House Microwave Library*,
turn to the back of this book

Electronic Filter Analysis and Synthesis

Michael G. Ellis, Sr.

Artech House
Boston • London

Library of Congress Cataloging-in-Publication Data
Ellis, Michael, 1950-
Electronic filter analysis and synthesis/Michael Ellis.
Includes bibliographical references and index.
ISBN 0-89006-616-7
1. Electric filters–Mathematical models. 2. Electric filters–Design and construction. I. Title
TK7872.F5E45 1994 94-7654
621.3815'324–dc20 CIP

A catalogue record for this book is available from the British Library

685 Canton Street
Norwood, MA 02062

International Standard Book Number: 0-89006-616-7
Library of Congress Catalog Card Number: 94-7654

10 9 8 7 6 5 4 3 2 1

To Mary Kay
and our children
Michael, David, and Sara

Table of Contents

Preface

This text presents the derivation of design techniques for electronic filters in a manner conducive for writing computer-aided design and analysis programs. It is assumed that the reader has an understanding of the basic procedures used to analyze electric circuits and systems, such as taught in an undergraduate network theory series. Some familiarity with numerical analysis and a computer language would be helpful. The text is written at a senior or graduate level for students and practicing engineers who desire to understand the theory behind filter design methodology. To a large extent, tabulation of resistor, inductor, and capacitor component values for filters has been avoided because the formulas given in the text allow the engineer to generate tables of component values for a large variety of circuit configurations. The prototype computer programs given in Appendix B are meant to supplement the derivations in the text by providing a starting point for those interested in programming more exotic designs. A complete set of compiled programs, which illustrate the methods given in this text, are also available from Artech House, Inc.

The book can be divided into four main sections, categorized by passive, active, digital, and switched capacitor realizations. Chapter 1 and Appendix A are intended to be a review of background material from elementary numerical analysis and network theory. Basic methods are given for the analysis of analog and digital filters with regard to magnitude, phase, group delay, impulse and step response.

Chapter 2 provides a foundation for understanding the low-pass Butterworth filter. Simple equations are given for determining L-C component values. The locations of the Butterworth poles are defined and formulas are derived for the order, attenuation, and response of Butterworth functions.

Chapters 3 through 5 derive and define similar concepts for low-pass Tchebyscheff, low-pass inverse Tchebyscheff filters, and low-pass elliptic filters. For the elliptic filter type, two computer programs are given in Appendix B as an aid for those interested in programming elliptic filter solutions.

Chapter 6 deals with the transformation of low-pass filters to other bandpass topologies. Impedance scaling, frequency scaling, duality, reciprocity, network transformations, two-port network parameters, lattice structures, and group delay equalizer topologies for passive filters are covered in detail.

The realization of unevenly terminated passive filters is treated in Chapter 7. The equation for input impedance is derived based solely on network theory and used to generate unequally terminated structures for Butterworth, Tchebyscheff, and elliptic filters. Both classical filter synthesis and the permutation method of Amstutz are discussed in the synthesis of elliptic configurations.

Chapter 8 provides over 16 active filter topologies, complete with derivations. Poles and zeros, determined mathematically from the methods given in Chapters 2–5 and Chapter 7, can be used to generate low-pass, bandpass, bandstop, and high-pass active circuits. Filter types include multiple feedback, voltage controlled voltage source, state-variable, and many other configurations. A method for optimization of the gain of each active filter section in a cascade is presented as a means to maximize dynamic range.

Chapter 9 introduces infinite impulse response digital filters using the bilinear transform to map the analog domain into the digital domain. The effects of frequency warping and aliasing are discussed.

In Chapter 10, the Fourier transform is presented as a design tool for finite impulse response (FIR) filters. Window types are given including triangular, Hamming, Hanning, Kaiser, and Blackman. A brief discussion of sigma-delta is given as it pertains to high-resolution analog-to-digital conversion.

The exchange method is presented in Chapter 11 as an optimization technique useful for determining a minimax polynomial that approximates a desired set of data points with an equiripple error. The Parks-McClellan algorithm is shown to be an extension of the exchange method applied to FIR digital filters to generate a minimax error in both the passband and the stopband.

Chapter 12 deals with switched capacitor filters. The concept of equivalent resistance in a switched capacitor is introduced. Charge conservation is used as an analysis tool for obtaining the z-domain transfer function. Several parasitic insensitive switched capacitor circuit topologies are given for use in low-pass, bandpass, bandstop, or high-pass circuits. Consideration is given to the effect of nonideal op-amp characteristics.

Equations shown in bold represent fundamental concepts from which other equations will be derived.

The author would like to express thanks to *RF Design Magazine* and IngSOFT Limited for their support in writing this text.

Chapter 1
Basic Network Theory

1.1 INTRODUCTION

Modern filter theory is based on many diverse but interrelated discoveries over the past two centuries. In 1785 Laplace [1] wrote the almost modern Laplace transform to solve certain differential equations with variable coefficients. Baron Jean-Baptiste-Joseph Fourier [2], as a prefect in Napoleon's army, invented a method for solving equations that described the conduction of heat in solid bodies, and it was published in 1807. Laplace was intrigued and obtained additional results that in turn inspired Fourier's discovery of his own transform. Many nineteenth century mathematicians did not accept Fourier's claim that any initial temperature distribution could be decomposed into a simple arithmetic sum consisting of a fundamental variation and higher frequency harmonics, and the theoretical basis of the Fourier transform was not formally published until 1822.

In 1829, Carl Gustav Jacob Jacobi derived and reported the main properties of a set of new functions useful for solving elliptic integrals. At that time, the elliptic functions were used to solve problems in finding the length of an arc of an ellipse and in determining the motion of a pendulum.

The first known record of sampling analog signals is found on pages 420–425 of the *Treatise on Electricity and Magnetism* (1873) by James Clark Maxwell [3] in his discussion on the equivalent resistance of a periodically switched capacitor.

Filter technology was officially born in 1915 when K. Wagner (Germany) and G. Campbell (United States), working independently, proposed the basic concept of the filter. Their results evolved from earlier work on loaded transmission lines and the classical theory of vibrating systems.

In 1923, Zobel [4] reported a practical method of designing image parameter filters with an unlimited number of reactances. This filter method is summarized in Figure 1.1 but will not be discussed further because it is a little used technique that gives the designer only a very loose control over the passband and stopband characteristics. However, it was the only known method until 1939 and the only practical method until computers became widespread in the mid-1950s. S. Darlington [5] in the United States and W. Cauer in Germany, both inspired by the work of Norton, produced a theory in 1939 that involved a set of problems relating to modern synthesis procedure. The filters proposed by Darlington and Cauer used the Jacobian elliptic integral to determine element values for very sharp roll-off filters with Tchebyscheff (or equiripple) behavior in both the passband and the stopband.

This behavior takes its name from the Russian mathematician who in the last century developed a powerful method of approximation by means of orthogonal polynomials. The elliptic filters of Darlington and Cauer should not be confused with Tchebyscheff filters because the latter exhibit equiripple behavior in either the passband or the stopband but not both. The importance of polynomial filter synthesis was not recognized immediately because of the extremely heavy burden required on computation. It was not until the mid-1950s that Cauer-Darlington (elliptic) filters came into widespread use.

The theory of sampling analog signals was also well developed by the late 1950s. In the 1960s several schemes were proposed that used switches and capacitors to simulate resistors in filters. It was shown that the filter transmittances had the important property of depending only on capacitor ratios.

The Fairchild uA 709 was introduced in 1965 at $70 and became the first integrated circuit (IC) operational amplifier (op-amp) manufactured with such high yields in volume production that it facilitated the implementation of active RC filters equaling in performance and usefulness its passive counterpart. Although the op-amp was a major component of the filter circuit and eliminated the need for bulky inductors, discrete resistors and capacitors were still required because IC technology did not permit resistors and capacitors to be integrated with a high degree of accuracy.

That same year, James W. Cooley of IBM's Thomas J. Watson Research Center and John W. Tukey [6] of the Bell Telephone Laboratories in Murray Hill, N.J., rediscovered a program known as the fast Fourier transform (FFT), which allowed a speed increase of up to several orders of magnitude in the calculation of the Fourier transform. The FFT has become the basis for the implementation of many digital filters because the computations can be done in real time by high-speed digital signal processor (DSP) chips.

The importance and significance of many of the ideas proposed for analog sampled data signal processing had to wait until technology provided the means of turning ideas into practical reality. In 1972, it was suggested that metal oxide semiconductor (MOS) technology would be applicable for the construction of integrated circuit analog sampled data filters using switched capacitors to simulate resistors in active filters. This suggestion was then followed by a rapid development in the implementation of analog signal processing integrated circuits. The switched capacitor techniques provided very high accuracy in the design of analog IC filters because the filter parameters depended only on the ratios of capacitors and this ratio can be controlled to a very high accuracy in the design of MOS integrated circuits.

The Zobel Image Parameter Filter Design Method

Let f_c be the cutoff frequency of the low-pass prototype in Hertz. Then

$\omega_c = 2\pi f_c$

R = termination resistance (equal at both ends)

$m = 0.6$

compute

$$L = \frac{2R}{\omega_c} \qquad C = \frac{2}{R \cdot \omega_c}$$

The filter components can now be calculated as

$$L_{a1} = \frac{(1 - m^2) \cdot L}{2 \cdot m} \qquad L_b = \frac{L}{2} \qquad L_{c1} = \frac{x \cdot L}{2} \qquad L_{a1} = \frac{(1 - m^2) \cdot L}{2 \cdot m}$$

$$C_a = \frac{m \cdot C}{2} \qquad C_b = C \qquad C_c = x \cdot C \qquad C_a = \frac{m \cdot C}{2}$$

$$L_{a2} = \frac{m \cdot L}{2} \qquad L_{c2} = \frac{(1 - x^2) \cdot L}{4 \cdot x} \qquad L_{a2} = \frac{m \cdot L}{2}$$

where $x = \sqrt{[1 - (\omega_c/\omega_\infty)^2]}$.

The frequency, ω_∞, must be specified by the designer in the stopband for each M-derived section. The total number of constant-K sections, the number of M-derived sections, and an ω_∞ for each M-derived section must be specified. The sections are then cascaded to create the entire filter.

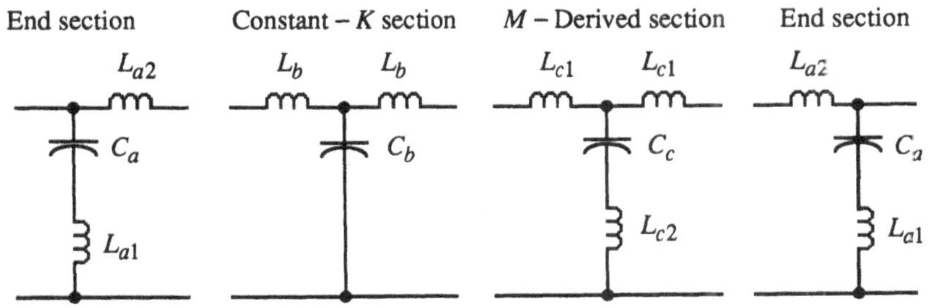

Figure 1.1 Image parameter filters were popular during the 1940s because they require very simple calculations to design.

This chapter reviews the fundamental theories essential for the design and analysis of electronic filters. The design methods derived later in this text are based on developments over the last 200 years and cover the majority of synthesis techniques in use today. The four major classes of polynomial filters are discussed in sufficient detail to allow practicing engineers flexibility in specifying design parameters. Algorithms for the more advanced designs are given as prototype programs in Appendix B. A complete set of computer aided design and analysis tools, corresponding to the methods used in this text, is also available from Artech House, Inc.

1.2 MAGNITUDE AND PHASE RESPONSE OF ANALOG TRANSFER FUNCTIONS

An analog voltage transfer function, $G(s)$, with real poles and zeroes, of the form

$$G(s) = \frac{(s - s_{01})(s - s_{02}) \ldots (s - s_{0M})}{(s - s_1)(s - s_2) \ldots (s - s_N)} \tag{1.1}$$

has a magnitude response as a function of frequency, ω, given by

$$|G(s)| = \frac{\sqrt{\omega^2 + s_{01}^2}\,\sqrt{\omega^2 + s_{02}^2} \ldots \sqrt{\omega^2 + s_{0M}^2}}{\sqrt{\omega^2 + s_1^2}\,\sqrt{\omega^2 + s_2^2} \ldots \sqrt{\omega^2 + s_N^2}} \tag{1.2}$$

If the poles (roots of the denominator) and zeroes (roots of the numerator) are complex, then $G(s)$ can be written as

$$G(s) = \frac{(s^2 + A_{01}s + B_{01})(s^2 + A_{02}s + B_{02}) \ldots (s^2 + A_{0M}s + B_{0M})}{(s^2 + A_1s + B_1)(s^2 + A_2s + B_2) \ldots (s^2 + A_Ns + B_N)} \tag{1.3}$$

with a magnitude response as a function of frequency, ω, as given by equation (1.4):

$$|G(s)| = \frac{\sqrt{(B_{01} - \omega^2)^2 + (A_{01}\omega)^2}\sqrt{(B_{02} - \omega^2)^2 + (A_{02}\omega)^2} \ldots \sqrt{(B_{0M} - \omega^2)^2 + (A_{0M}\omega)^2}}{\sqrt{(B_1 - \omega^2)^2 + (A_1\omega)^2}\sqrt{(B_2 - \omega^2)^2 + (A_2\omega)^2} \ldots \sqrt{(B_N - \omega^2)^2 + (A_N\omega)^2}} \tag{1.4}$$

In both cases the magnitude response was formed by taking the square root of the sum of the squares of the real part and the imaginary part of each section. Similarly the phase response is given by the arc tangent of the imaginary component divided by the real component. For equation (1.1), the phase response becomes

$$\angle G(s) = \tan^{-1}\left(\frac{\omega}{-s_{01}}\right) + \tan^{-1}\left(\frac{\omega}{-s_{02}}\right) + \ldots \tan^{-1}\left(\frac{\omega}{-s_{0M}}\right) - \tan^{-1}\left(\frac{\omega}{-s_1}\right) - \tan^{-1}\left(\frac{\omega}{-s_2}\right) - \ldots \tan^{-1}\left(\frac{\omega}{-s_N}\right) \quad (1.5)$$

whereas for equation (1.3), the phase response is

$$\angle G(s) = \tan^{-1}\left(\frac{A_{01}\omega}{B_{01} - \omega^2}\right) + \tan^{-1}\left(\frac{A_{02}\omega}{B_{02} - \omega^2}\right) + \ldots + \tan^{-1}\left(\frac{A_{0M}\omega}{B_{0M} - \omega^2}\right) - \tan^{-1}\left(\frac{A_1\omega}{B_1 - \omega^2}\right) - \tan^{-1}\left(\frac{A_2\omega}{B_2 - \omega^2}\right) - \ldots - \tan^{-1}\left(\frac{A_N\omega}{B_N - \omega^2}\right) \quad (1.6)$$

1.3 TIME DELAY AND GROUP DELAY OF ANALOG FILTERS

The phase response of a filter affects the transient response, which is of particular concern in pulse systems such as radar and digital communications. In these cases, the pulse rise time and overshoot are of interest because they indicate the degree of pulse distortion. If a filter is such that the output pulse is an exact replica of the input pulse, except for a time delay, then the transient response is ideal. The transfer function of this "ideal" filter is given by $G(s) = e^{-sT} = e^{-j\omega T}$, where T is the time delay in seconds. The phase of $G(s)$ is given by

$$\angle G(s) = \angle e^{-sT} = \angle[\cos(\omega T) - j\ \sin(\omega T)] = \tan^{-1}\left(\frac{-\sin(\omega T)}{\cos(\omega T)}\right) = -\omega T \quad (1.7)$$

From equation (1.7) the time delay of a linear phase filter is given by $-\phi/\omega$ seconds; however, in practice a constant group delay is often sufficient to achieve linear phase over a portion of the frequency band of interest. The group delay of a filter is defined by

$$\textbf{group delay} = -\frac{d\phi}{d\omega} \quad (1.8)$$

Using the relation

$$\frac{d}{du}\arctan(u) = \frac{1}{1 + u^2}\frac{du}{d\omega} \quad (1.9)$$

the expression for group delay based on equation (1.6) becomes

$$-\frac{A_{01}(B_{01}+\omega^2)}{(B_{01}-\omega^2)^2+(A_{01}\omega)^2}+\frac{A_1(B_1+\omega^2)}{(B_1-\omega^2)^2+(A_1\omega)^2}-\dots \tag{1.10}$$

1.4 TRANSIENT ANALYSIS USING LAPLACE TRANSFORMS

Laplace transforms are useful for analytically determining the impulse and step response of analog filters. The use of Table 1.1 in conjunction with partial fraction expansion (see Appendix A) provides the necessary information to find the impulse and step response of all analog filters with real or complex poles occurring with a multiplicity of not more than two.

Example 1.1

Find expressions for the magnitude, phase, group delay, impulse response, and step response for the filter transfer function given by

$$G(s)=\frac{0.1604485(s^2+2.0762144)}{s^3+0.9684197s^2+1.2422106s+0.666251}$$
$$=\frac{0.1604485(s^2+2.0762144)}{(s+0.6446584)(s^2+0.3236714s+1.0334946)}$$

Table 1.1
Laplace Transforms

$F(s)$	$f(t)$
1	$\delta(t)$, unit impulse at $t = 0$
$1/s$	$u(t)$, unit step at $t = 0$
$1/(s+a)$	$u(t)\ e^{-at}$
$\omega_o/(s^2+\omega_o^2)$	$u(t)\sin(\omega_o t)$
$s/(s^2+\omega_o^2)$	$u(t)\cos(\omega_o t)$
$\omega_o/[(s+a)^2+\omega_o^2]$	$u(t)\ e^{-at}\sin(\omega_o t)$
$(s+a)/[(s+a)^2+\omega_o^2]$	$u(t)\ e^{-at}\cos(\omega_o t)$
$1/s^2$	$u(t)t$
$1/(s+a)^2$	$u(t)t\ e^{-at}$
$1/[(s+a)^2+\omega_o^2]^2$	$u(t)\ e^{-at}[\sin(\omega_o t)-(\omega_o t)\cos(\omega_o t)]/(2\omega_o^3)$
$s\ F(s)$	$d[f(t)]/dt$
$K\ F(s)$	$K\ f(t)$
e^{-sT}	Delay of T sec

Solution

Magnitude response:

$$|G(s)| = \frac{0.1604485|(2.0762144 - \omega^2)|}{\sqrt{\omega^2 + (0.6446584)^2}\,\sqrt{(1.0334946 - \omega^2)^2 + (0.3236714\omega)^2}}$$

Phase response:

$$\angle G(s) = -\tan^{-1}\left(\frac{\omega}{0.6446584}\right) - \tan^{-1}\left(\frac{0.3236714\omega}{1.0334946 - \omega^2}\right), \qquad \text{for } \omega < 1.44091$$

$$\angle G(s) = -180 - \tan^{-1}\left(\frac{\omega}{0.6446584}\right) - \tan^{-1}\left(\frac{0.3236714\omega}{1.0334946 - \omega^2}\right), \qquad \text{otherwise}$$

Note: $\tan^{-1}(-a/b) \neq \tan^{-1}(a/-b)$, so be careful.

Group delay response:

$$\text{group delay} = \frac{0.6446584}{(0.6446584)^2 + \omega^2} + \frac{0.3236714(1.0334946 + \omega^2)}{(1.0334946 - \omega^2)^2 + (0.3236714\omega)^2}$$

Impulse response (from inverse Laplace transform table):

$$Vo(t) = \Im^{-1}\left(\frac{0.1604485(s^2 + 2.0762144)}{s^3 + 0.9684197s^2 + 1.2422106s + 0.666251}\right)$$

$$= \Im^{-1}\left(\frac{0.322329}{s + 0.6446582} + \frac{-0.1618806s}{s^2 + 0.3237614s + 1.033495}\right)$$

$$= 0.322329\ e^{-0.6446582t} - 0.1618806\ e^{-0.1618808t}\cos(1.003638\ t) + 0.02611021\ e^{-0.1618808t}\sin(1.003638\ t)$$

Step response (from inverse Laplace transform table):

$$Vo(t) = \Im^{-1}\left(\frac{1}{s}\,\frac{0.1604485(s^2 + 2.0762144)}{s^3 + 0.9684197s^2 + 1.2422106s + 0.666251}\right)$$

$$= \Im^{-1}\left(\frac{0.5}{s} + \frac{-0.5}{s + 0.6446582} + \frac{-0.1618805}{s^2 + 0.3237616s + 1.033495}\right)$$

$$= 0.5 - 0.5\ e^{-0.6446582t} - 0.1612937\ e^{-0.1618808t}\sin(1.003638\ t)$$

1.5 SAMPLED DATA SYSTEMS

Sampled data systems use samples of the input voltage at discrete times. Digital and switched capacitor filters use discrete time, continuous amplitude, signals as inputs. Discrete time signals are defined only for discrete values of the independent variable, time. Generally time is quantized uniformly; that is, $t = NT$, where T is the interval between time samples. Discrete time signals are represented mathematically as sequences of numbers whose amplitude may take on a continuum of values. Normally an analog-to-digital (A/D) converter is used to obtain a sequence by sampling a continuous-time waveform. The notation

$$\{h(n)\} \qquad N_1 \leq n \leq N_2$$

is often used to denote the various samples in a sequence. A digital impulse, or unit sample, is defined by the relation

$$u_o(n) = \begin{cases} 1 & n = 0 \\ 0 & n \neq 0 \end{cases} \tag{1.11}$$

in much the same way that the analog impulse, or Dirac delta, function $\delta(t)$ plays in continuous-time systems.

A discrete time system converts the input sequence into an output sequence as shown by the simple representation in Figure 1.2.

A linear system is defined as follows. If $x_1(n)$ and $x_2(n)$ are specific inputs to a linear system and $y_1(n)$ and $y_2(n)$ are the respective outputs, then if the sequence $ax_1(n) + bx_2(n)$ is applied at the input, the sequence $ay_1(n) + by_2(n)$ is obtained at the output, where a and b are arbitrary constants.

In a time-invariant system, if the input sequence $x(n)$ produces an output sequence $y(n)$, then the input sequence $x(n - n_o)$ produces the output sequence $y(n - n_o)$ for all n_o.

A linear, time-invariant (LTI) system is causal, or realizable, if the output at $n = n_o$ is dependent only on values of the input for $n \leq n_o$. This implies that the impulse response $h(n)$ is zero for $n < 0$. An LTI system is stable if every bounded input produces a bounded output.

The impulse response of a sampled data system, or digital filter, can be either of finite duration or infinite duration. A nonrecursive digital filter, in which previous output samples are not used to calculate the current output, is always a finite impulse response

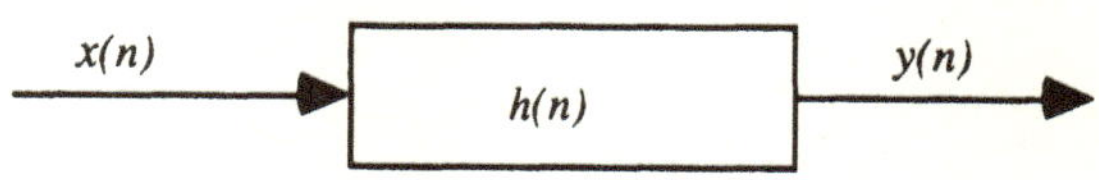

Figure 1.2 A discrete time system.

(FIR) filter. Infinite impulse response (IIR) filters are obtained most often by conformal mapping from the analog domain (s-domain) to the digital domain. FIR filters, however, have no analogy in the s-domain and can be obtained either by taking the inverse Fourier transform of the desired frequency response or by using a linear optimization method such as the simplex method or Remez exchange. Unlike analog filters, FIR filters can have perfectly constant group delay.

A general expression that models both FIR and IIR filters for an Mth order linear, constant coefficient difference equation of a causal system is of the form

$$y(n) = \sum_{i=0}^{M} b_i x(n-i) - \sum_{i=1}^{M} a_i y(n-i) \qquad n \geq 0 \tag{1.12}$$

where a_i and b_i characterize the appropriate system. A nonrecursive finite impulse response filter usually has $a_i = 0$ for all i. A recursive infinite impulse response filter necessarily has one, or more, nonzero values of a_i.

1.6 THE FOURIER TRANSFORM

A periodic function repeats itself every T sec and can be expressed as

$$f(t+T) = f(t) \qquad -\infty \leq t \leq \infty \tag{1.13}$$

where T is called the period of the function. As in the case of sinusoids,

$$f_o = \frac{1}{T} \tag{1.14}$$

is called the frequency of the function, measured in Hertz, and

$$\omega_o = 2\pi f_o = \frac{2\pi}{T} \tag{1.15}$$

is the angular frequency in radians per second. If

$$\int_{-T/2}^{T/2} |f(t)|dt \tag{1.16}$$

is finite, then the periodic function can be expanded into an infinite series of the form

$$f(t) = a_o + a_1 \cos(\omega_o t) + a_2 \cos(2\omega_o t) + \ldots + b_1 \sin(\omega_o t) + b_2 \sin(2\omega_o t) + \ldots \tag{1.17}$$

where the a and b terms are constants. The periodic function consisting of the sum of sine and cosine terms in (1.17) is called a *Fourier series.* Equation (1.17) can be rewritten as

$$f(t) = a_o + \sum_{m=1}^{\infty} (a_m \cos(m\omega_o t) + b_m \sin(m\omega_o t) \tag{1.18}$$

The coefficients for the Fourier series are given by

$$a_o = \frac{1}{T}\int_{-T/2}^{T/2} f(t)\, dt \tag{1.19}$$

$$a_m = \frac{2}{T}\int_{-T/2}^{T/2} f(t) \cos(m\omega_o t)\, dt \tag{1.20}$$

$$b_m = \frac{2}{T}\int_{-T/2}^{T/2} f(t) \sin(m\omega_o t)\, dt \tag{1.21}$$

The Fourier series is often given in exponential form as

$$f(t) = \sum_{k=-\infty}^{k=\infty} C_k\, e^{jk\omega_o t} \tag{1.22}$$

where

$$C_k = \frac{1}{T}\int_{-T/2}^{T/2} f(t)\, e^{-jk\omega_o t}\, dt \tag{1.23}$$

$$C_k = \frac{a_k - jb_k}{2} \tag{1.24}$$

The Fourier series can be extended for a *nonperiodic* signal to produce the Fourier integral pairs given by

$$F(j\omega) = \int_{-\infty}^{\infty} f(t)\, e^{-j\omega t}\, dt \tag{1.25}$$

$$f(t) = \frac{1}{2\pi}\int_{-\infty}^{\infty} F(j\omega)\, e^{j\omega t}\, d\omega \tag{1.26}$$

or

$$f(t) = \int_{-\infty}^{\infty} F(j\omega)\, e^{j2\pi ft}\, df \tag{1.27}$$

The inverse Fourier transform given in (1.26) may produce values of $f(t)$ that are complex and nonzero in negative time. The system frequency response is $F(j\omega)$ and the corresponding time impulse response is $f(t)$. When $f(t)$ must be real and also causal (zero in negative time) then equation (1.26) may be written as

$$f(t) = \frac{2}{\pi}\int_{0}^{\infty} R(\omega)\cos(\omega t)d\omega \tag{1.28}$$

where $R(\omega)$ is the real part of the frequency response. In evaluating equation (1.28) it is useful to recognize that

$$\frac{1}{2\pi}\int_{-\infty}^{\infty} \cos(\omega t)d\omega = \delta(t) = \text{the unit impulse} \tag{1.29}$$

Equation (1.29) can be proven rigorously only by means of distribution theory, which is beyond the scope of this chapter.

Example 1.2

Find the impulse response of the high-pass filter $s/(s + 1)$.

Solution

By the inverse Laplace transform method, $s/(s + 1)$ may be written as $1 - 1/(s + 1)$, which has an impulse response from Table 1.1 given by $\delta(t) - e^{-t}$.

By the inverse Fourier transform method, use $R(\omega) = \omega^2/(1 + \omega^2) = 1 - 1/(1 + \omega^2)$. The integration formula becomes

$$f(t) = \frac{1}{2\pi}\int_{-\infty}^{\infty}\left[1 - \frac{1}{1+\omega^2}\right]\cos(\omega t)d\omega$$

$$= \frac{1}{2\pi}\int_{-\infty}^{\infty}\cos(\omega t)d\omega - \frac{2}{\pi}\int_{0}^{\infty}\left[\frac{1}{1+\omega^2}\right]\cos(\omega t)d\omega$$

$$= \delta(t) - \frac{2}{\pi}\int_{0}^{\infty}\left[\frac{1}{1+\omega^2}\right]\cos(\omega t)d\omega$$

which can be integrated numerically for a given value of t.

It can be seen from example 1.2 that the numerical integration is valid only to some finite frequency on the assumption that $F(j\omega)$ is band limited, for example, is zero above some limiting frequency. If this is not the case as in a high-pass or bandstop filter, then an appropriate constant term must be subtracted from the real part of the system frequency response so that the remaining response at infinite frequency goes to zero, and the integration broken into two parts, using (1.29) to evaluate one of the two parts.

For sampled data systems, the Fourier transform relates the impulse response, $h(n)$, of a sampled data system to the frequency response, $H(e^{j\omega})$, of the system by the relations (assuming a normalized time, T, between samples of 1 second):

$$\mathbf{H(e^{j\omega}) = \sum_{n=-\infty}^{n=\infty} h(n)\, e^{-j\omega n}} \tag{1.30}$$

$$\mathbf{h(n) = \frac{1}{2\pi} \int_{-\pi}^{\pi} H(e^{j\omega})\, e^{j\omega n} d\omega} \tag{1.31}$$

The relations in (1.30) and (1.31) are valid for any sequence including an arbitrary input sequence, $x(n)$. The frequency components of an output sequence can be expressed by simple multiplication of $X(e^{j\omega})H(e^{j\omega})$ such that

$$Y(e^{j\omega}) = X(e^{j\omega})H(e^{j\omega})$$

Example 1.3

Find the frequency response of a sampled data system whose impulse response, $h(n)$, is given by

$$h(-1) = 1$$
$$h(0) = 2$$
$$h(1) = 1$$

Solution

From (1.30), using the relations $\cos(\omega) = \cos(-\omega)$ and $\cos(0) = 1$,

$$\begin{aligned} H(e^{j\omega}) &= \cos(\omega) + j\sin(\omega) + 2\cos(0) + j2\sin(0) + \cos(\omega) - j\sin(\omega) \\ &= 2\cos(\omega) + 2 \end{aligned}$$

Example 1.4

The frequency response, $H(e^{j\omega})$, of a sampled data system is given by $2\cos(\omega) + 2$. Find the impulse response.

Solution

Using $h(n) = \frac{1}{2\pi}\int_{-\pi}^{\pi} H(e^{j\omega})e^{j\omega n}d\omega$ yields

$$h(n) = \frac{1}{2\pi}\int_{-\pi}^{\pi} [2 + 2\cos(\omega)]e^{j\omega n}d\omega = \frac{1}{2\pi}\int_{-\pi}^{\pi} [2 + 2\cos(\omega)]\cos(\omega n)d\omega$$

$$= \frac{1}{2\pi}\int_{-\pi}^{\pi} 2\cos(\omega n)d\omega + \frac{1}{2\pi}\int_{-\pi}^{\pi} 2\cos(\omega)\cos(\omega n)d\omega$$

$$= \frac{1}{2\pi}\frac{2}{n}\sin(\omega n)\Big|_{-\pi}^{\pi} + \frac{1}{2\pi}\frac{1}{n-1}\sin[\omega(n-1)]\Big|_{-\pi}^{\pi} + \frac{1}{2\pi}\frac{1}{n+1}\sin[\omega(n+1)]\Big|_{-\pi}^{\pi}$$

$$= \frac{1}{2\pi}\frac{4}{n}\sin(\pi n) + \frac{1}{2\pi}\frac{2}{n-1}\sin[\pi(n-1)] + \frac{1}{2\pi}\frac{2}{n+1}\sin[\pi(n+1)]$$

Using L'Hôpital's rule, which states that if $f(x)$ and $g(x)$ are both 0, then $f(x)/g(x) = f'(x)/g'(x)$, the first term is 2 only for $n = 0$ and is 0 for all other values of n; the second term is 1 only for $n = 1$ and is 0 for all other values of n; the third term is 1 only for $n = -1$ and is 0 for all other values of n; therefore, $h(-1) = 1$, $h(0) = 2$, and $h(1) = 1$.

1.7 THE FAST FOURIER TRANSFORM

The fast Fourier transform (FFT) is an algorithm used for reducing the computational time required to compute a discrete Fourier transform (DFT) by removing the inefficiencies of the DFT process. The number of complex multiplications and additions in a DFT is on the order of N^2, where N is the number of data points. For a FFT, the time is proportional to $N \log_2 N$. Use of the FFT over the DFT can result in a gain in speed of several orders of magnitude. For an FFT, the value of N must be a power of 2 ($N = 2, 4, 8, 16, 32 \ldots$); however, most practical problems are sufficiently malleable so that no difficulties occur when the sequence must be artificially lengthened by augmentation with signal samples of zero value. The resulting spectrum will be an interpolated version of the nonaugmented spectrum.

Because the derivation of the FFT is too complicated to be presented here, program B1 in Appendix B is provided as a decimation-in-time, radix 2 in-place FFT adapted from Cooley, Lewis, and Welch. The following example is given for the use of this program.

Example 1.5

Use a 1024-point FFT to find the frequency response of a sampled data system with a sampling time $T = 0.0001$ sec and an impulse response, $h(n)$, is given by

$$h(0) = 1$$
$$h(1) = 2$$
$$h(2) = 1$$

Solution

We know from example 1.3 that the frequency response is $|H(e^{j\omega})| = 2\cos(\omega) + 2$, but this assumes a sampling time of 1 sec ($T = 1$). If $T = 1/10000$, then $|H(e^{j\omega T})| = 2\cos(\omega T) + 2$. The phase response can be obtained from $H(e^{j\omega T}) = \sum_{n=-\infty}^{n=\infty} h(nT)\, e^{-j\omega nT} = (1) + (2)\, e^{-j\omega T} + (1)\, e^{-2j\omega T}$ by rewriting it as $H(e^{j\omega T}) = (e^{-j\omega T})[e^{j\omega T} + 2 + e^{-j\omega T}]$. Therefore the expression for the phase angle becomes

$$\angle[e^{-j\omega T}H(e^{j\omega})] = \angle e^{-j\omega T} + \angle \sum_{n=-\infty}^{n=\infty} h(nT)\, e^{-j\omega nT}$$
$$= \angle e^{-j\omega T} + 0 = -\omega T \text{ radians or } 360 \cdot f \cdot T \text{ degrees}$$

Running the FFT program B1 using a sampling frequency of 10,000 Hz with three taps and coefficients of 1, 2, and 1 for each tap, respectively, provides the following response at the first three frequencies, which corresponds exactly to the analytical solution:

Frequency	*Magnitude*	*Phase*
0	4	0 degrees
9.765625	3.999962	0.35156 degrees
19.53125	3.999849	0.703125 degrees
and so forth		

The response for frequencies between those given by the FFT routine can be obtained by interpolation of the given magnitude and phase for the desired frequencies.

Equations (1.19), (1.20), and (1.21) can be used to develop numerical integration formulas for the DFT to provide the harmonics for a waveform of period T_p. The term, a_o, is the average value of $f(t)$ over one period. If the periodic waveform, $f(t)$, is divided into N segments such that $f(t)$ is approximately constant over any one segment then (1.20) can be written as a sum of integrals over each segment such that

$$a_m = \frac{2}{T_p}\sum_{i=1}^{N} f\left[\frac{-T_p}{2} + (i - 0.5)\frac{T_p}{N}\right]\left[\int_{\frac{-T_p}{2} + (i-1)\frac{T_p}{N}}^{\frac{-T_p}{2} + i\frac{T_p}{N}} \cos\left(m\frac{2\pi}{T_p}t\right)dt\right] \tag{1.32}$$

Notice that $f(t)$ has been taken outside the integral because it is assumed to be constant over the limits of integration having a value of $f(-T_p/2 + (i - 0.5)T_p/N)$. The constant value for $f(t)$ is taken as the midpoint of each segment. Performing the integration yields

$$a_m = \frac{1}{\pi m}\sum_{i=1}^{N} f\left(\frac{-T_p}{2} + (i - 0.5)\frac{T_p}{N}\right)\left[\sin\left(-m\pi + \frac{2\pi mi}{N}\right) - \sin\left(-m\pi + \frac{2\pi m(i-1)}{N}\right)\right] \tag{1.33}$$

or

$$a_m = \frac{1}{\pi m}(-1)^m \sum_{i=1}^{N} f\left(\frac{-T_p}{2} + (i - 0.5)\frac{T_p}{N}\right)\left[\sin\left(\frac{2\pi mi}{N}\right) - \sin\left(\frac{2\pi m(i-1)}{N}\right)\right] \tag{1.34}$$

The term $(-1)^m$ is due to the fact that the harmonics of the fundamental frequency start at $-T_p/2$. If the harmonic frequencies are referenced to $t = 0$, then the $(-1)^m$ term can be eliminated, providing

$$a_m = \frac{1}{\pi m}\sum_{i=1}^{N} f\left(\frac{-T_p}{2} + (i - 0.5)\frac{T_p}{N}\right)\left[\sin\left(\frac{2\pi mi}{N}\right) - \sin\left(\frac{2\pi m(i-1)}{N}\right)\right] \tag{1.35}$$

Similarly for b_m,

$$b_m = -\frac{1}{\pi m}\sum_{i=1}^{N} f\left(\frac{-T_p}{2} + (i - 0.5)\frac{T_p}{N}\right)\left[\cos\left(\frac{2\pi mi}{N}\right) - \cos\left(\frac{2\pi m(i-1)}{N}\right)\right] \tag{1.36}$$

Equations (1.35) and (1.36) can be evaluated directly or they can be further modified so that they can be evaluated as the sum of two FFTs.

1.8 THE z TRANSFORM

One of the most common techniques for representing and manipulating sequences is the z transform. Given a sampled data sequence $x(n)$, its z transform is defined as

$$H(z) = \sum_{n=-\infty}^{n=\infty} h(n)z^{-n} \tag{1.37}$$

where z is a complex variable. If the z transform is evaluated on a circle of unit radius, that is, $z = e^{j\omega}$, then

$$H(z)\Big|_{z=e^{j\omega}} = H(e^{j\omega}) = \sum_{n=-\infty}^{n=\infty} h(n)\, e^{-j\omega n} \tag{1.38}$$

which is the Fourier transform of the sampled data sequence.

It is common to use the equivalence

$$z = e^{j\omega}, \qquad -\pi \leq \omega \leq \pi$$

however, this implicitly assumes a sampling interval of 1 sec ($T = 1$). In general,

$$\begin{aligned} z &= e^{j\omega T}, \qquad -\pi \leq \omega T \leq \pi \\ z^{-1} &= e^{-j\omega T}, \qquad -\pi \leq \omega T \leq \pi \end{aligned}$$

If a sequence $h_1(n)$ has z transform $H_1(z)$, then the sequence $h_1(n - n_o)$ has the z transform $z^{-n_o}H_1(z)$ for all n_o. This property of the z transform is useful for converting between a difference equation representation of a sampled data system and its z transform representation. For example, the difference equation

$$y(n) = v(n) - b_1 y(n-1) - b_2 y(n-2) \tag{1.39}$$

has a z transform representation

$$Y(z) = V(z) - b_1 z^{-1} Y(z) - b_2 z^{-2} Y(z) \tag{1.40}$$

or

$$Y(z) = \frac{V(z)}{1 + b_1 z^{-1} + b_2 z^{-2}} \tag{1.41}$$

By substituting

$$z^{-1} = e^{-j\omega T} \tag{1.42}$$

where T is the time between samples, in seconds, the magnitude, phase, and group delay of sampled data systems can be determined by grouping the real and imaginary components of the numerator and denominator and applying the same methods used for analog transfer functions.

Example 1.6

Find expressions for the magnitude, phase, and group delay for the sampled data system with an arbitrary sample time, T, given by

$$\frac{V_o(z)}{V_{in}(z)} = H(z) = \frac{0.2992562z^{-1} + 0.2992562}{-0.4014877z^{-1} + 1} \; \frac{0.1512348z^{-2} + 0.09804408z^{-1} + 0.1512348}{0.4657942z^{-2} - z^{-1} + 0.9347194}$$

Solution

Magnitude response:

$$|H(z)| = |H(e^{j\omega})| = \frac{0.2992562\sqrt{[1 + \cos(\omega T)]^2 + [\sin(\omega T)]^2}}{\sqrt{[1 - 0.4014877\cos(\omega T)]^2 + [0.4014877\sin(\omega T)]^2}}$$

$$\frac{\sqrt{[0.15123 + 0.098044\cos(\omega T) + 0.15123\cos(2\omega T)]^2 + [0.098044\sin(\omega T) + 0.15123\sin(2\omega T)]^2}}{\sqrt{[0.93472 - \cos(\omega T) + 0.46579\cos(2\omega T)]^2 + [\sin(\omega T) - 0.46579\sin(2\omega T)]^2}}$$

Phase response:

$$\angle H(e^{j\omega}) = \tan^{-1}\left[\frac{-0.2992562\sin(\omega T)}{0.2992562 + 0.2992562\cos(\omega T)}\right] - \tan^{-1}\left[\frac{0.4014877\sin(\omega T)}{1 - 0.4014877\cos(\omega T)}\right]$$

$$+ \tan^{-1}\left[\frac{-0.09804408\sin(\omega T) - 0.1512348\sin(2\omega T)}{0.1512348 + 0.09804408\cos(\omega T) + 0.1512348\cos(2\omega T)}\right]$$

$$- \tan^{-1}\left[\frac{\sin(\omega T) - 0.4657942\sin(2\omega T)}{0.9347194 - \cos(\omega T) + 0.4657942\cos(2\omega T)}\right]$$

Group delay: for simplicity of notation, define

$$N_1 = -0.2992562 \sin(\omega T)$$

$$N_1' = -0.2992562\, T \cos(\omega T)$$

$$D_1 = 0.2992562 + 0.2992562 \cos(\omega T)$$

$$D_1' = -0.2992562\, T \sin(\omega T)$$

$$N_2 = 0.4014877 \sin(\omega T)$$

$$N_2' = 0.4014877\, T \cos(\omega T)$$

$$D_2 = 1 - 0.4014877 \cos(\omega)$$

$$D_2' = 0.4014877\, T \sin(\omega T)$$

$$N_3 = -0.09804408 \sin(\omega T) - 0.1512348 \sin(2\omega T)$$

$$N_3' = -0.09804408\, T \cos(\omega T) - 0.3024696\, T \cos(2\omega T)$$

$$D_3 = 0.1512348 + 0.09804408 \cos(\omega T) + 0.1512348 \cos(2\omega T)$$

$$D_3' = -0.09804408\, T \sin(\omega T) - 0.3024696\, T \sin(2\omega T)$$

$$N_4 = \sin(\omega T) - 0.4657942 \sin(2\omega T)$$

$$N_4' = T \cos(\omega T) - 0.9315884\, T \cos(2\omega T)$$

$$D_4 = 0.9347194 - \cos(\omega T) + 0.4657942 \cos(2\omega T)$$

$$D_4' = T \sin(\omega T) - 0.9315884\, T \sin(2\omega T)$$

Using equations (1.8) and (1.9), the group delay becomes

$$\text{Group delay} = -\frac{D_1N_1' - N_1D_1'}{D_1^2 + N_1^2} + \frac{D_2N_2' - N_2D_2'}{D_2^2 + N_2^2} - \frac{D_3N_3' - N_3D_3'}{D_3^2 + N_3^2} + \frac{D_4N_4' - N_4D_4'}{D_4^2 + N_4^2}$$

If $T = 0.0001$ sec, then at 1 kHz, $|H(e^{j\omega})| = 0.8912512$, $\angle H(e^{j\omega}) = -101.5°$, and the group delay is 378.0953 μsec.

A graphical interpretation of the frequency response for a z-domain transfer function is given in Figure 1.3. The z transform of a sequence can always be written as a ratio of a product of factors representing the zeros of $H(z)$ and a product of factors representing the poles of $H(z)$. Therefore, $H(z)$ assumes the form

$$H(z) = A\frac{\prod_{i=1}^{M} 1 - z_i z^{-1}}{\prod_{i=1}^{N} 1 - p_i z^{-1}} \tag{1.43}$$

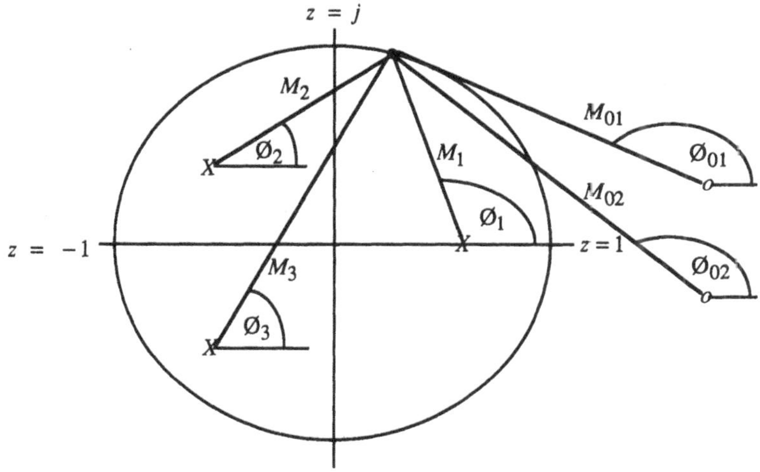

Figure 1.3 A graphical interpretation of the frequency response in the z domain.

Multiplying the numerator and denominator through by z yields

$$|H(e^{j\omega})| = |A|\frac{\prod_{i=1}^{M}|e^{j\omega} - z_i|}{\prod_{i=1}^{N}|e^{j\omega} - p_i|} \tag{1.44}$$

$$\angle H(e^{j\omega}) = \angle A + \sum_{i=1}^{M} \angle(e^{j\omega} - z_i) - \sum_{i=1}^{N} \angle(e^{j\omega} - p_i) \tag{1.45}$$

From the point $z = e^{j\omega}$ (on the unit circle) vectors are drawn to each of the poles and zeros. The magnitudes of the vectors are used to determine the magnitude of the system function at the given frequency, ω, and the angles are used to determine the phase. In Figure 1.3, there are three poles and two zeros and $A = 1$ so that

$$|H(e^{j\omega})| = \frac{M_{01}M_{02}}{M_1M_2M_3} \quad \text{and} \quad \angle H(e^{j\omega}) = \phi_{01} + \phi_{02} - (\phi_1 + \phi_2 + \phi_3)$$

To determine the entire system function (as ω varies from 0 to π) z moves counterclockwise around the unit circle beginning at $z = +1$ and ending at $z = -1$.

1.9 SENSITIVITIES AND MONTE CARLO ANALYSIS

The sensitivity of parameter F with respect to changes in h are defined as

$$S_h^F = \frac{h}{F}\frac{\partial F}{\partial h} \tag{1.46}$$

In network applications, the function F can be a network function, its pole, or zero; the Q and ω_o of the pole or zero, and so on, whereas the parameters h can be component values, the frequency variable s, the operating temperature, or humidity, and so forth. Sensitivity is a first-order measure and may not be meaningful when large parameter changes need to be considered. Instead a Monte Carlo analysis is usually carried out before the designs are put into volume production. A Monte Carlo analysis is a computer simulation of the response of a network to changes in component tolerance. A number of test circuits, typically 100 or more, are "manufactured" on the computer, with the component values picked at random from their respective statistical distributions. The circuits are then analyzed and the results tabulated. A statistical analysis is then performed on the results of the computer simulations to determine the viability of the design under actual production and field conditions.

Monte Carlo analysis requires that we pick the components from their probability density functions. For discrete components, manufacturers often screen the components and sell those that are closer to the mean values at higher prices. To better model the distribution of discrete components, uniform, bimodal, or split triangular distributions can be used. These distributions are shown in Figure 1.4.

When considering two or more components, the coupling between the component values also becomes important. The correlation coefficient relates a pair of random values. In the simplest case, generally valid in discrete realizations, the components are taken as independently distributed and have zero correlation. This assumption is not valid for

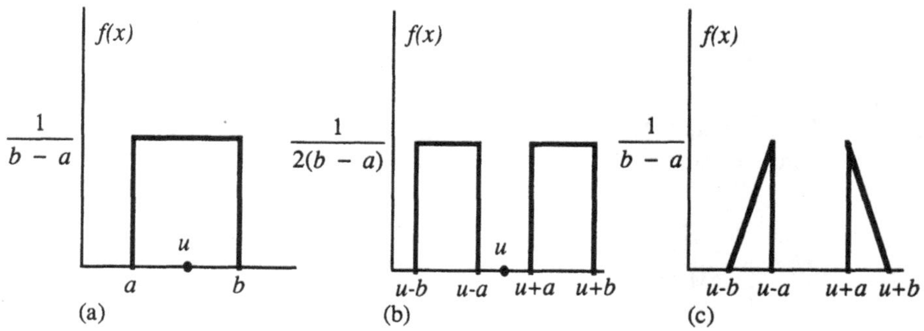

Figure 1.4 Typical component tolerance distributions [7], by permission of the publisher: (a) uniform distribution, (b) bimodal distribution, (c) split triangular distribution.

semiconductor device model parameter values or integrated circuit realizations. For components manufactured on the same semiconductor wafer, a multivariate normal distribution should be used to analyze component sensitivities.

1.10 POLYNOMIAL FILTERS

Polynomial filters generally refer to Butterworth, Tchebyscheff, inverse Tchebyscheff, and elliptic filters, because the poles and zeroes can be found directly from the roots of the denominator and numerator polynomials, respectively. The rest of this text will deal with the implementation of active, passive, digital, and switched capacitor polynomial filters. The term $G(s)$ will be used to denote the voltage transfer function as a ratio of polynomials in s. Because s is an imaginary variable, $G(s)$ will have a real component and an imaginary component at each frequency. The product $G(s)G(-s)$ is normally used to represent the squared magnitude of the voltage transfer function because $G(s)G(-s)$ is a real expression at all frequencies and is a function only of ω. The transfer function, $G(s)$, can be written in quadratic factored form as shown previously.

The term $G(s)$ is characterized by pole-zero values, magnitude response, phase response, group delay, impulse response, and step response. Each of the polynomial low-pass filters tries to approximate the ideal low-pass curve in Figure 1.5. For 1 dB of attenuation at cutoff frequency of 1000 Hz with 50 dB attenuation at 1100 Hz, a Butterworth polynomial would be of order $N = 68$, a Tchebyscheff polynomial would require an order of $N = 17$, and an elliptic filter would satisfy the requirements with an eighth-order filter.

All of the filters to be discussed in the next four chapters will be low-pass filters with a normalized cutoff frequency of 1 rad/sec, and equal source and load terminations (with the exception of the even-order Tchebyscheff filter). Impedance scaling, frequency scaling, and bandpass transformations will be discussed in later chapters. The characteristics for each of the polynomial filters are given in Table 1.2.

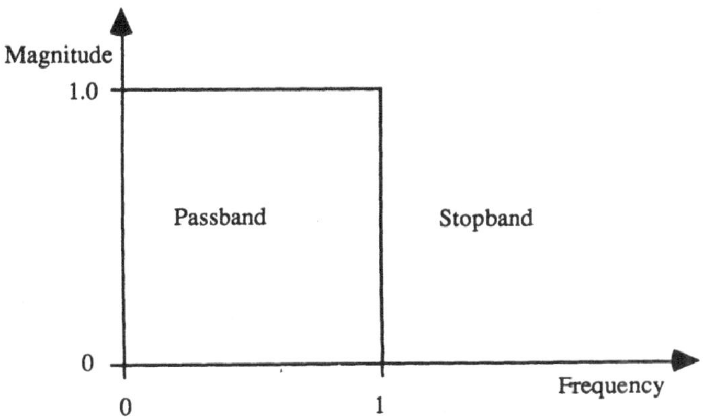

Figure 1.5 Ideal low-pass response.

Table 1.2
Polynomial Filters

Filter Type	*Characteristics*
Butterworth	Maximally flat response in passband and stopband, sharp attenuation response
Tchebyscheff	Ripple in passband, maximally flat in stopband, sharper roll-off than Butterworth
Inverse Tchebyscheff	Maximally flat in passband, ripple in stopband, exact same attenuation response as Tchebyscheff
Elliptic	Ripple in passband and stopband, roll-off sharper than Tchebyscheff

Figure 1.6 shows the bandshape responses for Butterworth, Tchebyscheff, inverse Tchebyscheff, and elliptic filters, each with exactly 3.0103 dB of attenuation at a cutoff frequency of 1 rad/sec. Each plotted filter response is a ninth-order filter with 60 dB attenuation in the stopband. A 3 dB passband ripple was used for all four cases, but could have been designed for any desired ripple. Both the Butterworth and the inverse Tchebyscheff filters are maximally flat in the passband. Although the inverse Tchebyscheff response never exceeds –60 dB in the stopband, at very high frequencies the Butterworth filter will give greater attenuation. The initial frequency where the Tchebyscheff and inverse Tchebyscheff filters first obtain 60 dB of attenuation is exactly the same. A Butterworth or Tchebyscheff low-pass passive implementation would require nine reactive elements (same as the filter order). The inverse Tchebyscheff and elliptic filter would both require 13 reactive elements for a passive realization (filter order times 1.5).

In Figure 1.7, the group delay is plotted for the same filters given in Figure 1.6. The filters with the sharpest roll-off have the highest group delay. It can be shown that for a minimum phase transfer function, the magnitude response uniquely determines the phase response (and the group delay) and vice versa. The Hilbert transform is used to relate magnitude and phase. If α and β are defined by

$$H(j\omega) = e^{\alpha + j\beta} \tag{1.47}$$

then

$$\alpha = \ln|H(j\omega)|, \qquad \beta = \angle H(j\omega)$$

where α is called the *attenuation function,* and β is the *phase function.* The attenuation and phase are related by [8]

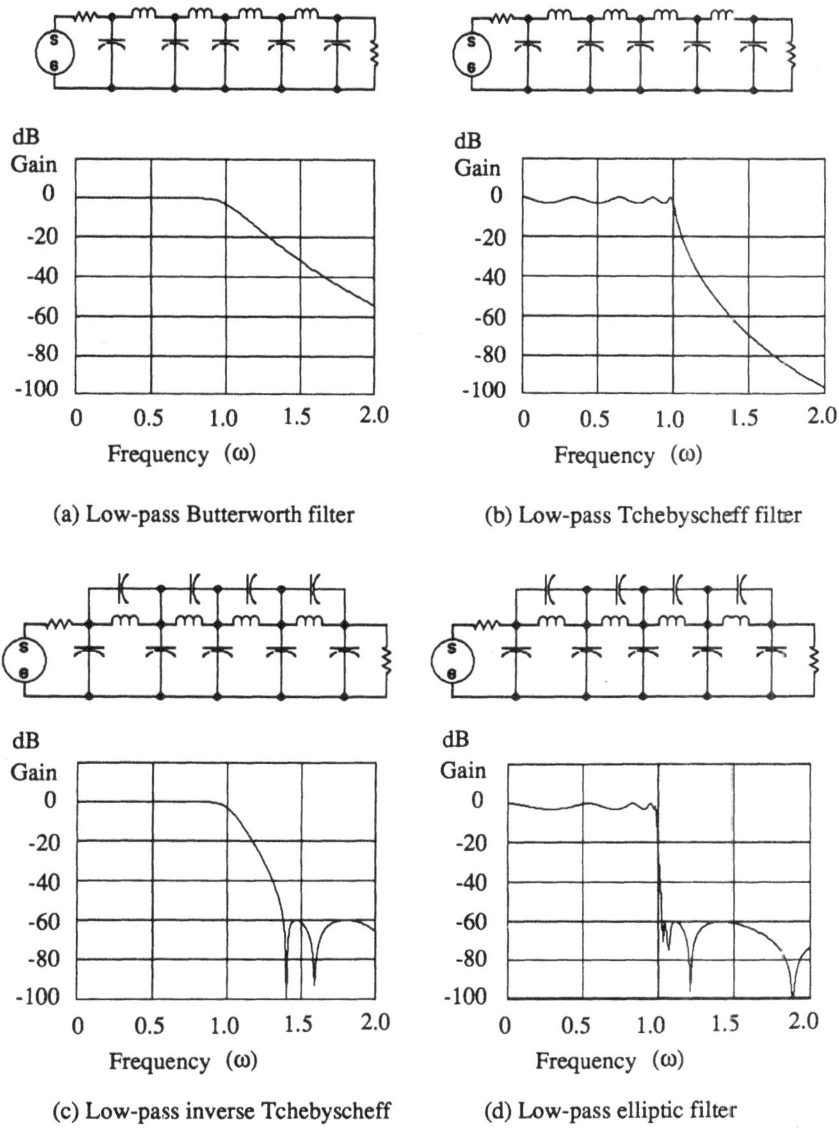

Figure 1.6 A comparison of four bandshape responses each of ninth order with 3.0103 dB of attenuation at the passband edge and 60 dB, or more, attenuation in the stopband.

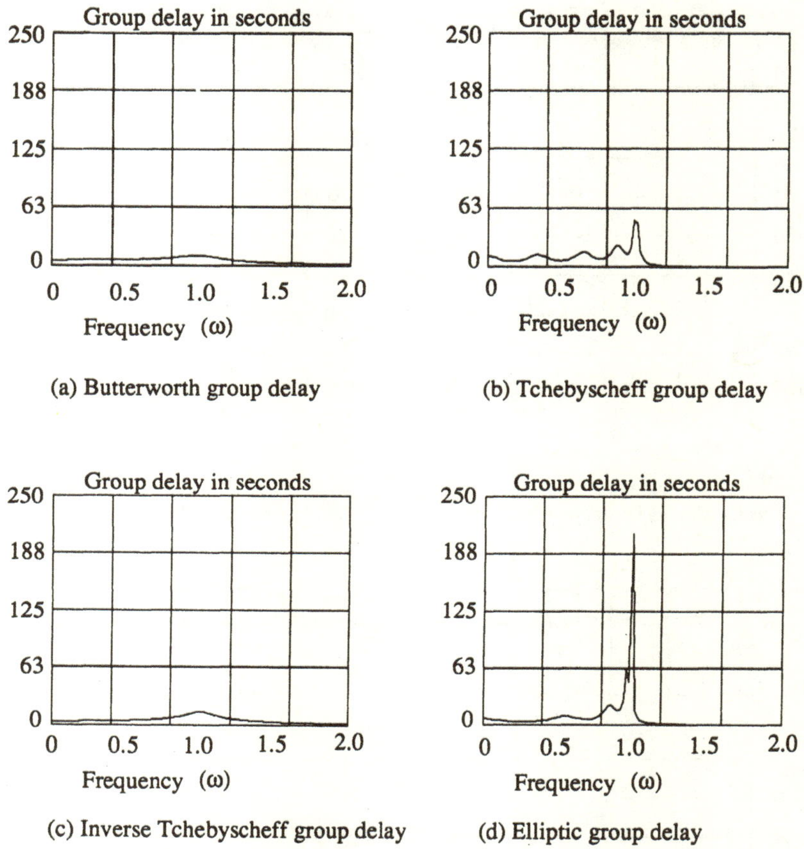

Figure 1.7 A comparison of four group delay responses for the filters in Figure 1.6.

$$\beta(\omega) = -\frac{2\omega}{\pi}\int_0^\infty \frac{\alpha(\xi)d\xi}{\omega^2 - \xi^2} \tag{1.48}$$

$$\alpha(\omega) = \alpha(0) + \frac{2\omega^2}{\pi}\int_0^\infty \frac{\beta(\xi)d\xi}{\xi(\omega^2 - \xi^2)} \tag{1.49}$$

Thus the magnitude of a minimum phase transfer function completely determines the phase and the phase determines the amplitude within a constant multiplier.

If an all-pass function in the continuous time s-domain is given by

$$G(s) = \frac{X(-s)}{X(s)} \tag{1.50}$$

then the group delay function can be expressed as

$$\text{Group delay} = \frac{2 \text{ Real part of } \left\{ X(s)\dfrac{dX(-s)}{ds} \right\}}{X(s)X(-s)} = \frac{N(s)}{D(s)} \tag{1.51}$$

As the denominator of (1.51) is a magnitude-squared function, it can be designed by standard optimization techniques. Equation (1.51) is often used as the starting point in the design of group delay equalizers using linearized approximation methods [9]. The mechanism of such an algorithm is dependent on the similarity between the desired group delay function and the denominator magnitude function of (1.51). The numerator, $N(s)$, is observed to be a smooth function over the approximating region, which warps the delay function of the denominator. The $N(s)$ is usually taken as unity, although $D(s)$ is first computed, then $N(s)$ is formed by Hurwitz factorization of $D(s)$ and the process is repeated until the group delay converges to an equiripple value.

REFERENCES

[1] Nahin, Paul J., "Behind the Laplace Transform," *IEEE Spectrum,* March 1991, pp. 60–61.

[2] Bracewell, Ronald N., "The Fourier Transform," *Scientific American,* June 1989, pp. 86–95.

[3] Maxwell, James Clerk, *A Treatise on Electricity and Magnetism,* Oxford London: University Press, 1873; reprinted, London: Lowe and Brydone Printers, 1946.

[4] Zobel, O., "Theory and Design of Electric Wave Filters," *Bell System Tech. J.*, January 1923.

[5] Darlington, S., "Synthesis of Reactance-Four-Poles Which Produce Prescribed Insertion Loss Characteristics," *J. Math. Phys.,* Vol. 30, September 1939, pp. 257–353.

[6] Cooley, J. W., and Tukey, J. W., "An Algorithm for the Machine Computation of Complex Fourier Series," *Math. Comp.,* Vol. 19, April 1965, pp. 297–301.

[7] Vlach, Jiri, and Singhal, Kishore, *Computer Methods for Circuit Analysis and Design,* New York: Van Nostrand and Reinhold, 1983.

[8] Guillemin, E. A., *Theory of Linear Physical Systems,* New York: John Wiley & Sons, 1963. pp. 554–558.

[9] Henderson, R. K., Ping, Li, and Sewell, J. I., "Extended Remez Algorithms for Filter Amplitude and Group Delay Approximation," *IEE Proc.-G,* Vol. 138, No. 3, June 1991.

PROBLEMS

1.1 Find the values for the ten reactive elements of an image parameter low-pass filter with a cutoff frequency of 1 rad/sec, one constant-K section, one M-derived section, and with a notch at $\omega = 1.25$ rad/sec in the M-derived section. Place the sections in the same order as given in Figure 1.1

1.2 Show that the group delay contribution of a pole at $s = -\alpha + j\beta$ is given by $\alpha/[\alpha^2 + (\omega - \beta)^2]$.

1.3 Show that the impulse response of $1/(s^5 + 3s^4 + 5s^3 + 5s^2 + 3s + 1)$ is given by $e^{-t} + 0.5773503\ e^{-0.5t} \sin(0.8660254\ t) - e^{-0.5t} \cos(0.8660254\ t) - 0.5773503\ t\ e^{-0.5t} \sin(0.8660254\ t) + 0.3849002\ e^{-0.5t}[\sin(0.8660254\ t) - 0.8660254\ t \cos(0.8660254\ t)]$

1.4 For the bandpass function given by

$$G(s) = \frac{6.362849(0.01877977s^2 + 39.725544)}{(s + 6.362849)(s^2 + 6.243356s + 39.72554)}$$

determine as a function of frequency
(a) the magnitude response,
(b) the phase response,
(c) the group delay,
(d) the impulse response, and
(e) the step response.

1.5 Find the response to an impulse and a step function for the sampled data system given by $1 + 2z^{-1} - z^{-2} + 3z^{-3}$.

1.6 Using equations (1.35) and (1.36), find the first ten harmonics of a square wave alternating between −1 and 1 volt, using only one sample point to represent the middle of each half cycle ($N = 2$, $m = 1, 2, \ldots, 10$).

1.7 Perform the numerical integration in example 1.2 and show that the solution is equal to $\delta(t) - e^{-t}$.

1.8 Derive the transfer function of the circuit in Figure 1.8 as a function of the component values given by $R_1 = 1$, $R_2 = 1$, $C_1 = C_3 = 1.551209$ farads, $C_2 = 0.7687431$ farads, and $L_1 = 0.6265367$ henries. Write a computer program that performs a Monte Carlo simulation of the magnitude response assuming that each component has a ±10% tolerance with a uniform distribution centered on the theoretical component values given above.

1.9 Derive equation (1.20) from (1.18) given that $[\cos(x)]^2 = 0.5 + 0.5\cos(2x)$.

1.10 Use equation (1.48) to show that an ideal low-pass filter (with infinite attenuation in the stopband) is physically unrealizable because it would have infinite delay.

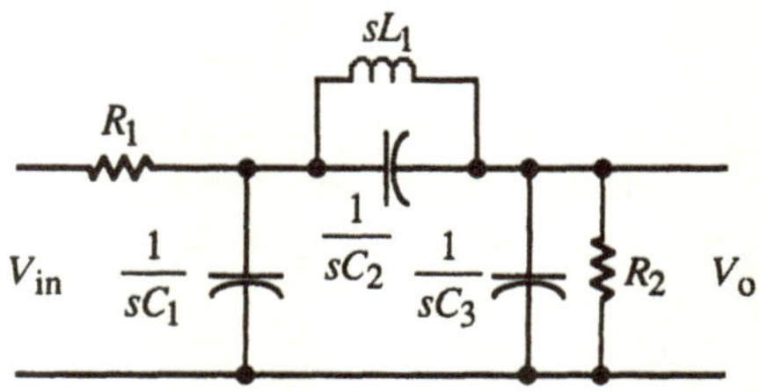

Figure 1.8 Passive bandpass circuit.

Chapter 2
The Low-Pass Butterworth Filter

2.1 THE BUTTERWORTH FUNCTION

The Butterworth filter tries to approximate the ideal low-pass response of Figure 1.5 with the equation $1/(1 + \omega^{2N})$ because

$$\frac{1}{1 + \omega^{2N}} \approx 1 \text{ for large } N \quad \text{and} \quad 0 < \omega < 1$$

$$\frac{1}{1 + \omega^{2N}} \approx 0 \text{ for large } N \quad \text{and} \quad \omega > 1 \tag{2.1}$$

2.2 POLE LOCATIONS

The pole locations of a Butterworth low-pass filter with a 3.0103 dB cutoff frequency, ω_c, of 1 rad/sec are equally spaced about the unit circle on the $j\omega$ plane as shown in Figure 2.1 for a fifth order low pass. These pole positions can be obtained by substituting $\omega = -js$ into equation (2.1) and taking the roots of the resulting polynomial in s that lie in the left half-plane. The pole locations can also be obtained [1] directly from the equation

$$-\sin\frac{(2K - 1)\pi}{2N} + j\cos\frac{(2K - 1)\pi}{2N}, \qquad K = 1, 2, \ldots, N \tag{2.2}$$

where N is the order of the filter. These are the lefthand poles of the unit circle. All attenuation characteristics for the Butterworth low-pass filter can be computed directly by

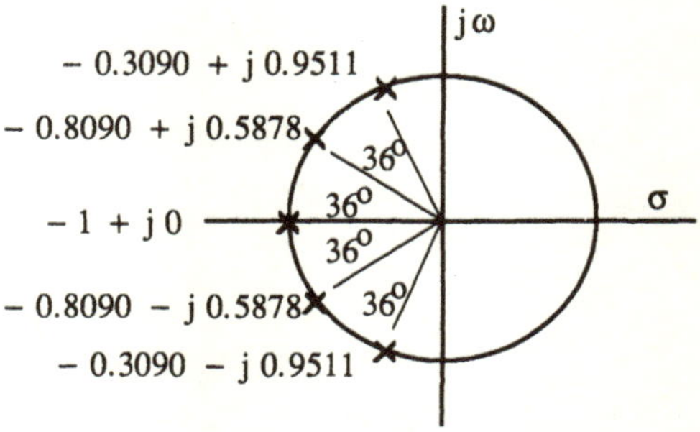

Figure 2.1 Pole locations for a fifth-order low-pass Butterworth filter are marked by the five X's and are equally spaced on the unit circle in the lefthand plane.

$$G(s)G(-s) = \frac{1}{1 + \omega^{2N}} = [\text{Mag}]^2 \text{ for } 0 < \omega < \infty \tag{2.3}$$

where ω rad/sec = $2\pi f$ in Hertz, $s = j\omega$, and $G(s)$ is the voltage transfer function of the Butterworth filter.

2.3 CALCULATION OF FILTER ORDER

The required filter order for an attenuation of R_s dB at ω_x (where $\omega_x > 1$) can be obtained from the basic Butterworth equation, given by

$$\frac{1}{1 + \omega_x^{2N}} = 10^{(-R_s/10)} \tag{2.4}$$

where $R_s = -10 \log[\text{Mag}]^2$ dB.

Butterworth filters do not necessarily have to be specified by their 3 dB cutoff frequency. If an attenuation of R_p dB is desired at the cutoff frequency, ω_c, of 1 rad/sec, then a frequency scale factor, C, can be introduced such that

$$\frac{1}{1 + \left(\dfrac{\omega_c}{C}\right)^{2N}} = 10^{-R_p/10} \tag{2.5}$$

The attenuation at ω_x can likewise be expressed as

$$\frac{1}{1 + \left(\dfrac{\omega_x}{C}\right)^{2N}} = 10^{-R_s/10} \tag{2.6}$$

where ω_x is the stopband frequency greater than ω_c, where the desired attenuation, R_s, is achieved. The required low-pass Butterworth filter order, N, can be found by solving equations (2.5) and (2.6) simultaneously to eliminate the variable C. For example, if an attenuation of 1 dB (R_p = 1) is required at the passband edge frequency of 1 rad/sec (ω_c = 1), and a stopband attenuation of 20 dB (R_s = 20) is required at the stopband edge frequency of 2 rad/sec (ω_x = 2) then the voltage magnitude at the cutoff frequency is given by

$$1 \text{ dB} = -20 \log[\text{Mag}], \qquad \text{or Mag} = 0.89125 \text{ (at } \omega = 1 \text{ rad/sec)} \tag{2.7}$$

At the cutoff frequency, the magnitude (voltage gain) is given by

$$20 \text{ dB} = -20 \log[\text{Mag}], \qquad \text{or Mag} = 0.1 \text{ (at } \omega = 2 \text{ rad/sec)} \tag{2.8}$$

Because R_p, R_s, and ω_c are known, the filter order, N, can be found from

$$\omega_x^{2N} = \frac{10^{R_s/10} - 1}{10^{R_p/10} - 1} \tag{2.9}$$

In general,

$$N = \frac{0.5 \cdot \log\left[\dfrac{10^{R_s/10} - 1}{10^{R_p/10} - 1}\right]}{\log\left[\dfrac{\omega_x}{\omega_c}\right]} \tag{2.10}$$

Example 2.1

If $N = 5$, calculate the gain at $\omega = 2$ rad/sec for a Butterworth filter with a 3 dB cutoff frequency of 1 rad/sec.

Solution

Substituting $N = 5$ and $\omega = 2$ into the equation for Butterworth attenuation, the voltage gain is

$$\frac{1}{1 + 2^{(2 \cdot 5)}} = 0.0009756 = [\text{Mag}]^2 = -30.1 \text{ dB}, \qquad \text{or}$$

$$V_o/V_{in} = [\text{Mag}] = 0.0312345$$

2.4 BUTTERWORTH ELEMENT VALUES

A simple expression [1] exists for calculating the element values of a Butterworth low-pass filter with equal 1 Ω source and load resistors. The equation

$$g_k = 2 \sin \frac{(2K - 1)\pi}{2N}, \qquad K = 1, 2, \ldots, N \tag{2.11}$$

can be applied to the four low-pass prototype topologies in Figure 2.2, where g_k represents the inductance in henries, or the capacitance in farads, depending on the component. This equation is valid only for 1 Ω terminating resistors and a 3 dB cutoff frequency, ω_c, of 1 rad/sec. Any other terminating resistors and cutoff frequencies can be realized by impedance scaling and frequency scaling the low-pass prototype.

Table 2.1 gives the component values for the Butterworth filters shown in Figure 2.2 as derived from equation (2.11). Inductor values are in henries, and capacitor values are in farads. These values are valid only for source and load terminations of 1 Ω and assume a 3 dB cutoff frequency of 1 rad/sec. The source and load terminations can be increased, or scaled, by increasing the *impedance* of all resistive, inductive, and capacitive elements by the same factor, which implies increasing inductor values and decreasing capacitor values. The cutoff frequency can be increased, or scaled, by any arbitrary constant by decreasing inductor values and decreasing capacitor values by the same constant. Resistors are not affected in frequency scaling because they are frequency independent devices.

Example 2.2

Derive the transfer function for a third-order Butterworth low-pass filter.

Solution

For $N = 3$, the Butterworth response is given by

$$G(s)G(-s) = \frac{1}{1 + \omega^6} = \frac{1}{1 - s^6} = \frac{1}{s^3 + 2s^2 + 2s + 1} \frac{1}{-s^3 + 2s^2 - 2s + 1}$$

The voltage transfer function is given by

$$G(s) = \frac{1}{s^3 + 2s^2 + 2s + 1} = \frac{1}{1 + s} \frac{1}{s^2 + s + 1}$$

There are three zeroes at infinity, indicating the order of the numerator is three less than the order of the denominator, and the poles are located at

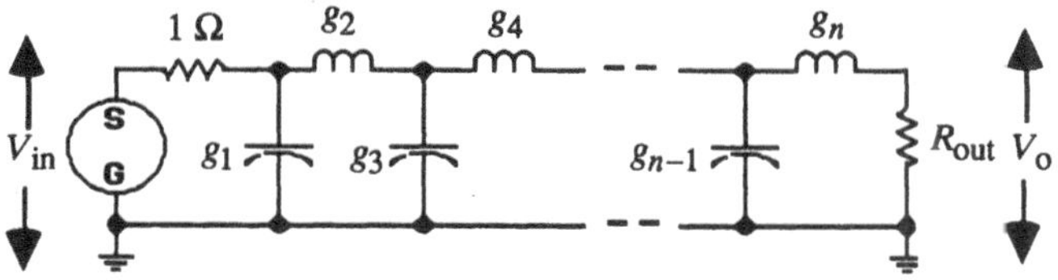

(a) Shunt capacitor configuration with N even

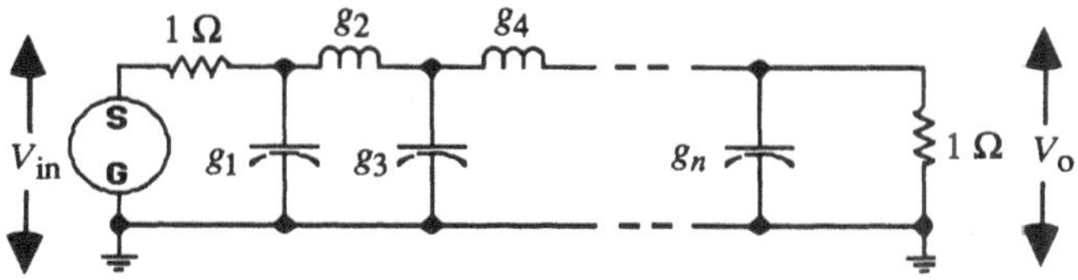

(b) Shunt capacitor configuration with N odd

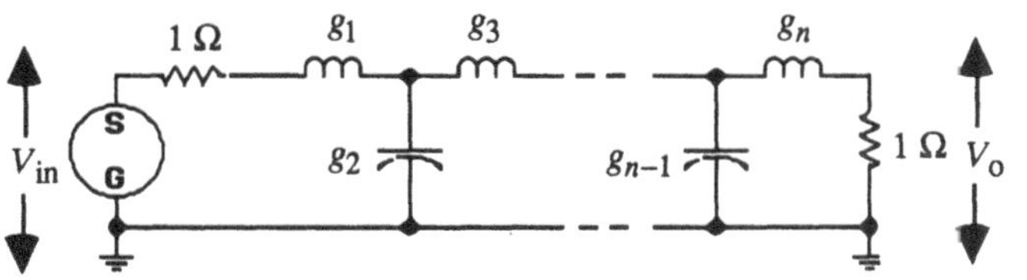

(c) Series inductor configuration with N odd

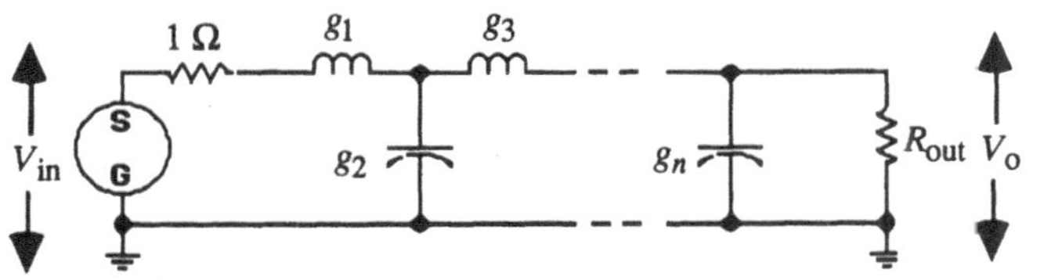

(d) Series inductor configuration with N even.

Figure 2.2 Passive Butterworth and Tchebyscheff low-pass prototype topologies with equal terminating resistors. Capacitor values are in farads. Inductor values are in henries. $R_{out} = 1\ \Omega$ if the filter is Butterworth.

$$s = -1 + j0$$

$$s = -0.500000 \pm j0.866025$$

Alternatively, the poles could be found by applying a polynomial root finder to the denominator polynomial $1 - s^6$ and using only those roots with negative real coefficients.

Table 2.1
Butterworth Component Values in Farads and Henries

N	g_1	g_2	g_3	g_4	g_5	g_6	g_7	g_8	g_9	g_{10}
1	2.000									
2	1.414	1.414								
3	1.000	2.000	1.000							
4	0.765	1.848	1.848	0.765						
5	0.618	1.618	2.000	1.618	0.618					
6	0.517	1.414	1.932	1.932	1.414	0.517				
7	0.445	1.247	1.802	2.000	1.802	1.247	0.445			
8	0.390	1.111	1.663	1.962	1.962	1.663	1.111	0.390		
9	0.347	1.000	1.532	1.879	2.000	1.879	1.532	1.000	0.347	
10	0.313	0.908	1.414	1.782	1.975	1.975	1.782	1.414	0.908	0.313

REFERENCE

[1] Williams, Arthur B., and Taylor, Fred J. *Electronic Filter Design Handbook*, New York: McGraw-Hill, 1988, pp. 2–34.

PROBLEMS

2.1 Find the degree, N, for a normalized Butterworth filter with 3 dB attenuation at $\omega = 1$ and 80 dB attenuation at $\omega = 2$.

2.2 Find the degree, N, for a Butterworth filter with 1 dB attenuation at $\omega = 1$ and 80 dB attenuation at $\omega = 2$.

2.3 Find the pole locations for a normalized tenth-order Butterworth low pass.

2.4 Find the transfer function, $G(s)$, for a normalized fourth-order Butterworth low pass.

2.5 Find the component values for a 12th-order normalized passive Butterworth low pass.

2.6 Plot the magnitude response, phase response, group delay, impulse, and step response for a tenth-order Butterworth low pass.

2.7 Perform a worst case analysis of the magnitude response of a tenth-order passive Butterworth low pass using a Monte Carlo simulation for

(a) a 10% uniformly distributed variation in the real part of each pole,

(b) a 10% uniformly distributed variation in the value of the inductors and capacitors.

Use at least 100 Monte Carlo runs. What is the maximum deviation at $\omega = 1$?

Chapter 3
The Low-Pass Tchebyscheff Filter

3.1 THE TCHEBYSCHEFF FUNCTION

The Tchebyscheff filter tries to approximate the ideal low-pass response of Figure 1.5 with the equation

$$\frac{1}{1 + \epsilon^2[T_N(\omega)]^2} \approx 1 \qquad \text{for } 0 < \omega \leq 1$$

$$\frac{1}{1 + \epsilon^2[T_N(\omega)]^2} \approx 0 \qquad \text{for } \omega > 1 \tag{3.1}$$

where $T_N(\omega)$ is a Tchebyscheff polynomial of order N. The term $T_N(\omega)$ has the characteristic such that

$$-1 \leq T_N(\omega) \leq 1 \qquad \text{for } -1 \leq \omega \leq 1$$

$$|T_N(\omega)| > 1 \qquad \text{for } |\omega| > 1 \tag{3.2}$$

The equiripple character of the Tchebyscheff polynomials, as shown in Figure 3.1 makes them highly desirable in filter approximation. The equation

$$\boldsymbol{G(s)G(-s) = \frac{1}{1 + \epsilon^2[T_N(\omega)]^2} = [\text{Mag}]^2} \tag{3.3}$$

defines a Tchebyscheff low-pass filter of order N, where $T_N(\omega)$ is given by

$$\boldsymbol{T_N(\omega) = \cos[N \arccos(\omega)] \qquad \text{for } 0 \leq \omega \leq 1} \tag{3.4}$$

$$T_N(\omega) = \cosh[N\, \text{arccosh}(\omega)] \qquad \text{for } \omega > 1 \tag{3.5}$$

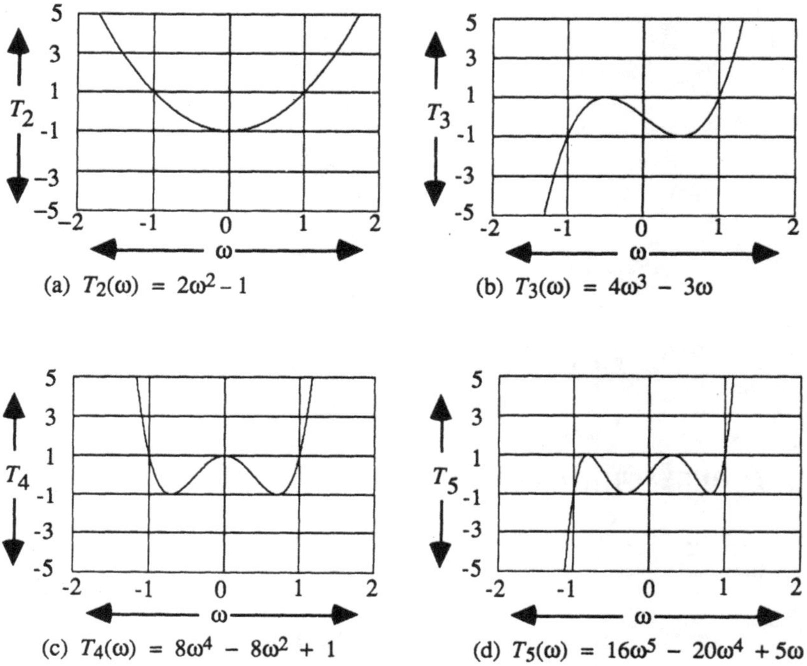

Figure 3.1 Tchebyscheff polynomials, $T_N(\omega)$, for N = 2, 3, 4, and 5.

where

$$
\begin{aligned}
\boldsymbol{T_0(\omega)} &= \mathbf{1} \\
\boldsymbol{T_1(\omega)} &= \boldsymbol{\omega} \\
T_2(\omega) &= 2\omega^2 - 1 \\
T_3(\omega) &= 4\omega^3 - 3\omega \\
T_4(\omega) &= 8\omega^4 - 8\omega^2 + 1 \\
T_5(\omega) &= 16\omega^5 - 20\omega^3 + 5\omega \\
T_6(\omega) &= 32\omega^6 - 48\omega^4 + 18\omega^2 - 1 \\
T_7(\omega) &= 64\omega^7 - 112\omega^5 + 56\omega^3 - 7\omega \\
T_8(\omega) &= 128\omega^8 - 256\omega^6 + 160\omega^4 - 32\omega^2 + 1 \\
T_9(\omega) &= 256\omega^9 - 576\omega^7 + 432\omega^5 - 120\omega^3 + 9\omega \\
T_{10}(\omega) &= 512\omega^{10} - 1280\omega^8 + 1120\omega^6 - 400\omega^4 + 50\omega^2 - 1, \qquad \text{and so on}
\end{aligned}
\tag{3.6}
$$

The recursive formula

$$\boldsymbol{T_{N+1}(\omega) = 2\omega T_N(\omega) - T_{N-1}(\omega)} \tag{3.7}$$

will generate Tchebyscheff polynomials of any order given $T_0(\omega)$ and $T_1(\omega)$.

It is easy to show that the trigonometric equation (3.4) produces the polynomials shown in (3.6). If

$$T_N(\omega) = \cos[N \arccos(\omega)] \tag{3.8}$$

then let

$$\arccos(\omega) = \phi, \quad \text{or} \quad \omega = \cos(\phi) \tag{3.9}$$

and the Tchebyscheff polynomials can be written, using trigonometric expansion, as

$$T_1(\omega) = \cos(\phi) = \omega$$

$$T_2(\omega) = \cos(2\phi) = 2\cos^2(\phi) - 1 = 2\omega^2 - 1 \tag{3.10}$$

$$T_3(\omega) = \cos(3\phi) = 2\cos(\phi)[\cos(2\phi)] - \cos(\phi) = 4\omega^3 - 3\omega, \qquad \text{and so on}$$

3.2 POLE LOCATIONS

The passband ripple, R_p, in dB can be related to ϵ by (3.3). Because $T_N(\omega)^2$ varies from 0 to 1, the maximum ripple occurs when $T_N(\omega)^2$ is equal to 1. Because the ripple, R_p, in dB is equal to $-10 \log[\text{Mag}]^2$, ϵ can be found by solving the equation

$$\frac{1}{1 + \epsilon^2} = 10^{-R_p/10} \tag{3.11}$$

or

$$\epsilon = \sqrt{10^{(0.1 \cdot R_p)} - 1} \tag{3.12}$$

where R_p is specified as a positive number. Once the variable ϵ has been found, the Tchebyscheff filter function, $G(s)$, can be found by substituting the appropriate Tchebyscheff polynomial for $T_N(\omega)$ into (3.3), expanding the denominator, making the substitution $\omega = -js$, and using a polynomial root finder to select only those roots of the denominator with negative real parts.

Pole locations for Tchebyscheff low-pass filters fall on an ellipse around the origin as shown in Figure 3.2.

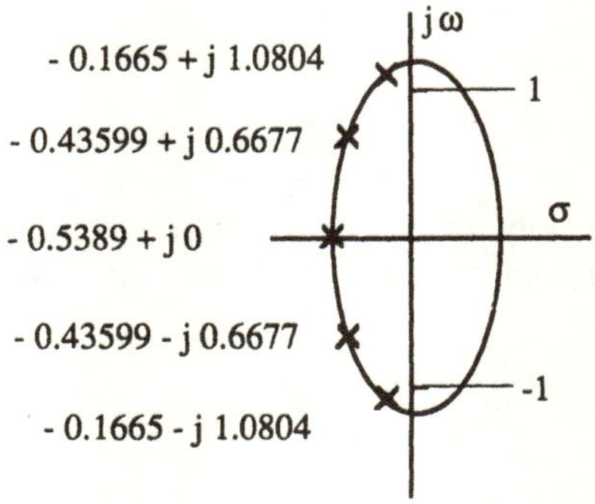

Figure 3.2 Pole locations for a Tchebyscheff 5th-order lowpass, with 0.1 dB passband ripple and a normalized cutoff frequency of 1 rad/sec.

There is an alternate procedure [1] that allows the poles of a Tchebyscheff filter to be found from the poles of a Butterworth filter. First define α such that

$$\alpha = \frac{1}{\epsilon} + \sqrt{1 + \frac{1}{\epsilon^2}} \tag{3.13}$$

The equations that transform the Butterworth poles to Tchebyscheff poles are given by

$$A = 0.5 \cdot \alpha^{(1/N)} - 0.5 \cdot \alpha^{(-1/N)}$$

$$B = 0.5 \cdot \alpha^{(1/N)} + 0.5 \cdot \alpha^{(-1/N)} \tag{3.14}$$

The real part of each Butterworth pole is multiplied by A, and the imaginary part of each Butterworth pole is multiplied by B.

The required filter order, N, to achieve a certain attenuation in the stopband can be computed if the passband ripple and the desired stopband attenuation are given. From Euler's equation

$$e^{jx} = \cos(x) + j\sin(x) \tag{3.15}$$

$$e^{-jx} = \cos(x) - j\sin(x) \tag{3.16}$$

Adding equations (3.15) and (3.16) and dividing by 2 yields

$$\frac{e^{jx} + e^{-jx}}{2} = \cos(x) \tag{3.17}$$

Now let $\cos(x) = \omega$,

$$\frac{e^{jx} + e^{-jx}}{2} = \omega \tag{3.18}$$

or

$$e^{jx} + e^{-jx} = 2\omega \tag{3.19}$$

Multiplying (3.19) through by e^{jx} yields

$$e^{2jx} - 2\omega e^{jx} + 1 = 0 \tag{3.20}$$

which is a quadratic equation in e^{jx}. Using the quadratic formula and solving for e^{jx},

$$e^{jx} = \omega + \sqrt{\omega^2 - 1} \tag{3.21}$$

or

$$x = -j\ \ln(\omega + \sqrt{\omega^2 - 1}), \qquad \omega > 1 \tag{3.22}$$

Since $\cos(x) = \omega$,

$$\mathbf{arccos(\boldsymbol{\omega}) = -j\ ln(\boldsymbol{\omega} + \sqrt{\boldsymbol{\omega}^2 - 1})}, \qquad \omega > 1 \tag{3.23}$$

which provides the necessary equation needed to calculate the required filter order, N, because ω is outside the passband.

3.3 CALCULATION OF FILTER ORDER

Case 1

Given the passband ripple and stopband attenuation at frequency ω_x, the required filter order, N, can be obtained from (3.3) and (3.4) to yield

$$N = \frac{\arccos\left(\sqrt{\dfrac{1 - [\mathrm{Mag}]^2}{[\mathrm{Mag}]^2\epsilon^2}}\right)}{\arccos(\omega_x)} \tag{3.24}$$

Using the results of (3.23),

$$N = \frac{\ln\left(\sqrt{\frac{1 - [\mathrm{Mag}]^2}{[\mathrm{Mag}]^2\epsilon^2}} + \sqrt{\left[\frac{1 - [\mathrm{Mag}]^2}{[\mathrm{Mag}]^2\epsilon^2}\right] - 1}\right)}{\ln\left[\omega_x + \sqrt{\omega_x^2 - 1}\right]} \tag{3.25}$$

Example 3.1

Find the order, N, to realize a Tchebyscheff low-pass filter with 0.1 dB bandpass ripple, cutoff frequency of 1 rad/sec, and 50 dB attenuation at 1.1 rad/sec.

Solution

Because $[\mathrm{Mag}]^2 = 10^{-50/10} = 0.00001$, ϵ^2 can be calculated from equation (3.12) as

$$\epsilon^2 = 10^{0.01} - 1 = 0.0232929$$

$$\sqrt{\frac{1 - [\mathrm{Mag}]^2}{[\mathrm{Mag}]^2\epsilon^2}} = 2071.99$$

$$N = \frac{\ln\left[2071.99 + \sqrt{(2071.99)^2 - 1}\right]}{\ln\left(1.1 + \sqrt{1.1^2 - 1}\right)} = 18.78$$

Choose N to be the next largest integer, or $N = 19$.

Case 2

Given the passband ripple, the filter order, N, and the required attenuation of the filter in the stopband, the first frequency, ω_x, where that attenuation occurs can be found by solving (3.3) for ω_x such that

$$\omega_x = \cos\left[\frac{1}{N}\arccos\left(\sqrt{\frac{1 - [\mathrm{Mag}]^2}{[\mathrm{Mag}]^2\epsilon^2}}\right)\right] \tag{3.26}$$

Let

$$k = \sqrt{\frac{1 - [\text{Mag}]^2}{[\text{Mag}]^2\epsilon^2}}$$

Using (3.17) and (3.23) and rearranging terms yields

$$\omega_x = \frac{e^{(1/N)\ln(k+\sqrt{k^2-1})} + e^{-(1/N)\ln(k+\sqrt{k^2-1})}}{2} \tag{3.27}$$

Example 3.2

A Tchebyscheff lowpass filter of 19th order has 0.1 dB bandpass ripple, cutoff frequency of 1 rad/sec. Find the first stopband frequency where the attenuation is 50 dB.

Solution

Because $[\text{Mag}]^2 = 10^{-50/10} = 0.00001$, ϵ^2 can be calculated from (3.12) as

$$\epsilon^2 = 10^{0.01} - 1 = 0.0232929$$

$$\sqrt{\frac{1 - [\text{Mag}]^2}{[\text{Mag}]^2\epsilon^2}} = 2071.99$$

$$\omega_x = \frac{e^{(1/19)\ln(4143.98)} + e^{-(1/19)\ln(4143.98)}}{2} = 1.097 \text{ rad/sec}$$

Case 3

Given the passband ripple, the filter order, N, and a frequency, ω_x, in the stopband, the attenuation at ω_x can be found by substituting (3.4) into (3.3) so that

$$\frac{1}{1 + \epsilon^2\{\cos[N \arccos(\omega)]\}^2} = [\text{Mag}]^2 \tag{3.28}$$

Using (3.17) and (3.23) yields

$$\frac{1}{1 + \epsilon^2\left[\dfrac{e^{N\ln(\omega_x+\sqrt{\omega_x^2-1})} + e^{-N\ln(\omega_x+\sqrt{\omega_x^2-1})}}{2}\right]^2} = [\text{Mag}]^2 \tag{3.29}$$

The dB attenuation at ω_x will be

$$\text{dB attenuation at } \omega_x = -20 \log[\text{Mag}] = -10 \log[\text{Mag}]^2 \tag{3.30}$$

Example 3.3

Find the magnitude response of a Tchebyscheff low-pass filter of 19th order with 0.1 dB bandpass ripple and a cutoff frequency of 1 rad/sec at a frequency of 1.1 rad/sec.

Solution

ϵ^2 can be calculated from equation (3.12) as

$$\epsilon^2 = 10^{0.01} - 1 = 0.0232929$$

Direct substitution into (3.29) yields

$$\frac{1}{1 + 0.0232929\left[\dfrac{e^{19\ln(1.1+\sqrt{1.1^2-1})} + e^{-19\ln(1.1+\sqrt{1.1^2-1})}}{2}\right]^2} = [\text{Mag}]^2 = 0.00000821$$

or −50.86 dB.

3.4 TCHEBYSCHEFF ELEMENT VALUES

A closed form set of equations exists for computing the element values of a passive Tchebyscheff low-pass filter and are given in Table 3.1.

Note that the source and load terminations are 1 Ω except in the case of an even-order Tchebyscheff filter, where the value of the load termination is given by R_{out}. Component values for the topologies in Figure 2.2 are given by the values of $g(I)$ in Table 3.1, with L–C values listed in Table 3.2.

REFERENCE

[1] Geffe, Philip R., "Designers' Guide to Active Bandpass Filters (Part 2)," *EDN Magazine*, March 5, 1974, p. 40.

Table 3.1
Tchebyscheff Element Values (reference Figure 2.2a for even order or Figure 2.2b or 2.2c for odd order)

Start: For $I = 1$ to N — N is the Tchebyscheff filter order

$$X_1 = 10^{R_s/10} - 1$$

R_s is passband ripple and is a positive number

$$X_2 = \sqrt{X_1}$$

$$\beta = \frac{1}{N}\ln\left[\frac{1}{X_2} + \sqrt{\left(\frac{1}{X_2}\right)^2 + 1}\right]$$

$$\phi = (2I - 1)\frac{\pi}{2N}$$

$$A_k = \sin(\phi)$$

$$\sinh = 0.5\ e^{\beta} - 0.5\ e^{-\beta}$$

$$B_k = \sinh^2 + \sin^2\left(I\frac{\pi}{N}\right)$$

If $I = 1$ then $g(I) = 2\,\dfrac{A_k}{\sinh}$

If $I > 1$ then $g(I) = 4\,A_k A_{km1}\dfrac{1}{g(I-1)\cdot B_{km1}}$

$$A_{km1} = A_k$$

$$B_{km1} = B_k$$

Loop: Next I

$$\text{Res} = \sqrt{X_1 + 1}$$

If N is even, then $R_{out} = 2\ \text{Res}^2 - 2\ \text{Res}\sqrt{\text{Res}^2 - 1} - 1$

If N is odd, then $R_{out} = 1$

PROBLEMS

3.1 Find the degree, N, of a normalized Tchebyscheff filter with 1 dB passband ripple at $\omega = 1$ and 80 dB attenuation at $\omega = 2$.

3.2 Find $T_{11}(\omega)$, $T_{12}(\omega)$, $\log_{10}(-3)$, $\arccos(10)$.

3.3 Find the element values for a normalized Tchebyscheff 12th-order low pass with 0.5 dB passband ripple.

Table 3.2
Tchebyscheff Component Values for Passband Ripples (reference Figure 2.2a for even order, or Figure 2.2b or 2.2c for odd order)

N	R_{out}	g_1	g_2	g_3	g_4	g_5	g_6	g_7	g_8	g_9	g_{10}
1 dB											
1	1.000	1.018									
2	0.376	1.822	0.685								
3	1.000	2.024	0.994	2.024							
4	0.376	2.099	1.064	2.831	0.789						
5	1.000	2.135	1.091	3.001	1.091	2.135					
6	0.376	2.155	1.104	3.063	1.152	2.937	0.810				
7	1.000	2.167	1.112	3.094	1.174	3.094	1.112	2.167			
8	0.376	2.174	1.116	3.111	1.184	3.149	1.170	2.969	0.818		
9	1.000	2.180	1.119	3.121	1.190	3.175	1.190	3.121	1.119	2.180	
10	0.376	2.184	1.121	3.129	1.193	3.189	1.199	3.174	1.176	2.982	0.821
0.1 dB											
1	1.000	0.305									
2	0.738	0.843	0.622								
3	1.000	1.032	1.147	1.032							
4	0.738	1.109	1.306	1.770	0.818						
5	1.000	1.147	1.371	1.975	1.371	1.147					
6	0.738	1.168	1.404	2.056	1.517	1.903	0.862				
7	1.000	1.181	1.423	2.097	1.573	2.097	1.423	1.181			
8	0.738	1.190	1.435	2.120	1.601	2.170	1.564	1.944	0.878		
9	1.000	1.196	1.443	2.135	1.617	2.205	1.617	2.135	1.443	1.196	
10	0.738	1.200	1.448	2.144	1.627	2.225	1.642	2.205	1.582	1.963	0.885

3.4 Find the Q and resonant frequency of each quadratic section of a 12th-order Tchebyscheff low pass with 1.0 dB passband ripple.

3.5 A normalized 12th-order Tchebyscheff low pass has 0.001 dB passband ripple. Find the stopband frequency where the attenuation is 50 dB.

3.6 Find the order, N, of a normalized Tchebyscheff low pass with 3 dB passband ripple that has the same stopband attenuation as a 10th-order Butterworth filter at $\omega = 2$.

3.7 Perform a worst case analysis of the magnitude response of a 10th-order passive Tchebyscheff low pass with 1.0 dB passband ripple using a Monte Carlo simulation for

(a) a 10% uniformly distributed variation in the real part of each pole,

(b) a 10% uniformly distributed variation in the value of the inductors and capacitors.

Use at least 100 Monte Carlo runs. What is the maximum deviation at $\omega = 1$?

Chapter 4
The Low-Pass Inverse Tchebyscheff Filter

4.1 THE INVERSE TCHEBYSCHEFF FUNCTION

The inverse Tchebyscheff filter is sometimes called a Tchebyscheff Type II filter. It differs from a normal, or Type I, Tchebyscheff in that the passband is maximally flat, with ripple in the stopband. The voltage transfer function for the inverse Tchebyscheff is given by

$$G(s)G(-s) = \frac{\epsilon^2\left[T_N\left(\frac{1}{\omega}\right)\right]^2}{1 + \epsilon^2\left[T_N\left(\frac{1}{\omega}\right)\right]^2} = [\text{Mag}]^2 \tag{4.1}$$

The passband and stopband are defined for equation (4.1) as

Passband: $0 \leq \omega \leq 1/\omega_x$, where $1/\omega_x < 1$

Stopband: $1 \leq \omega \leq \infty$

The minimum stopband attenuation, R_s, occurs when $T_N^2\ (1/\omega) = 1$; therefore,

$$\frac{\epsilon^2}{1 + \epsilon^2} = [\text{Mag}]^2 = 10^{-R_s/10} \tag{4.2}$$

or

$$\epsilon^2 = \frac{10^{-R_s/10}}{1 - 10^{-R_s/10}} \tag{4.3}$$

where R_s is given as a positive number (usually in the range of 40 dB to 80 dB). Using (4.1) to solve for $1/\omega_x$, where R_p is a positive number that represents the passband ripple (usually in the range of 0.001 dB to 3 dB)

$$\frac{\epsilon^2[T_N(\omega_x)]^2}{1 + \epsilon^2[T_N(\omega_x)]^2} = 10^{-R_p/10} \tag{4.4}$$

$$[T_N(\omega_x)]^2 = \frac{\frac{1}{\epsilon^2}10^{-R_p/10}}{1 - 10^{-R_p/10}} \tag{4.5}$$

From equation (3.8)

$$\cos[N \arccos(\omega_x)] = \sqrt{\frac{\frac{1}{\epsilon^2}10^{-R_p/10}}{1 - 10^{-R_p/10}}} \tag{4.6}$$

Let

$$K = \sqrt{\frac{\frac{1}{\epsilon^2}10^{-R_p/10}}{1 - 10^{-R_p/10}}} \tag{4.7}$$

using equations (3.17) and (3.23)

$$\omega_x = \frac{e^{(1/N)\ln(K+\sqrt{K^2-1})} + e^{-(1/N)\ln(K+\sqrt{K^2-1})}}{2} \tag{4.8}$$

Equation (4.1) can now be frequency scaled such that

$$\boldsymbol{G(s)G(-s) = \frac{\epsilon^2\left[T_N\left(\frac{\omega_x}{\omega}\right)\right]^2}{1 + \epsilon^2\left[T_N\left(\frac{\omega_x}{\omega}\right)\right]^2} = [\mathrm{Mag}]^2} \tag{4.9}$$

where passband and stopband are redefined for

Passband: $0 \leq \omega \leq 1$

Stopband: $\omega_x \leq \omega \leq \infty, \qquad$ where $\omega_x > 1$

4.2 POLE AND ZERO LOCATIONS

The poles and zeroes of the inverse Tchebyscheff function, $G(s)$, can now be found by expanding the numerator and denominator of equation (4.9) and applying a polynomial root finder. There is a closed form method [1] of finding the poles and zeroes that avoids the numerical instabilities associated with iterative root finding techniques. By analogy with the Tchebyscheff Type I case, first define α such that

$$\alpha = \frac{1}{\epsilon} + \sqrt{1 + \frac{1}{\epsilon^2}} \tag{4.10}$$

The equations that transform the Butterworth poles to inverse Tchebyscheff poles are given by

$$A = 0.5 \cdot \alpha^{(1/N)} - 0.5 \cdot \alpha^{(-1/N)}$$

$$B = 0.5 \cdot \alpha^{(1/N)} + 0.5 \cdot \alpha^{(-1/N)} \tag{4.11}$$

The real part of each Butterworth pole is multiplied by A, and the imaginary part of each Butterworth pole is multiplied by B. If $g_R(I)$ denotes the real part of a Butterworth pole, and $g_I(I)$ denotes the imaginary part, then from (2.2),

$$g_R(I) = -\sin\frac{(2I - 1)\pi}{2N}, \qquad I = 1, 2, \ldots, N \tag{4.12}$$

$$g_I(I) = \cos\frac{(2I - 1)\pi}{2N}, \qquad I = 1, 2, \ldots, N \tag{4.13}$$

Multiplying by the constants A and B, the intermediate products become

$$g'_R(I) = A \cdot g_R(I), \qquad I = 1, 2, \ldots, N \tag{4.14}$$

$$g'_I(I) = B \cdot g_I(I), \qquad I = 1, 2, \ldots, N \tag{4.15}$$

The analogy with the Tchebyscheff Type I case departs at this point because the denominator of (4.1) is a polynomial in $1/\omega$ (instead of ω) and has an implied scale factor of ω_x. It is necessary to take the complex reciprocal of the intermediate values given in (4.14) and (4.15) and frequency scale by ω_x. The final poles for the inverse Tchebyscheff filter become,

$$P_R(I) = \text{Real pole } (I) = \omega_x \frac{g'_R(I)}{g'_R(I)^2 + g'_I(I)^2}, \qquad I = 1, 2, \ldots, N \tag{4.16}$$

$$P_I(I) = \text{Imaginary pole } (I) = \omega_x \frac{-g'_I(I)}{g'_R(I)^2 + g'_I(I)^2}, \qquad I = 1, 2, \ldots, N \tag{4.17}$$

The zeroes can be found by realizing that the numerator of (4.1) is a Tchebyscheff polynomial. The zero locations of a Tchebyscheff polynomial are real and given by

$$\mathbf{X(I) = \cos\frac{(2I - 1)\pi}{2N}, \qquad I = 1, 2, \ldots, N} \tag{4.18}$$

Frequency scaling by ω_x and taking the reciprocals of the values of $X(I)$ to obtain the final zeroes of $G(s)$ yields

$$Z_I(I) = \text{Imaginary zero } (I) = \omega_x \frac{1}{X(I)}, \qquad I = 1, 2, \ldots, N \qquad \text{for } N \text{ even} \tag{4.19}$$

$$Z_I(I) = \text{Imaginary zero } (I) = \omega_x \frac{1}{X(I)},$$

$$I = 1, 2, \ldots \frac{N+1}{2} - 1, \frac{N+1}{2} + 1 \ldots, N, \tag{4.20}$$

for N odd

Because the zeroes lie on the $j\omega$ axis, their real parts, $Z_R(I)$, are zero. If N is odd, the middle term in (4.20) is skipped because an odd-order inverse Tchebyscheff low-pass filter always has one zero at infinity (one less zero than pole).

4.3 CALCULATION OF FILTER ORDER

Given a normalized inverse Tchebyscheff with cutoff frequency, $\omega_c = 1$ rad/sec, a maximum passband attenuation of R_p, a minimum stopband attenuation of R_s, and the first frequency in the stopband, ω_x at which R_s occurs, the computation of the required filter order, N, can be obtained by solving (4.1) for N and is analogous to the Tchebyscheff Type I case. Define

$$K = \sqrt{\frac{\frac{1}{\epsilon^2}[\text{Mag}]^2}{1 - [\text{Mag}]^2}} \tag{4.21}$$

where ϵ^2 is defined in (4.3). The magnitude must be taken in the passband because the stopband ripple, R_s, was used to define ϵ^2; therefore,

$$[\text{Mag}]^2 = 10^{-R_p/10} \tag{4.22}$$

Then

$$N = \frac{\ln\left(K + \sqrt{K^2 - 1}\right)}{\ln\left[\frac{\omega_x}{\omega_c} + \sqrt{\left(\frac{\omega_x}{\omega_c}\right)^2 - 1}\right]} \tag{4.23}$$

with $\omega_c = 1$ in this derivation. It can be shown that (4.23) for calculating N is identical to (3.25) for calculating N for the Tchebyscheff Type I case. The frequency in the stopband, ω_x, where the desired stopband attenuation is first obtained and given by (4.8), is also exactly the same as (3.27) for the Tchebyscheff Type I case. Figure 4.1 illustrates this graphically.

Example 4.1

Find the transfer function, $G(s)$, for a fifth-order inverse Tchebyscheff low-pass filter with 1 dB passband attenuation, cutoff frequency of 1 rad/sec, and a minimum attenuation of 50 dB in the stopband. What is the amplitude response at 2 rad/sec?

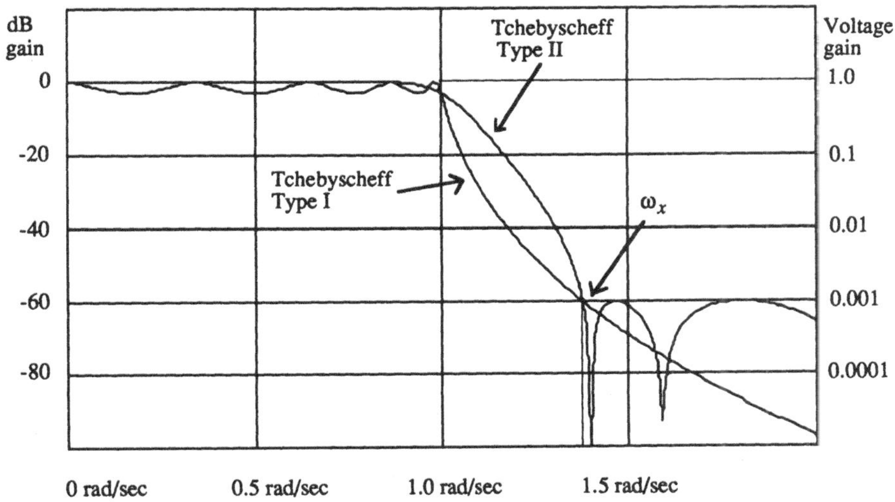

Figure 4.1 Tchebyscheff and inverse Tchebyscheff ninth-order low-pass filters.

Solution

The function, $G(s)$, can be found two ways. In Method 1, solving for the missing parameters,

$$\epsilon^2 = 0.00001 \qquad \text{[from equation (4.3)]}$$

$$K = 621.456 \qquad \text{[from equation (4.7)]}$$

$$\omega_x = 2.201037 \qquad \text{[from equation (4.8)]}$$

Substituting into equation (4.9) yields

$$G(s)G(-s) = \frac{0.00001\left[16\left(\frac{2.20137}{\omega}\right)^5 - 20\left(\frac{2.20137}{\omega}\right)^3 + 5\left(\frac{2.20137}{\omega}\right)\right]^2}{1 + 0.00001\left[16\left(\frac{2.20137}{\omega}\right)^5 - 20\left(\frac{2.20137}{\omega}\right)^3 + 5\left(\frac{2.20137}{\omega}\right)\right]^2} \tag{4.24}$$

or

$$G(s)G(-s) = \frac{0.0012115\omega^8 - 0.046968\omega^6 + 0.6373035\omega^4 - 3.5296159\omega^2 + 6.8418986}{\omega^{10} + 0.0012115\omega^8 - 0.046968\omega^6 + 0.6373035\omega^4 - 3.5296159\omega^2 + 6.8418986}$$

Substituting $\omega = s/\mathrm{j}$,

$$G(s)G(-s) = \frac{0.0012115s^8 + 0.046968s^6 + 0.6373035s^4 + 3.5296159s^2 + 6.8418986}{-s^{10} + 0.0012115s^8 + 0.046968s^6 + 0.6373035s^4 + 3.5296159s^2 + 6.8418986}$$

The poles and zeroes of $G(s)G(-s)$ (Figure 3.2) are

Poles	Zeroes
−0.306777 + j1.09900390	0 + j2.30938925
−0.306777 − j1.09900390	0 − j2.30938925
−0.945087 + j0.79925575	0 + j3.73131205
−0.945087 − j0.79925575	0 − j3.73131205
−1.311428 + j0	0 + j2.30938925
0.306777 − j1.09900390	0 − j2.30938925
0.306777 + j1.09900390	0 + j3.73131205
0.945087 − j0.79925575	0 − j3.73131205
0.945087 + j0.79925575	
1.311428 + j0	

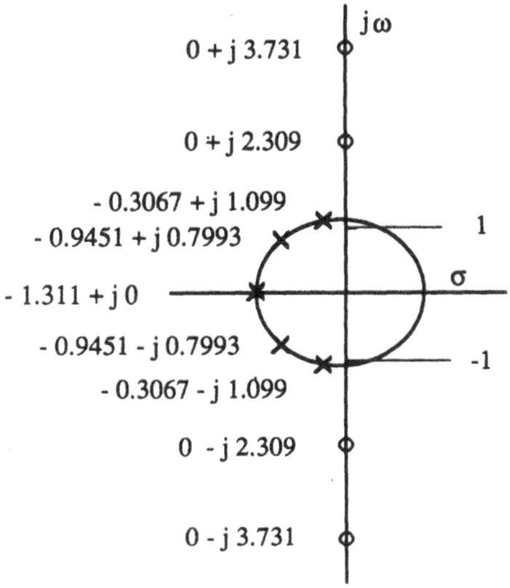

Figure 4.2 Pole-zero diagram for example 4.1.

Taking the poles with negative real parts, and one set of four zeroes, $G(s)$ in factored form becomes

$$G(s) = \frac{0.0352258(s^2 + 5.3332785)(s^2 + 13.922689)}{(s^2 + 0.613554s + 1.3019217)(s^2 + 1.89017s + 1.532)(s + 1.3114)} \tag{4.25}$$

or

$$G(s) = \frac{0.0352258(s^4 + 19.255968s^2 + 74.253578)}{(s^5 + 3.812925s^4 + 7.27414s^3 + 8.6380788s^2 + 6.4543767s + 2.61545)} \tag{4.26}$$

The constant, 0.0352258, must be inserted to make the d.c. gain unity. Equation (4.25) is the preferred form for solving for magnitude, phase, and group delay. The polynomial form given by (4.26) is numerically unstable for larger values of N and should not be used for response calculations.

In Method 2 (the preferred way), using equation (4.7) and (4.10) through (4.20) the poles and zeroes are determined to be

Poles	Zeroes
−0.306777 + j1.09900390	0 + j2.30938925
−0.306777 − j1.09900390	0 − j2.30938925
−0.945087 + j0.79925575	0 + j3.73131205
−0.945087 − j0.79925575	0 − j3.73131205
−1.311428 + j0	

from which

$$G(s) = \frac{0.0352258(s^2 + 5.3332785)(s^2 + 13.922689)}{(s^2 + 0.613554s + 1.3019217)(s^2 + 1.89017s + 1.532)(s + 1.3114)}$$

The amplitude response at $\omega = 2$ can be computed four ways. The preferred method (Case 1) substitutes $\omega = 2$ into equation (4.25). The magnitude becomes

$$|G(s)| = \frac{0.0352258\ (5.3332785 - 2^2)(13.922689 - 2^2)}{\sqrt{(1.3019217 - 2^2)^2 + (0.613554 \cdot 2)^2}\ \sqrt{(1.532 - 2^2)^2 + (1.89017 \cdot 2)^2}\ \sqrt{(1.3114)^2 + (2)^2}}$$

or

$$|G(s)| = 0.01456 = -36.7 \text{ dB}$$

In Case 2, substitution of $\omega = 2$ into equation (4.26) yields

$$|G(s)| = \frac{0.0352258\ (2^4 - 19.255968 \cdot 2^2 + 74.253578)}{\sqrt{(2^5 - 7.27414 \cdot 2^3 + 6.4543767 \cdot 2)^2 + (3.812925 \cdot 2^4 - 8.6380788 \cdot 2^2 + 2.61545)^2}}$$

or $|G(s)| = 0.01456 = -36.7$ dB; however, this method is unstable and yields inaccurate answers for larger values of N.

In Case 3, substituting $\omega = 2$ into equation (4.9) yields

$$G(s)G(-s) = \frac{\epsilon^2\left[T_N\left(\frac{\omega_x}{\omega}\right)\right]^2}{1 + \epsilon^2\left[T_N\left(\frac{\omega_x}{\omega}\right)\right]^2} = [\text{Mag}]^2$$

or

$$G(s)G(-s) = \frac{0.00001\left[16\left(\frac{2.20137}{\omega}\right)^5 - 20\left(\frac{2.20137}{\omega}\right)^3 + 5\left(\frac{2.20137}{\omega}\right)\right]^2}{1 + 0.00001\left[16\left(\frac{2.20137}{\omega}\right)^5 - 20\left(\frac{2.20137}{\omega}\right)^3 + 5\left(\frac{2.20137}{\omega}\right)\right]^2}$$

$$|G(s)| = \sqrt{|G(s)G(-s)|} = 0.01456 = -36.7 \text{ dB}$$

In Case 4, using (3.17), (3.23), and (4.9),

$$T_N\left(\frac{\omega_x}{\omega}\right) = \cos\left[N \arccos\left(\frac{\omega_x}{\omega}\right)\right] = 4.649$$

the amplitude response becomes

$$|G(s)| = \sqrt{|G(s)G(-s)|} = 0.01456 = -36.7 \text{ dB}$$

REFERENCE

[1] Rabiner, Lawrence R., and Gold, Bernard, *Theory and Application of Digital Signal Processing*, Englewood Cliffs, NJ: Prentice-Hall, 1975, pp. 232–234.

PROBLEMS

4.1 Find the poles and zeros of a tenth-order Tchebyscheff Type II normalized low-pass filter with 1 dB attenuation at $\omega = 1$ and 60 dB attenuation in the stopband.

4.2 For the filter in problem 4.1, find an expression for
(a) the magnitude response,
(b) the log magnitude response,
(c) the phase response,
(d) the group delay, and
(e) the Q and resonant frequency of each quadratic section.

4.3 Compare the attenuation at $\omega = 10$ of the low pass in problem 4.1 with
(a) a tenth-order Tchebyscheff Type I normalized low pass with 1 dB passband ripple,
(b) a tenth-order normalized Butterworth low pass.

4.4 A normalized 12th-order Tchebyscheff Type II low pass has 0.001 dB passband attenuation at $\omega = 1$ and 50 dB minimum attenuation in the stopband. Find the first stopband frequency where the attenuation is 50 dB.

4.5 Find the transfer function, $G(s)$, of a third-order Tchebyscheff Type II low-pass filter with a maximum of 1 dB passband attenuation and 20 dB minimum stopband attenuation such that no energy is transmitted at $\omega = 2$.

4.6 (a) Find the passband frequencies for unity transmission gain of a 12th-order Tchebyscheff Type II low pass with 0.001 dB maximum passband attenuation at $\omega = 1$ rad/sec and 50 dB minimum stopband attenuation.
(b) Find the stopband frequencies where no energy is transmitted.

4.7 Show that the magnitude response of an even-order normalized Tchebyscheff Type II low pass with 40 dB minimum stopband attenuation approaches a voltage gain of 0.01 as ω approaches infinity.

4.8 Show that equations (3.25) and (4.23) are identical. Use the facts that the definitions for ϵ^2 and [Mag] are different for the Tchebyscheff Type I and Type II cases.

Chapter 5
The Low-Pass Elliptic Filter

5.1 THE TCHEBYSCHEFF RATIONAL FUNCTION

The poles and zeros of an elliptic filter cannot be found by an analytical formula. Elliptic filters are defined by the equation

$$G(s)G(-s) = \frac{1}{1 + \epsilon^2[R_N(\omega)]^2} \tag{5.1}$$

where

$$R_N(\omega) = B_1\omega \prod_{i=1}^{(N-1)/2} \frac{\omega_{oi}^2 - \omega^2}{\omega_{zi}^2 - \omega^2} \qquad \text{for } N \text{ odd} \tag{5.2}$$

$$R_N(\omega) = B_2 \prod_{i=1}^{N/2} \frac{\omega_{oi}^2 - \omega^2}{\omega_{zi}^2 - \omega^2} \qquad \text{for } N \text{ even} \tag{5.3}$$

with

$$\omega_{zi} = \frac{\omega_x}{\omega_{oi}} \tag{5.4}$$

where ω_x is the first frequency greater than 1 rad/sec where the filter meets the prescribed stopband loss. For example, $R_N(\omega)$ for a fifth-order elliptic filter can be written as

$$R_5(\omega) = \frac{B_1\omega(\omega_{o1}^2 - \omega^2)(\omega_{o2}^2 - \omega^2)}{(\omega_{z1}^2 - \omega^2)(\omega_{z2}^2 - \omega^2)} \tag{5.5}$$

The function $R_N(\omega)$ is called the *Tchebyscheff rational function* or sometimes the *Zolotarev function.* If the zeros, ω_{oi}, are chosen so the passband has equiripple error, then the stopband will automatically have equiripple error. At $\omega = 1$, B_1 is computed by scaling $R_N(\omega)$ such that $R_N(1) = 1$. A typical Tchebyscheff rational function is plotted in Figure 5.1.

5.2 CALCULATION OF FILTER ORDER

The most common method used for finding coefficients of the Tchebyscheff rational function was researched by Cauer in 1931 and uses the elliptic function, $sn(u)$. It is summarized as follows. Define [1]

$$L^2 = \frac{10^{0.1 \cdot R_s} - 1}{10^{0.1 \cdot R_p} - 1} \tag{5.6}$$

where R_s is the minimum attenuation in the stopband, and R_p is the maximum attenuation in the passband, both expressed as positive numbers. The required filter order, N, can be computed by [2]

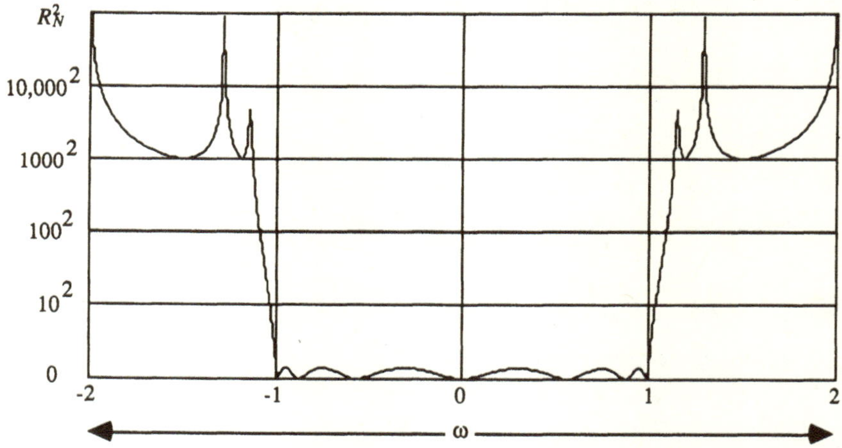

Figure 5.1 This function generates a seventh-order elliptic filter with 3.0103 dB passband ripple and 60 dB minimum attenuation in the stopband starting with $\omega = 1.133$ rad/sec with a characteristic equation given by

$$R_7^2(\omega) = \left[\frac{179.23\omega([0.56634]^2 - \omega^2)([0.87897]^2 - \omega^2)([0.988702]^2 - \omega^2)}{([2.00054]^2 - \omega^2)([1.28899]^2 - \omega^2)([1.145946]^2 - \omega^2)}\right]^2$$

$$N = \frac{K\left(\frac{1}{\omega_x}\right)K'\left(\frac{1}{L}\right)}{K'\left(\frac{1}{\omega_x}\right)K\left(\frac{1}{L}\right)} \tag{5.7}$$

where $K(k)$ is defined as

$$K(k) = \int_0^{\pi/2} (1 - k^2 \sin^2(x))^{-1/2} dx \tag{5.8}$$

and $K'(k)$ is defined as

$$K'(k) = \int_0^{\pi/2} [1 - (1 - k^2) \sin^2(x)]^{-1/2} dx \tag{5.9}$$

Equations (5.8) and (5.9) are complete elliptic integrals of the first kind. In general, if

$$K(z) = \int_0^{\pi/2} (1 - z^2 \sin^2(x))^{-1/2} dx \tag{5.10}$$

then taking a Taylor series about $z = 0$ yields

$$K(z) = \frac{\pi}{2}\left\{1 + \left(\frac{1}{2}\right)^2 z^2 + \left(\frac{1 \cdot 3}{2 \cdot 4}\right)^2 z^4 + \left(\frac{1 \cdot 3 \cdot 5}{2 \cdot 4 \cdot 6}\right)^2 z^6 + \ldots\right\} \tag{5.11}$$

The convergence of equation (5.11) is very slow. However, the trapezoidal rule will give more accurate integration results. The numerical integration algorithm, program B2 in Appendix B, for calculating N, given passband ripple, stopband attenuation, and stopband cutoff frequency, is a vast improvement in speed over both equation (5.11) and most numerical integration schemes. If N is known and the stopband cutoff frequency is unknown, then the algorithm used in program B2 can easily be modified to iterate the stopband cutoff frequency till the specified value of N is found.

5.3 POLE AND ZERO LOCATIONS

Once N is determined, U can be calculated as

$$U = \frac{(2I-1)\cdot K\left(\dfrac{1}{\omega_x}\right)}{N} \qquad \text{for } N \text{ even and } I = 1, 2, \ldots, N/2 \tag{5.12}$$

$$U = \frac{(2I)\cdot K\left(\dfrac{1}{\omega_x}\right)}{N} \qquad \text{for } N \text{ odd and } I = 1, 2, \ldots, (N-1)/2 \tag{5.13}$$

Now ϕ must be determined by

$$U = \int_0^{\phi} [1 - (1/\omega_x)^2 \sin^2(x)]^{-1/2} dx \tag{5.14}$$

Then $sn(U) = \sin(\phi) = \omega_{oi}$ and $\omega_{xi} = \omega_x/\omega_{oi}$. The equations for the Tchebyscheff rational function become

$$R_N(\omega) = B_1\omega \prod_{I=1}^{(N-1)/2} \frac{\omega^2 - sn^2(U)}{\omega^2 - \omega_x/sn^2(U)} \qquad \text{for } n \text{ odd} \tag{5.15}$$

$$R_N(\omega) = B_2 \prod_{I=1}^{N/2} \frac{\omega^2 - sn^2(U)}{\omega^2 - \omega_x/sn^2(U)} \qquad \text{for } n \text{ even} \tag{5.16}$$

with U as defined previously.

Whittaker and Watson [3] give two numerical methods for evaluating $sn(U)$. If q is defined such that

$$q = e^p, \qquad \text{with } p = -\pi \frac{K\left[\sqrt{1-\left(\dfrac{1}{\omega_x}\right)^2}\right]}{K\left[\dfrac{1}{\omega_x}\right]} \tag{5.17}$$

$$U = \frac{2xK\left[\dfrac{1}{\omega_x}\right]}{\pi} \tag{5.18}$$

then $sn(U)$ can be expressed as an infinite summation such that

$$sn(U) = \frac{2\pi}{\frac{1}{\omega_x}K\left[\frac{1}{\omega_x}\right]}\left[\frac{q^{1/2}\sin(x)}{1-q} + \frac{q^{3/2}\sin(3x)}{1-q^3} + \frac{q^{5/2}\sin(5x)}{1-q^5} + \ldots\right] \tag{5.19}$$

or with an infinite product formulation given by

$$sn(U) = 2q^{1/4}\left[\frac{1}{\omega_x}\right]^{-1/2}\sin(x)\prod_{n=1}^{\infty}\frac{1 - 2q^{2n}\cos(2x) + q^{4n}}{1 - 2q^{2n-1}\cos(2x) + q^{4n-2}} \tag{5.20}$$

The constant B in equations (5.15) and (5.16) is easily found once the poles and zeroes are known, because $R_N(\omega) = 1$ at $\omega = 1$. The voltage transfer function, $G(s)$, is determined by rationalizing as in equation (5.16) for N even:

$$G(s)G(-s) = \frac{[(\omega_{x1}^2 - \omega^2)(\omega_{x2}^2 - \omega^2)\ldots]^2}{[(\omega_{x1}^2 - \omega^2)(\omega_{x2}^2 - \omega^2)\ldots]^2 + \epsilon^2[B_1(\omega_{o1}^2 - \omega^2)(\omega_{o2}^2 - \omega^2)\ldots]^2} \tag{5.21}$$

Because $s = j\omega$, substitute $\omega = s/j$ for each ω, ($\omega^2 = -s^2$), which yields

$$G(s)G(-s) = \frac{[(\omega_{x1}^2 + s^2)(\omega_{x2}^2 + s^2)\ldots]^2}{[(\omega_{x1}^2 + s^2)(\omega_{x2}^2 + s^2)\ldots]^2 + \epsilon^2[B_1(\omega_{o1}^2 + s^2)(\omega_{o2}^2 + s^2)\ldots]^2} \tag{5.22}$$

The two halves of the denominator must be multiplied out and combined into a single polynomial. A polynomial root finder can then be employed to find the lefthand poles corresponding to $G(s)$. The denominator of (5.22) has roots that are in close proximity to each other and poses a problem for most iterative root finders. Bairstow's method, which is an adaptation of Newton's method, finds the roots by solving for the quadratic factors. The computer program B3 implements Bairstow's method and can be used to find the poles of ill-conditioned polynomials using a transformation variable that will be discussed in detail in Chapter 7.

Example 5.1

Find the voltage transfer function, $G(s)$, for an elliptic low-pass filter with 3.0103 dB passband ripple, 60 dB stopband attenuation, and a stopband frequency of 1.507 rad/sec.

Solution

From equation (5.7),

$$N = \frac{\displaystyle\int_0^{\pi/2} \left(1 - \frac{1}{\omega_x^2}\sin^2(x)\right)^{-1/2} dx}{\displaystyle\int_0^{\pi/2} \left[1 - \left(1 - \frac{1}{\omega_x^2}\right)\sin^2(x)\right]^{-1/2} dx} \; \frac{\displaystyle\int_0^{\pi/2} \left[1 - \left(1 - \frac{1}{L^2}\right)\sin^2(x)\right]^{-1/2} dx}{\displaystyle\int_0^{\pi/2} \left(1 - \frac{1}{L^2}\sin^2(x)\right)^{-1/2} dx} \tag{5.23}$$

where $\omega_x = 1.507$. Solving (5.23) numerically yields $N = 5$. From equation (3.12), $\epsilon^2 = 1$; therefore, equation (5.1) becomes

$$G(s)G(-s) = \frac{1}{1 + \epsilon^2\left[\dfrac{B_1\omega(\omega_{o1}^2 - \omega^2)(\omega_{o2}^2 - \omega^2)}{(\omega_{z1}^2 - \omega^2)(\omega_{z2}^2 - \omega^2)}\right]^2} \tag{5.24}$$

Using the computer program B2 to find the missing coefficients, or equations (5.6) through (5.20),

$$G(s)G(-s) = \frac{1}{1 + \left[\dfrac{152.93\omega(0.64258^2 - \omega^2)(0.963007^2 - \omega^2)}{(2.3452^2 - \omega^2)(1.5649^2 - \omega^2)}\right]^2}$$

Substituting $s = j\omega$ and combining terms,

$$G(s)G(-s) = \frac{(s^2 + 5.5)^2(s^2 + 2.4489)^2}{-23387.58s^{10} - 62691.24s^8 - 59908.38s^6 - 23916.29s^4 - 3215.22s^2 + 181.41}$$

Let $u = s^2$, then the denominator polynomial becomes $-23387.58u^5 - 62691.24u^4 - 59908.38u^3 - 23916.29u^2 - 3215.22u + 181.4112$. Using a polynomial root finder, the roots of the denominator are

$$u - 0.041986$$
$$u^2 + 0.823446u + 0.20375$$
$$u^2 + 1.8990077u + 0.9081868$$

or

$$(s - 0.2049049)(s + 0.2049049)$$
$$(s^2 + 0.2823721s + 0.4513869)(s^2 - 0.2823721s + 0.4513869)$$
$$(s^2 + 0.0834811s + 0.952988)(s^2 - 0.0834811s + 0.952988)$$

Then

$$G(s) = \frac{(1/152.93)(s^2 + 5.5)(s^2 + 2.4489)}{(s + 0.2049049)(s^2 + 0.2823721s + 0.4513869)(s^2 + 0.0834811s + 0.952988)}$$

The resulting elliptic poles and zeros are plotted in Figure 5.2.

In Example 5.1, the polynomial substitution of $u = s^2$ was essential in preparation for the root finder in order to reduce the order of the polynomial by 2, thus improving the speed and accuracy of the root finding process. The polynomials in s were recovered by equating like terms and recognizing that for $u^2 + 1.8990077u + 0.9081868 = (s^2 + As + B)(s^2 - As + B)$, $B = \sqrt{0.9081868}$, and $2B - A^2 = 1.8990077$.

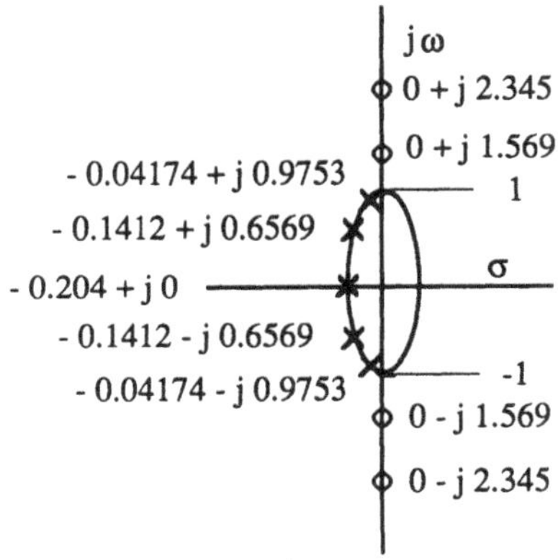

Figure 5.2 Elliptic poles and zeros for the fifth-order elliptic lowpass in Example 5.1.

REFERENCES

[1] Daniels, Richard W., *Approximation Methods for Electronic Filter Design,* New York: McGraw-Hill, 1975, p. 55.

[2] Zverev, Anatol I., *Handbook of Filter Synthesis,* New York: John Wiley & Sons, 1967, pp. 107–122.

[3] Whittaker, E. T., and Watson, G. N., *Modern Analysis,* Cambridge: Cambridge University Press, 1963, pp. 508–510.

PROBLEMS

5.1 Use program B2 iteratively to find the first stopband frequency of a normalized filter such that the theoretical order is exactly 4.00 with a 3.0103 dB passband ripple and 60 dB minimum stopband attenuation. Repeat for $N = 5.00$, 6.00, and 7.00.

5.2 Use equation (5.7) to find the order, N, required for a normalized low-pass elliptic filter with 1 dB passband ripple and 80 dB attenuation, or more, at a stopband edge frequency of $\omega = 2$ rad/sec.

5.3 Find the Tchebyscheff rational function for the filter in problem (5.2) by using equation (5.19). Repeat using equation (5.20).

5.4 Find the transfer function, $G(s)$, for the filter in problem (5.3) by applying a polynomial root finder to the denominator of equation (5.22).

5.5 For the elliptic low-pass filter based on the Tchebyscheff rational function given in Figure 5.1, identify the passband frequencies where the gain is exactly unity. What are the stopband frequencies such that the gain is exactly zero? Derive the expression for group delay and evaluate it at the stopband frequencies.

Chapter 6
Bandpass Transformations

6.1 IMPEDANCE SCALING

In the previous prototype filter designs, the source and load resistances were unity (except for equal-ripple passband filters with even N, which have nonunity load resistance). A source resistance of R_o can be obtained by multiplying the impedances of the prototype design by R_o. Then, letting primes denote impedance scaled quantities, the new filter component values are given by

$$L' = R_o \cdot L \tag{6.1a}$$

$$C' = \frac{C}{R_o} \tag{6.1b}$$

$$R_s' = R_o \cdot R_s \tag{6.1c}$$

$$R_L' = R_o \cdot R_L \tag{6.1d}$$

where L, C, R_s, and R_L are the component values for the original prototype.

6.2 FREQUENCY SCALING

To change the cutoff frequency of a low-pass prototype from unity to ω_c requires that all frequency-dependent elements be scaled by a factor $1/\omega_c$, which is accomplished by replacing ω by ω/ω_c. Substituting ω/ω_c into the reactance expression for the inductors and capacitors yields

$$L's = \frac{Ls}{\omega_c}, \quad \text{or} \quad L' = \frac{L}{\omega_c}, \quad \text{and} \quad \frac{1}{C's} = \frac{1}{C\frac{s}{\omega_c}}, \quad \text{or} \quad C' = \frac{C}{\omega_c} \tag{6.2}$$

The combination of impedance and frequency scaling can be applied to any low-pass prototype to set its cutoff frequency and its source and termination resistances.

Example 6.1

Design a Tchebyscheff Type I fifth-order low-pass prototype with 1 dB passband ripple, cutoff frequency of 1 MHz, and 100 Ω source and termination resistances.

Solution

From Table 3.2 the normalized component values are

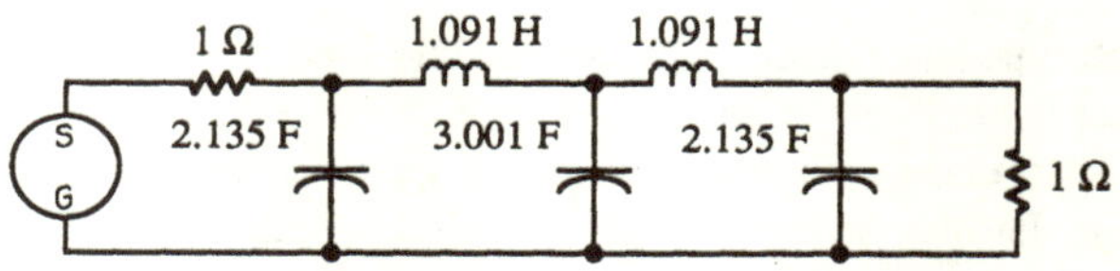

The impedance scale factor is 100, and the frequency scale factor is 6,283,200 (from $f =$ 0.159155 Hz to $f =$ 1,000,000 Hz). So the final network becomes

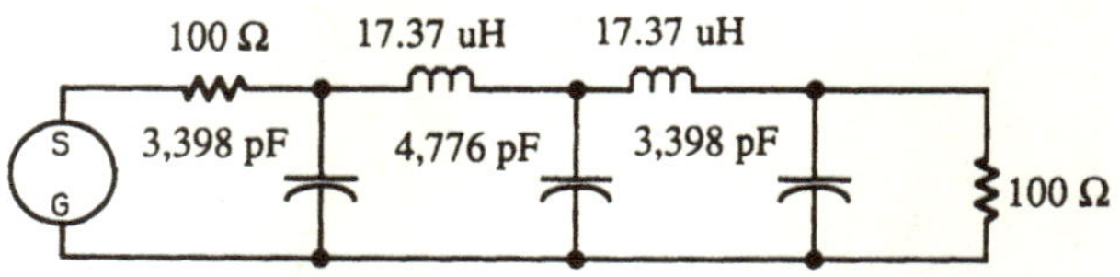

with 1 $pF = 1 \times 10^{-12}$ farads, and 1 uH $= 1 \times 10^{-6}$ henries.

6.3 DUALITY AND RECIPROCITY

A network and its dual have identical response characteristics and identical voltage transfer functions, $G(s)$, although the expression for input impedance will be drastically different. Any normalized ladder type passive network can be transformed into its dual by implementing the following rules.

1. Convert every series branch into a shunt branch and every shunt branch into a series branch.

2. Convert circuit branch elements in series to elements in parallel and vice versa.
3. Transform each inductor into a capacitor and vice versa. The values remain unchanged; that is, 2 H becomes 2 F.
4. Replace each resistance with a conductance; that is 2 Ω becomes 2 mhos or 1/2 Ω.
5. Change a voltage source into a current source and vice versa.

The theorem of reciprocity states that if a voltage located at one point of a linear network produces a current at any other point, the same voltage acting at the second point results in the same current at the first point. Alternately, if a current source at one point of a linear network results in a voltage measured at a different point, the same current source at the second point produces the same voltage at the first point. As a result, the response of an LC filter is the same regardless of which direction the signal flows in, except for a constant multiplier. It is not unusual to turn a filter schematic completely around with regard to its driving source, provided that the source and load resistive terminations are also interchanged.

The laws of duality and reciprocity are used to manipulate a filter to satisfy termination requirements or to force a desired configuration.

Example 6.2

The circuit in Figure 6.1 is a third-order elliptic bandpass. Find its dual but preserve the 100 Ω termination resistances.

Solution

The circuit dual prototype is shown in Figure 6.2a. However, the current source may be converted to a voltage source using Norton's equivalent, and the impedances scaled by 10,000 to recover the 100 Ω source and termination resistances to produce the final circuit.

6.4 BARTLETT'S BISECTION THEOREM

Bartlett's bisection theorem states that if a symmetrical network is bisected and one half is impedance scaled including the termination, the response shape will not change. All

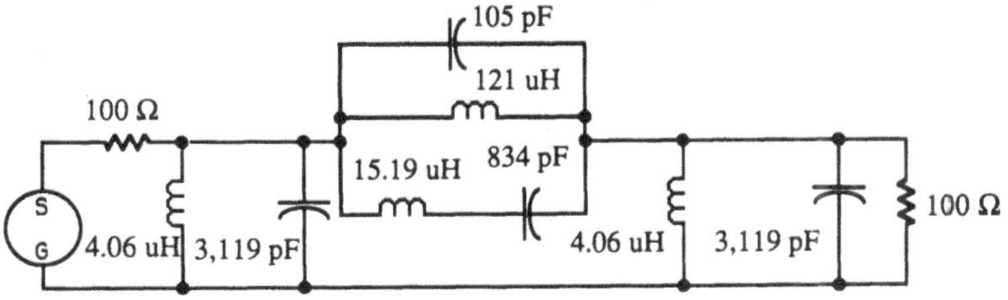

Figure 6.1 Bandpass third-order elliptic filter with 1 dB passband ripple, 50 dB stopband attenuation and a bandpass region of 1 MHz to 2 MHz.

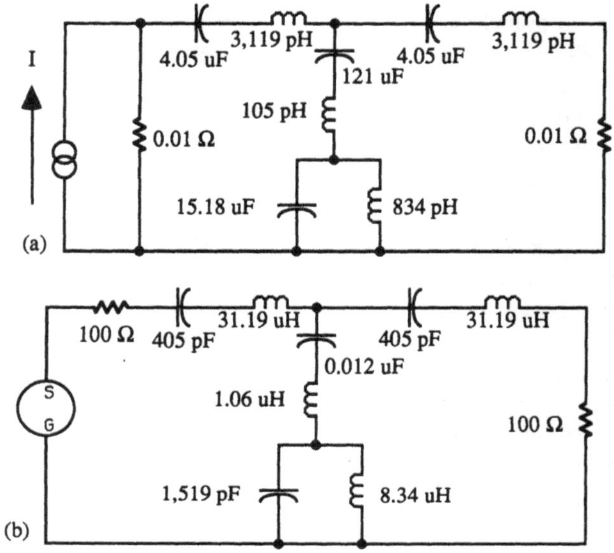

Figure 6.2 Dual of Figure 6.1 with 100 source and load resistances.

odd-order Butterworth and odd-order Tchebyscheff filters satisfy this requirement, and also third-order inverse Tchebyscheff and third-order elliptic filters. The symmetry must include not only the topology but also the component values; therefore, higher order inverse Tchebyscheff and elliptic filters are excluded. This extremely powerful result allows many passive filters to be modifieɑ for unequal source and load resistances.

Example 6.3

Modify the circuit in Figure 6.1 for a termination resistance of 1000 Ω.

Solution

The bisected circuit becomes

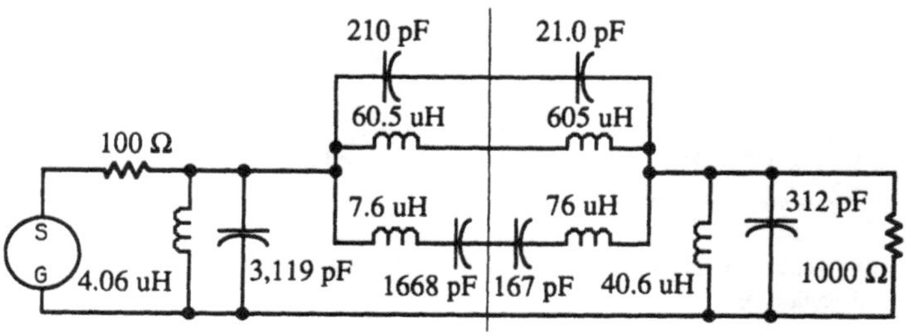

Combining the series inductors and capacitors yields

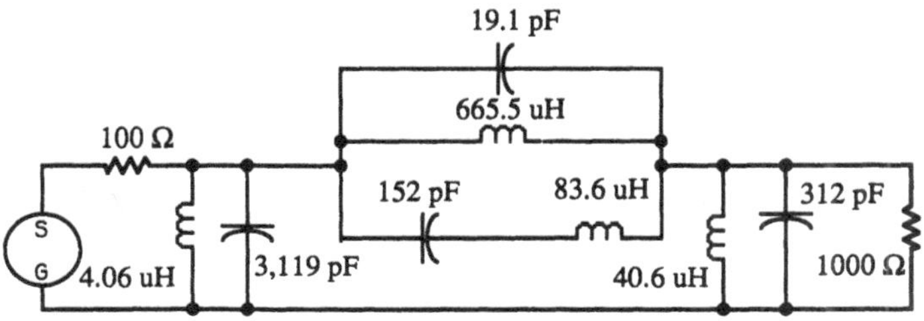

6.5 LOW-PASS TO BANDPASS TRANSFORMATIONS

Bandpass transformations are based on replacing the variable s with another variable from Table 6.1. The arrows in Table 6.1 should be read as "s gets replaced by $f(s)$." Whereas equation (6.3d) maps the low-pass cutoff frequency into ω_L and ω_u, (6.3e) has been provided because it is usually preferable to map ω_x into ω_L and ω_u for the bandstop filter case, because the desired bandstop attenuation is normally required throughout the range of ω_L to ω_u.

Table 6.1

Normalized Low-Pass to Unnormalized Bandpass Transformations

$s \quad f(s)$		
$s \rightarrow \dfrac{s}{\omega_u}$	converts 1 rad/sec low pass to ω_u rad/sec low pass by mapping the range $0 < \omega < \omega_u$ into $0 < \omega < 1$	(6.3a)
$s \rightarrow \dfrac{\omega_u}{s}$	converts 1 rad/sec low pass to ω_u rad/sec high pass by mapping the range $\omega_u < \omega < \infty$ into $0 < \omega < 1$	(6.3b)
$s \rightarrow \dfrac{s^2 + (\omega_u \cdot \omega_L)}{s \cdot (\omega_u - \omega_L)}$	converts 1 rad/sec *low pass to bandpass* by mapping the range $\omega_L < \omega < \omega_u$ into $-1 < \omega < 1$	(6.3c)
$s \rightarrow \dfrac{s \cdot (\omega_u - \omega_L)}{s^2 + (\omega_u \cdot \omega_L)}$	converts 1 rad/sec *low pass to bandstop* by mapping the range $\omega_L < \omega < \omega_u$ into $1 < \omega < \infty$	(6.3d)
$s \rightarrow \dfrac{\omega_x \cdot s \cdot (\omega_u - \omega_L)}{s^2 + (\omega_u \cdot \omega_L)}$	converts 1 rad/sec *low pass to bandstop* by mapping the range $\omega_L < \omega < \omega_u$ into $\omega_x < \omega < \infty$	(6.3e)

where ω_L = lower frequency cutoff in rad,
ω_u = upper frequency cutoff in rad,
ω_x = first frequency at which the low-pass prototype meets the prescribed bandstop loss.

The transforms can be applied either directly to the voltage transfer function, $G(s)$, or to the components values of the low-pass prototype (see Table 6.2). These low-pass-to-bandpass and low-pass-to-bandstop transforms do not provide for transition regions between the passband and the stopband that are exactly symmetrical. Although other transforms can easily be derived (see Table 6.2), equations (6.3a) through (6.3e) represent the ones used most often.

Table 6.2

Summary of Prototype Filter Transformations (from [1], by permission of the publisher)

Normalized Low-Pass Values	*Unnormalized High-Pass Values*	*Unnormalized Bandpass Values*	*Unnormalized Bandstop Values*

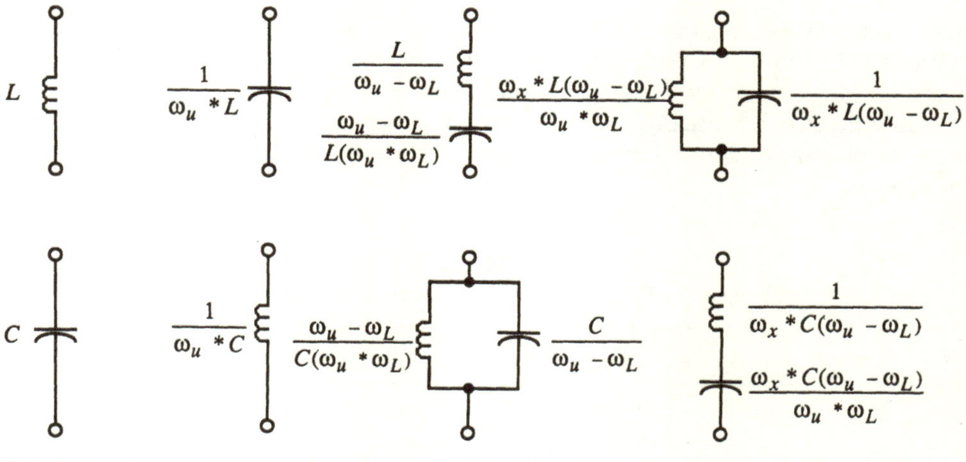

6.6 CALCULATING THE PASSBAND FILTER ORDER

Calculation of the order, N, for the bandpass, bandstop, and high-pass cases is done by using equation (6.3) to map the transition region between the passband and the stopband into the equivalent low-pass case. For the bandpass and the bandstop filter, the order is determined by the side that has the steepest skirt. A high-pass filter, cascaded with a low-pass filter, would be a better design for a bandpass or bandstop filter, if a highly unsymmetrical response is desired. In Table 6.3, ω_x is calculated from the passband requirements, then N can be calculated and rounded to the next highest integer. The bandstop transformation in Table 6.2 requires ω_x to be calculated based on the integer value of N, for a normalized low-pass prototype with a cutoff of 1 rad/sec.

For the Butterworth filter,

Table 6.3
Calculation of N

Bandpass	*Bandstop*	*High Pass*
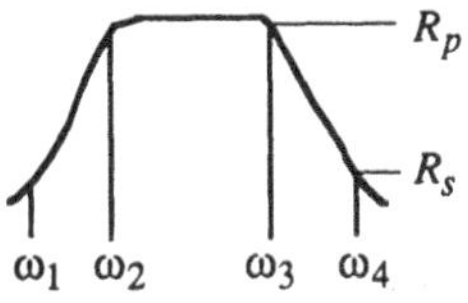	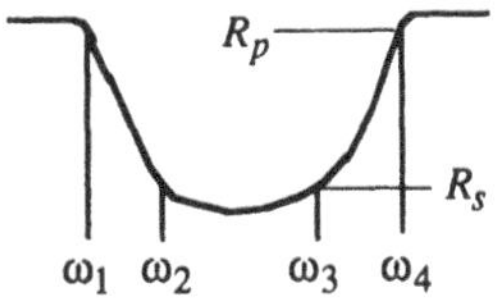	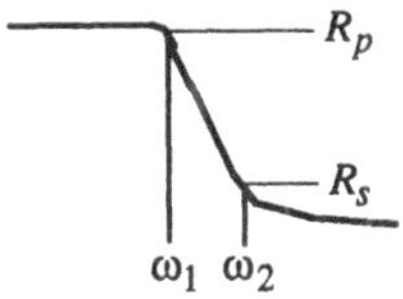
ω_1 = lower stopband edge freq.	ω_1 = lower passband edge freq.	ω_2 = stopband edge freq.
ω_2 = lower passband edge freq.	ω_2 = lower stopband edge freq.	ω_1 = passband edge freq.
ω_3 = upper passband edge freq.	ω_3 = upper stopband edge freq.	R_p = max. passband atten.
ω_4 = upper stopband edge freq.	ω_4 = upper passband edge freq.	R_s = min. stopband atten.
R_p = max. passband atten.	R_p = max. passband atten.	
R_s = min. stopband atten.	R_s = min. stopband atten.	
From equation (6.3c)	From equation (6.3d)	From equation (6.3b)
$\omega_{x1} = \left\lvert \frac{\omega_2 \cdot \omega_3 - \omega_1 \cdot \omega_1}{\omega_1 \cdot \omega_3 - \omega_1 \cdot \omega_2} \right\rvert$	$\omega_{x1} = \left\lvert \frac{\omega_2 \cdot \omega_3 - \omega_1 \cdot \omega_1}{\omega_1 \cdot \omega_3 - \omega_1 \cdot \omega_2} \right\rvert$	$\omega_x = \frac{\omega_2}{\omega_1}$
$\omega_{x2} = \left\lvert \frac{\omega_2 \cdot \omega_3 - \omega_4 \cdot \omega_4}{\omega_4 \cdot \omega_3 - \omega_4 \cdot \omega_2} \right\rvert$	$\omega_{x2} = \left\lvert \frac{\omega_2 \cdot \omega_3 - \omega_4 \cdot \omega_4}{\omega_4 \cdot \omega_3 - \omega_4 \cdot \omega_2} \right\rvert$	
$\omega_x = \omega_{x1}$ or ω_{x2}, whichever is smaller	$\omega_x = \omega_{x1}$ or ω_{x2}, whichever is smaller	

$$N = \frac{0.5 \cdot \log\left[\dfrac{10^{R_s/10} - 1}{10^{R_p/10} - 1}\right]}{\log[\omega_x]}$$

For the Tchebyscheff Type I and Type II filters,

$$\epsilon = \sqrt{10^{(0.1 \cdot R_p)} - 1}$$

$$[\mathrm{Mag}]^2 = 10^{-R_s/10}$$

$$N = \frac{\ln\left(\sqrt{\dfrac{1 - [\mathrm{Mag}]^2}{[\mathrm{Mag}]^2 \epsilon^2}} + \sqrt{\left(\dfrac{1 - [\mathrm{Mag}]^2}{[\mathrm{Mag}]^2 \epsilon^2}\right) - 1}\right)}{\ln(\omega_x + \sqrt{\omega_x^2 - 1})}$$

For elliptic filters,

$$L^2 = \frac{10^{0.1 \cdot R_s} - 1}{10^{0.1 \cdot R_p} - 1}$$

$$N = \frac{K\left(\frac{1}{\omega_x}\right) K'\left(\frac{1}{L}\right)}{K'\left(\frac{1}{\omega_x}\right) K\left(\frac{1}{L}\right)}$$

where

$$K(k) = \int_0^{\pi/2} [1 - k^2 \sin^2(x)]^{-1/2} dx$$

$$K'(k) = \int_0^{\pi/2} [1 - (1 - k^2) \sin^2(x)]^{-1/2} dx$$

The filter order, N, always refers to the order of the low-pass prototype, even though the total number of poles on the bandpass, or bandstop, filter is doubled.

6.7 STAR-DELTA TRANSFORMATIONS

The star-delta transformation is used primarily to modify reactive networks in order to achieve a different configuration of elements or to alter component values to achieve a more realizable design. The general star-delta transform, from which the special cases are derived, is shown in Figure 6.3; some of the special cases are shown in Figure 6.4.

6.8 NORTON'S TRANSFORMATIONS

Norton's transformation, shown in Figure 6.5, is useful in the elimination of redundant elements or for the modification of element values. By the use of this transformation, the impedances of the given network are modified on one side of the transformation such that all inductances and resistances are to be scaled by the factor $1/K^2$ and all the capacitors are scaled by K^2. With the aid of several transformations performed on various parts of a network, it is possible to eliminate several redundant elements but finally end with equal source and load terminations, if desired.

Given the bandpass filter in Figure 6.6, it is desired to economize the network by eliminating a redundant capacitor. The transformation of Figure 6.6 will be inserted in the position of the line AA', resulting in the schematic of Figure 6.7.

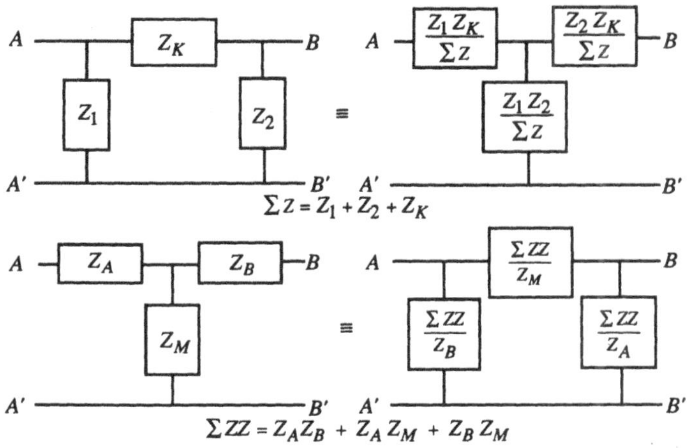

Figure 6.3 Star-delta generic transforms.

Note that it is possible to eliminate either the capacitor resonating with L_1, or with L_2, or neither of these capacitors could be eliminated but the output impedance could be modified.

To eliminate the capacitor resonating with L_1 in Figure 6.6, set

$$C_1 + (1 - K)C_2 = 0 \tag{6.4}$$

Solving for K,

$$K = \frac{C_1 + C_2}{C_2} \tag{6.5}$$

In this case, K will be greater than 1; and the transformation will be a step-down transformer; the impedance R/K^2 will be less than R. When the elimination of the capacitor resonating with L_2 is desired in Figure 6.6, set

$$K(K - 1)C_2 + K^2C_3 = 0 \tag{6.6}$$

Solving for K,

$$K = \frac{C_2}{C_2 + C_3} \tag{6.7}$$

Now K will be less than unity, and the resulting transformation will be a step-up transformer. The impedance R/K^2 will be greater than R. One word of caution must be observed.

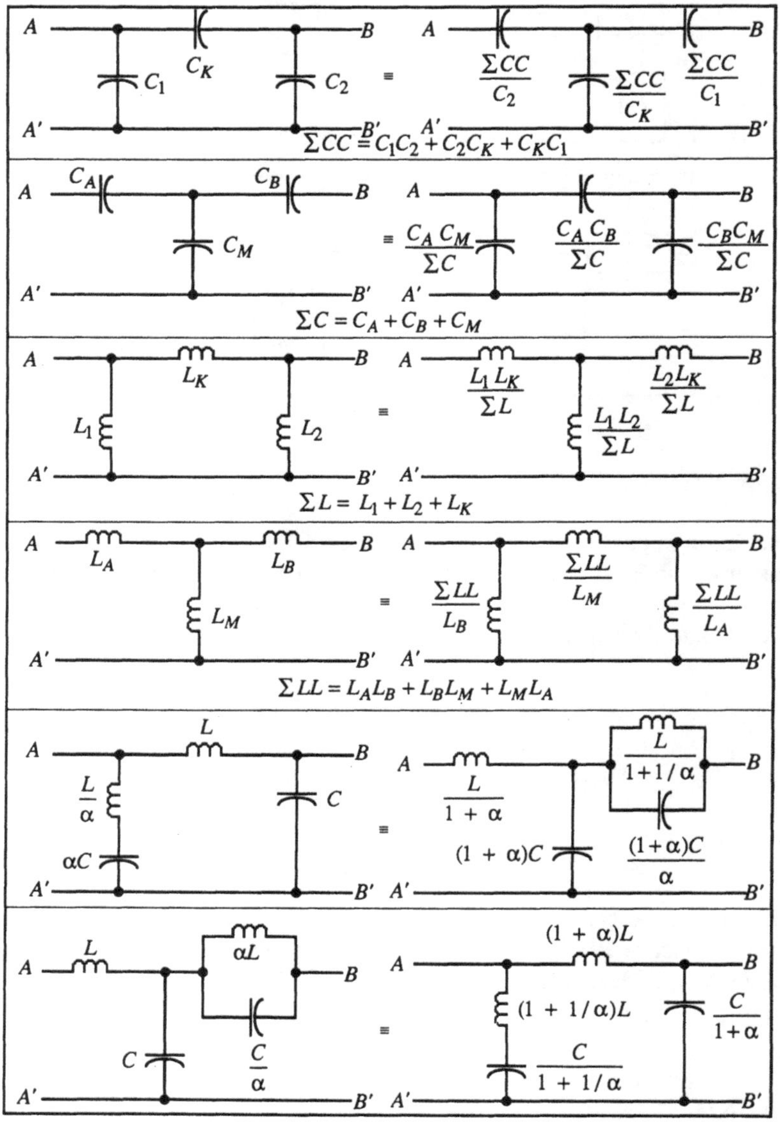

Figure 6.4 Common star-delta equivalents.

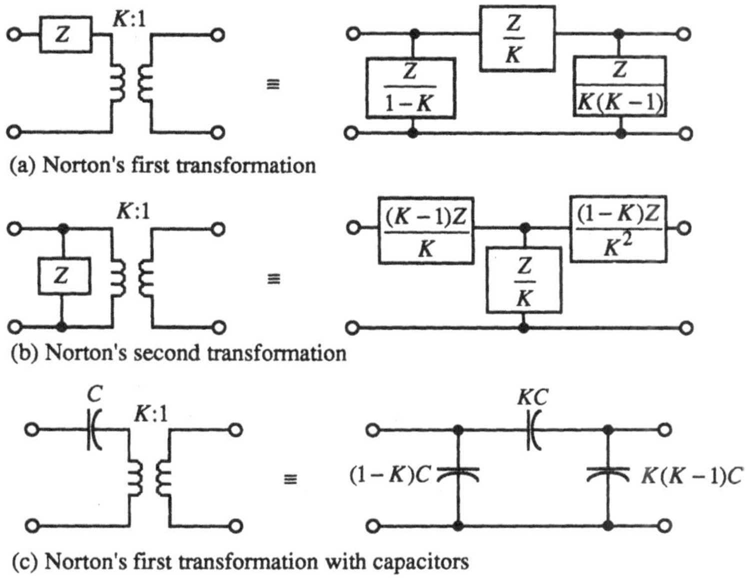

Figure 6.5 Norton's transformation.

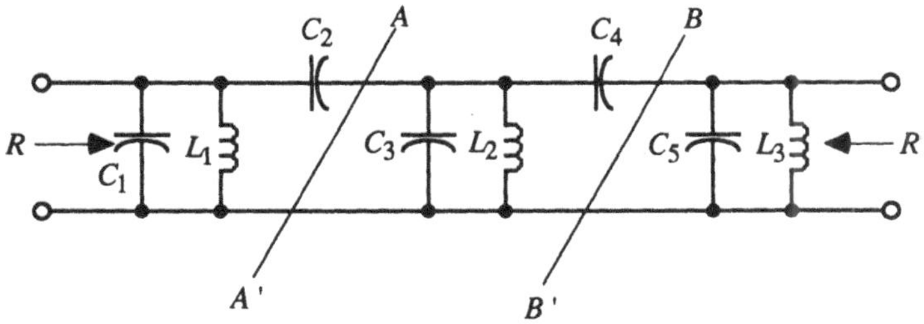

Figure 6.6 Bandpass filter to be transformed.

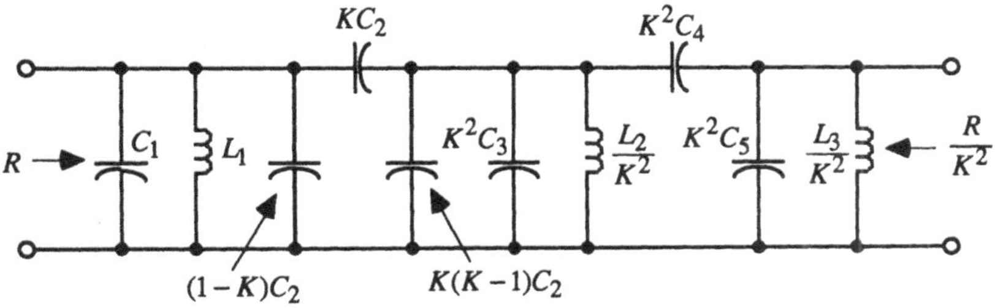

Figure 6.7 Bandpass filter with the transformation inserted.

If K were very small, then the voltage rise will be equal to $1/K$, and the components must be carefully chosen to withstand the high potential across the reactive elements.

It is also possible to eliminate either C_3 or C_5 from Figure 6.6 by inserting the same transformation at the line BB'. Also, it would be possible to bring the output impedance of the filter back to the value of R, if a transformation at BB' were made with $K_1K_2 = 1$, where K_1 is the transformation ratio performed at AA' and K_2 is the transformation ratio performed at BB'.

The element values for the bandpass in Figure 6.8 are obtained using the equations in Table 6.2. By inspection of these formulas, it is apparent that, at narrow bandwidths, $C_2 \gg C_1$ and $L_2 \ll L_1$, making the filter difficult to realize. By the use of the ideal transformer, however, this difficulty can be overcome.

Two ideal transformers will be inserted, such that the input and output impedances of the filter will be equal; that is, $K_1K_2 = 1$. The first transformation will be inserted at the dashed line AA' in Figure 6.8.

Norton's first transformation with capacitors is inserted into Figure 6.9(a). The capacitor, $K_1(K_1 - 1)C_1$, must be negative, making $K_1 < 1$. By a unique choice of K, it is possible to eliminate the shunt capacitor entirely. Here, we will set $K_1 = 0.5$ and $K_2 = 2$.

The value of C_x, the tuned circuit capacitor in Figure 6.9(b), is changed by the transformation

$$C_x = K_1^2C_2 + K_1(K_1 - 1)C_1$$

$$C_x = 0.25C_2 - 0.25C_1 = 0.25(C_2 - C_1)$$

The second transformer is to be inserted at a similar position on the right side between the coil and capacitor. This is shown in Figure 6.9(b). The value of K_2 to be used has been dictated: $K_2 = 2$. This will make the capacitor $0.25(1 - K_2)C_1$ negative.

The new element values are shown in Figure 6.9(c) where

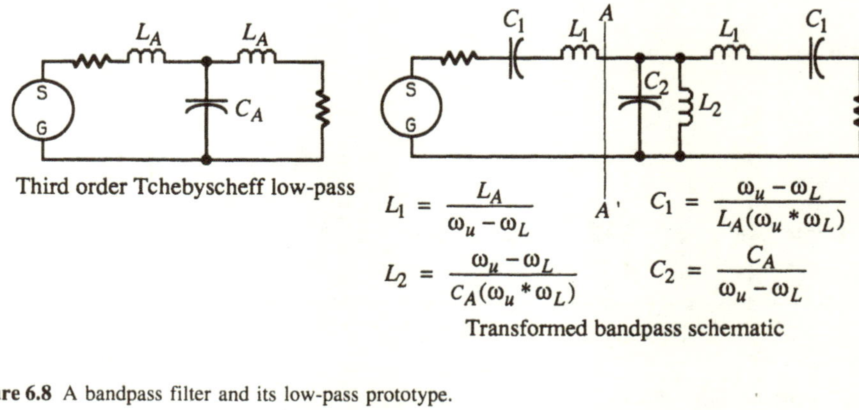

Figure 6.8 A bandpass filter and its low-pass prototype.

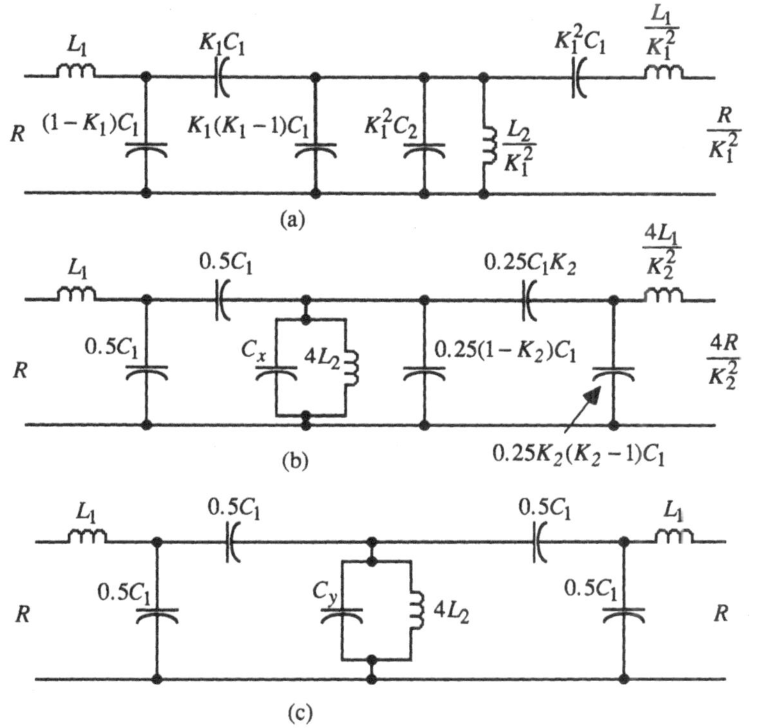

Figure 6.9 Transformation of a bandpass filter to realize practical component values.

$$C_y = 0.25C_2 - 0.25C_1 - 0.25C_1$$
$$C_y = 0.25(C_2 - 2C_1)$$

6.9 NONCANONIC TWO-TERMINAL NETWORKS*

Noncanonic network transformations are those that do not use the minimum number of components to realize a particular impedance function. This type of transformation is often applied when the inductance is rather large. One basic and very useful three-element network, illustrated in Figure 6.10(a), consists of a series resonant circuit in shunt with a capacitor. Figure 6.10(b) is its equivalent circuit. Two conditions of equivalence are imposed. The driving point impedance of the networks must be identical and the basic configuration must be similar. Figure 6.10(b) was derived by extracting a common capaci-

*The following discussion is taken from Anatol I. Zverev, *Handbook of Filter Synthesis*, New York: John Wiley & Sons, © copyright 1967, by permission of the publisher.

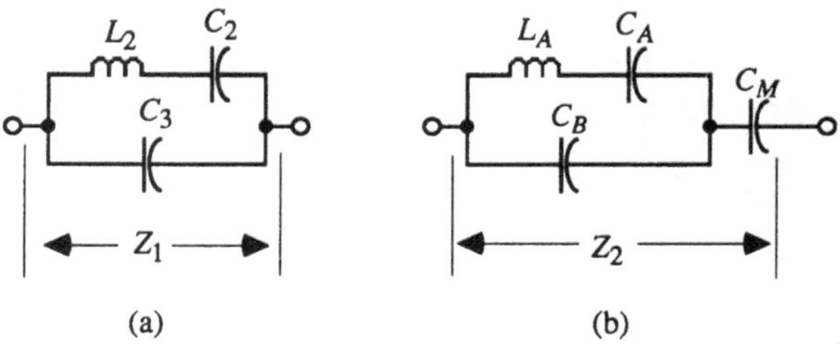

Figure 6.10 Noncanonic transformation.

tance from the network and putting it in a different position. This transformation is often applied when the inductance L_2 in Figure 6.10(a) is too large. The inductance L_A in Figure 6.10(b) will be considerably smaller.

Three conditions are required:

1. $Z_1 = Z_2 = Z$.
2. L_A must be less than L_2.
3. The basic configuration must be preserved.

Setting $Z_1 = Z_2$ and solving for L_A, C_A, C_B, and C_M, the following are obtained:

$$L_A = \left(1 - \frac{C_3}{C_M}\right)^2 L_2 \qquad \text{with } C_M = KC_3 \tag{6.8}$$

$$C_A = \frac{C_2C_M^2}{(C_M - C_3)(C_M - C_3 - C_2)} \tag{6.9}$$

$$C_B = \frac{C_3C_M}{(C_M - C_3)} \tag{6.10}$$

For an example, consider an application in crystal filter design, where the value of L_A can easily be produced by the crystal manufacturer. Then the ratio of L_A/L_2 is known. The term L_2 is the value of equivalent inductance and found to be of impractical magnitude. Using this ratio, (6.8) through (6.10) can be represented in the following forms:

$$a = \frac{L_A}{L_2} \tag{6.11}$$

$$K = \frac{1 + \sqrt{a}}{1 - a} \tag{6.12}$$

$$C_M = KC_3 \tag{6.13}$$

$$C_A = \frac{K^2C_2C_3}{(K - 1)[(K - 1)C_3 - C_2]} \tag{6.14}$$

$$C_B = \frac{KC_3}{K - 1} \tag{6.15}$$

The resonant frequency of L_A and C_A is designated f_A. With the aid of equation (6.14),

$$f_A = \frac{1}{2\pi K}\sqrt{\frac{(K-1)}{L_A}\left[\frac{K-1}{C_2} - \frac{1}{C_3}\right]} = \frac{1}{2\pi\sqrt{L_A C_A}} \tag{6.16}$$

This frequency, f_A, represents the principal ordering information for the crystal manufacturer. The physical dimensions of both crystals in series and lattice arms can always, by this method, be brought to the same practical values. The series L_A–C_A portion of the crystal equivalent circuit in Figure 6.10(b) is referred to as the *motional* arm of the circuit and arises from the mechanical crystal vibrations.

Example 6.4

Given that the two critical frequencies of a lattice network are

$$f_2 = 500.330 \text{ KHz}$$
$$f_3 = 502.585 \text{ KHz}$$

Calculated motional inductance of one crystal is L_2 = 15.92 H and calculated motional capacitance C_2 = 0.006356 pF. Interelectrode capacitance C_3 = 0.593529 pF. Equivalent motional inductance of the second crystal is L_A = 1.5535 H. The 15.92 H motional inductance is impractical. Use the noncanonic transformation to modify the reactance components and reduce this inductance to the same value as the second crystal (1.5535H).

Solution

A crystal with an equivalent motional inductance of 1.5535 H can easily be manufactured. Therefore L_A = 1.5535 H. From equation (6.11),

$$a = \frac{L_A}{L_2} = \frac{1.5535}{15.92} = 0.09758$$
$$\sqrt{a} = 0.312378$$

From equations (6.12) through (6.15),

$$K = \frac{1 + \sqrt{a}}{1 - a} = 1.4543$$

$$C_M = KC_3 = 1.4543 \cdot 0.593529 = 0.86317 \text{ pF}$$

$$C_A = \frac{K^2 C_2 C_3}{(K - 1)[(K - 1)C_3 - C_2]} = 0.0667037 \text{ pF}$$

$$C_B = \frac{KC_3}{K - 1} = 1.9 \text{ pF}$$

$$f_A = \frac{1}{2\pi\sqrt{L_A C_A}} = 494.413 \text{ KHz}$$

The frequency f_A is very important because it determines the series resonance of the crystal to be ordered.

6.10 CANONIC TWO-TERMINAL NETWORKS*

A well-designed filter should satisfy not only the stated attenuation requirements but should also be easily manufactured and aligned. Desirable features for passive filters designs include several factors:

1. The number of components, especially inductors, should be kept to a minimum.
2. The ratio of the largest to the smallest inductance should be minimized.
3. The ratio of the largest to the smallest capacitor should be minimized.
4. It is desirable to design for a capacitor in parallel with every shunt inductor in the ladder network to absorb the distributed capacitance of the inductor.
5. A capacitor is desired from the "hot" node of the series-resonant circuit in the series arm of the filter to ground to avoid problems with parasitic capacitances.
6. Proper distribution of tuning circuits is needed to minimize the sensitivity of the network with respect to terminal mismatches.

If the filter has been designed by means of some arbitrary method of synthesis, then there are numerous methods of modifying the schematic to gain some of these conditions. Norton's transformation, Bartlett's bisection theorem, and the star-delta transform can provide a variety of noncanonic network structures. Table 6.4 is included to minimize the time and effort for making network transformations. These are canonic networks because they use a minimum number of inductors and capacitors.

*The following discussion is taken from Anatol I. Zverev, *Handbook of Filter Synthesis,* New York: John Wiley & Sons, © copyright 1967, by permission of the publisher.

Table 6.4
Canonic Transformations (from [2], by permission of the publisher)

1

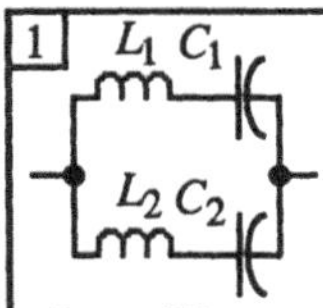

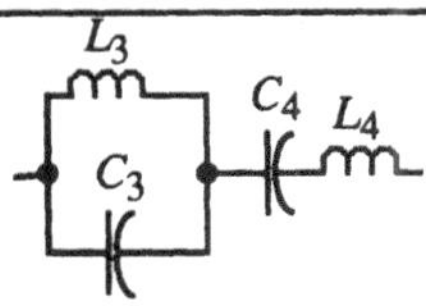

$L_1 = W$
$L_2 = X$
$C_1 = Y$
$C_2 = Z$
$D = C_4$
$E = L_3C_3C_4$
$A = L_3L_4C_3C_4$
$B = L_3C_3 + L_4C_4 + L_3C_4$

$$C_3 = \frac{C_1C_2(C_1+C_2)(L_1+L_2)^2}{(L_1C_1-L_2C_2)^2}$$

$$L_3 = \frac{(L_1C_1-L_2C_2)^2}{(L_1+L_2)(C_1+C_2)^2}$$

$$C_4 = C_1+C_2$$

$$L_4 = \frac{L_1L_2}{L_1+L_2}$$

2

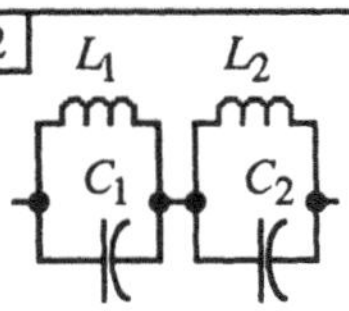

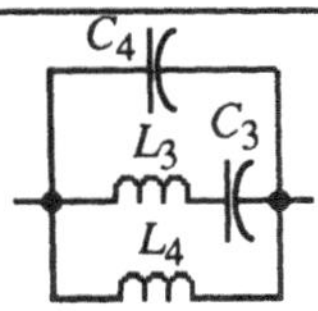

$C_1 = W$
$C_2 = X$
$L_1 = Y$
$L_2 = Z$
$A = L_3L_4C_3C_4$
$B = C_4L_4 + C_3L_3 + L_4C_3$
$D = L_4$
$E = L_3L_4C_3$

$$L_3 = \frac{L_1L_2(L_1+L_2)(C_1+C_2)^2}{(L_1C_1-L_2C_2)^2}$$

$$C_3 = \frac{(L_1C2-L_2C_2)^2}{(C_1+C_2)(L_1+L_2)^2}$$

$$C_4 = \frac{C_1C_2}{(C_1+C_2)}$$

$$L_4 = L_1+L_2$$

$$W = \frac{A(A-P^2)}{P(AD-PE)} \qquad Y = \frac{AD-PE}{A-P^2}$$

$$X = \frac{A-P^2}{E-PD} \qquad Z = \frac{P(E-PD)}{A-P^2}$$

$$\text{where } P = \frac{B+\sqrt{B^2-4A}}{2}$$

6.11 BANDPASS DESIGN TO ABSORB PARASITIC CAPACITANCE

One limitation of the design methods discussed so far is that they result in networks that are sometimes sensitive to stray capacitance. This occurs because some of the network nodes are not connected to ground through a shunt capacitor.

In Figure 6.11, a normalized elliptic low-pass filter is used as a prototype for a bandpass design. The desired bandpass edge frequencies are 95 MHz and 105 MHz. A bandpass transformation is applied via the relations in Table 6.2 with $\omega_L = 5.969 \times 10^8$, and $\omega_u = 6.5973 \times 10^8$. The canonical transform in Table 6.4 was used to further reduce the circuit as shown in Figure 6.11(c). Norton's transform, as shown in Figure 6.12, is used in the final configuration to absorb any parasitic capacitances at the two nodes in Figure 6.11(c) that have no direct shunt to ground [3].

The transformation applied to Figure 6.11(c) uses $K = 2$ and is applied immediately after the L–C section consisting of a 81.30 nH inductor and a 45.26 pF capacitor. The first transformation functions as a step-down transformer and all the impedances after it must be impedance scaled by K^2. A second transformation is applied in the opposite direction with $K = 2$ after the L–C section consisting of 27.49 nH inductor and 117.08 pF capacitor. The negative shunt components that result from the transformation are combined in parallel with the positive shunt inductors and capacitors to obtain realizable component values. The total capacitance of two capacitors in parallel is the sum of the individual capacitors whereas the total inductance of two noncoupled inductors in parallel is the product of the inductances divided by the sum of the inductances.

The use of Norton's transformation as shown in Figure 6.12 is the basis for many exotic types of network transformations. The values of K could have been chosen so that only two additional shunt capacitors, or two additional shunt inductors, would have resulted in the final configuration. In this case, the transformations usually will not maintain equal source and termination resistances.

6.12 GENERAL NETWORK EQUATIONS

In Figure 6.13, a four-terminal network is shown. According to Kirchhoff's laws, the following equations can be written.

$$\begin{aligned} V_1 &= I_1 Z_{11} + I_2 Z_{12} \\ V_2 &= I_1 Z_{21} + I_2 Z_{22} \end{aligned} \tag{6.17}$$

or

$$\begin{aligned} I_1 &= V_1 Y_{11} + V_2 Y_{12} \\ I_2 &= V_1 Y_{21} + V_2 Y_{22} \end{aligned} \tag{6.18}$$

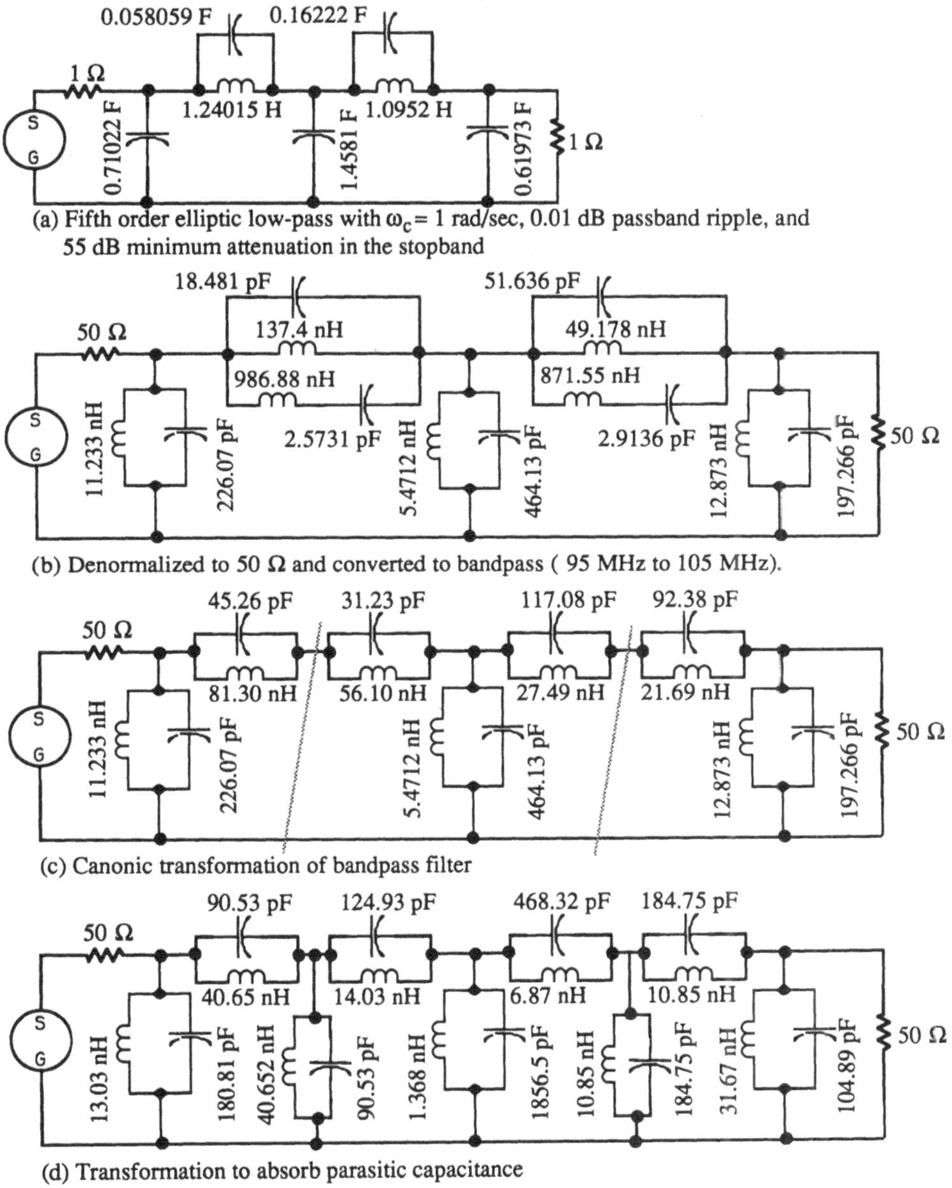

Figure 6.11 Use of Norton's transformation to absorb parasitic capacitance.

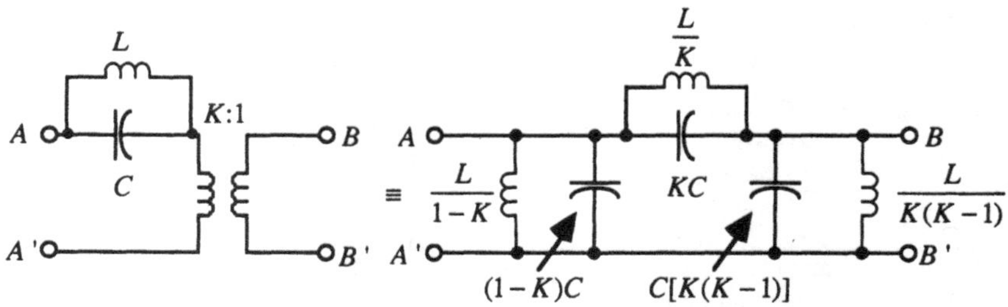

Figure 6.12 Equivalent Norton's transformation to generate parallel *L–C* sections.

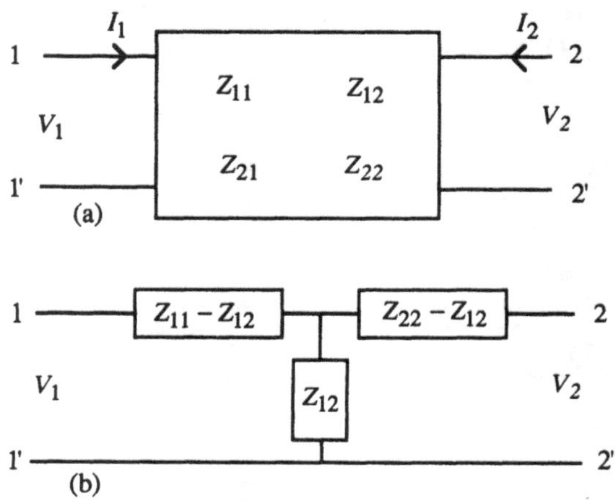

Figure 6.13 (a) Four-terminal network and (b) its *T*-equivalent.

The first pair of equations is called the *impedance equations,* and the second pair is called the *admittance equations.* From the impedance equations, it follows that when $I_2 = 0$ (a secondary open circuit),

$$Z_{11} = \frac{V_1}{I_1} \quad \text{primary open-circuit input impedance} \tag{6.19}$$

$$Z_{21} = \frac{V_2}{I_1} \quad \text{transfer open-circuit impedance}$$

When $I_1 = 0$,

$$Z_{22} = \frac{V_2}{I_2} \quad \text{secondary open-circuit input impedance} \tag{6.20}$$

$$Z_{12} = \frac{V_1}{I_2} \quad \text{transfer open-circuit impedance}$$

From the admittance equations, it follows that, when $V_2 = 0$ (a secondary short circuit),

$$Y_{11} = \frac{I_1}{V_1} \quad \text{primary short-circuit admittance} \tag{6.21}$$

$$Y_{21} = \frac{I_2}{V_1} \quad \text{transfer short-circuit admittance}$$

and when $V_1 = 0$ (a primary short circuit),

$$Y_{22} = \frac{I_2}{V_2} \quad \text{secondary short-circuit admittance} \tag{6.22}$$

$$Y_{12} = \frac{I_1}{V_2} \quad \text{transfer short-circuit admittance}$$

From the theory of reciprocity, $Z_{12} = Z_{21}$ and $Y_{12} = Y_{21}$ for all passive networks regardless of symmetry.

A different representation of network values is given by

$$\begin{aligned} V_1 &= V_2A + I_2B \\ I_1 &= V_2C + I_2D \end{aligned} \tag{6.23}$$

with a chain matrix given by

$$\|M\| = \begin{Vmatrix} A & B \\ C & D \end{Vmatrix}$$

The determinant of the chain matrices $|M| = |AD - BC|$ has, for all reversible networks, the value 1. (Nonreversible networks are those transmitting energy differently in different directions, and they contain amplifiers.) The *ABCD*, or chain, matrix, has its utility in allowing for the easy computation of network functions. For Figure 6.14, the voltage transfer function can be written as

$$\frac{V_3}{V_1} = \begin{bmatrix} A_1 & B_1 \\ C_1 & D_1 \end{bmatrix} \begin{bmatrix} A_2 & B_2 \\ C_2 & D_2 \end{bmatrix} \tag{6.24}$$

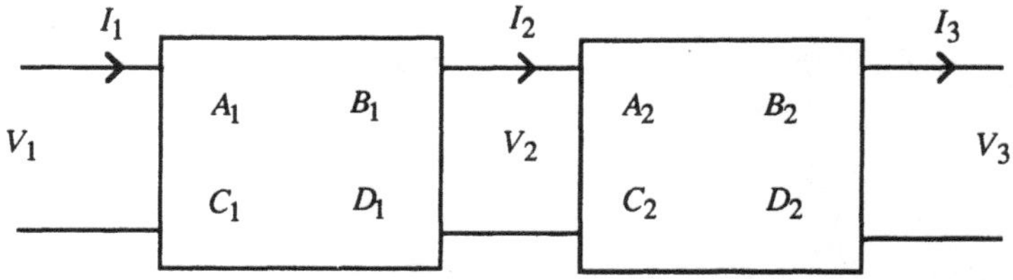

Figure 6.14 Transfer function representation using *ABCD* matrices.

A library of *ABCD* matrices for elementary two-port networks can be applied in building-block fashion to more complicated networks that consists of cascades of these simpler two-ports. Table 6.5 lists a number of two-port networks and their *ABCD* matrices.

Table 6.5
The *ABCD* Parameters of Common Two-Port Networks (from [1], by permission of the publisher)

Circuit	*ABCD Parameter*	
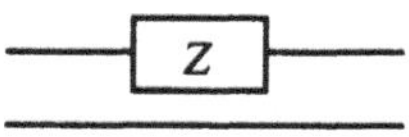	$A = 1$ $C = 0$	$B = Z$ $D = 1$
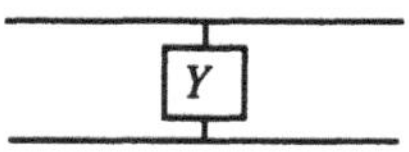	$A = 1$ $C = Y$	$B = 0$ $D = 1$
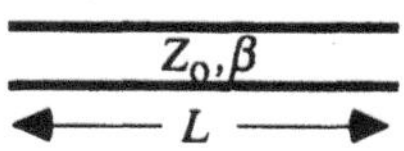	$A = \cos(\beta L)$ $C = j\,Y_o \sin(\beta L)$	$B = j\,Z_o \sin(\beta L)$ $D = \cos(\beta L)$
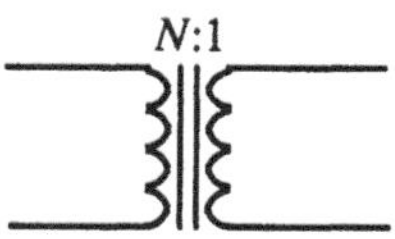	$A = N$ $C = 0$	$B = 0$ $D = \frac{1}{N}$
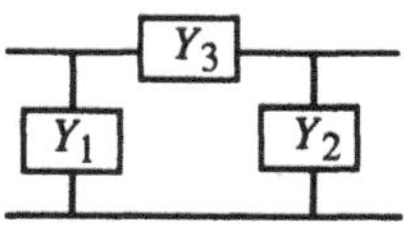	$A = 1 + \frac{Y_2}{Y_3}$ $C = Y_1 + Y_2 + \frac{Y_1 Y_2}{Y_3}$	$B = \frac{1}{Y_3}$ $D = 1 + \frac{Y_1}{Y_3}$
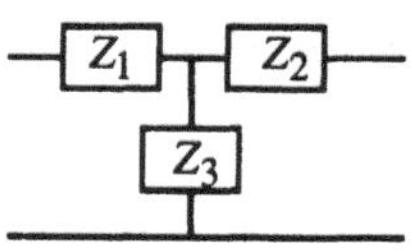	$A = 1 + \frac{Z_1}{Z_3}$ $C = \frac{1}{Z_3}$	$B = Z_1 + Z_2 + \frac{Z_1 Z_2}{Z_3}$ $D = 1 + \frac{Z_2}{Z_3}$

Like the impedance or admittance matrix for an N-port network, the scattering matrix [1] provides a complete description of the network as seen at its ports. Whereas the previous methods relate the total voltages and currents at the ports, the scattering matrix relates the voltage waves incident on the ports to those reflected from the ports. For some components and circuits, the scattering parameters can be calculated using network analysis techniques. Otherwise, the scattering parameters can be measured directly with a vector network analyzer. Once the scattering parameters of the network are known, conversion to other matrix parameters can be performed, if needed. Scattering parameters are popular because in RF and microwave circuits, the impedance or admittance parameters cannot be easily measured.

For the two-port network in Figure 6.15, the following equations can be written:

$$\begin{aligned} V_1^- &= S_{11}V_1^+ + S_{12}V_2^+ \\ V_2^- &= S_{21}V_1^+ + S_{22}V_2^+ \end{aligned} \tag{6.25}$$

where V_1^+ represents the incident voltage at port 1 and V_1^- is reflected voltage at port 1. The voltage standing wave ratio (VSWR) is given by

$$\text{VSWR} = \frac{|V^+| + |V^-|}{|V^+| - |V^-|} \tag{6.26}$$

For a transmission line with characteristic impedance Z_o and load impedance Z_L, the voltage reflection coefficient, Γ, at the load is

$$\Gamma = \frac{V_o^-}{V_o^+} = \frac{Z_L - Z_o}{Z_L + Z_o} = \frac{\text{VSWR} - 1}{\text{VSWR} + 1} \tag{6.27a}$$

The S parameters are easily extendable to N-port networks. If V_N^+ is the amplitude of the voltage wave incident on port N, and V_N^- is the amplitude of the voltage wave reflected from port N, the scattering matrix, or $[S]$ matrix, is defined in relation to these incident and reflected voltage waves as

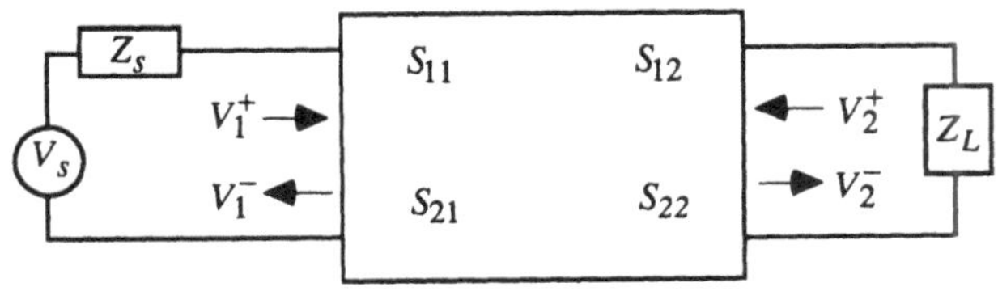

Figure 6.15 Two-port S parameter network.

$$\begin{bmatrix} V_1^- \\ V_2^- \\ \cdot \\ \cdot \\ \cdot \\ V_N^- \end{bmatrix} = \begin{bmatrix} S_{11} & S_{12} & \ldots & S_{1N} \\ S_{21} & S_{22} & \ldots & S_{2N} \\ \cdot & & & \\ \cdot & & & \\ \cdot & & & \\ S_{N1} & S_{N2} & \ldots & S_{NN} \end{bmatrix} = \begin{bmatrix} V_1^+ \\ V_2^+ \\ \cdot \\ \cdot \\ \cdot \\ V_N^+ \end{bmatrix}$$

or

$$[V^-] = [S][V^+] \tag{6.27b}$$

The term S_{ii} is the reflection coefficient (Γ_{ii}) seen looking into port i when all other ports are terminated in matched loads, and S_{ij} is the transmission coefficient ($T_{ij} = 1 + \Gamma_{ij}$) from port j to port i when all other ports are terminated in matched loads. Useful conversions for two-port networks are given in Table 6.6.

For interconnected two-port networks, the overall impedance matrix can be found by adding the individual impedance matrices of the networks that are connected in series. The overall admittance for networks connected in parallel can be found by adding the admittance matrices of the individual networks connected in shunt. As an example, a bridged-T circuit is shown in Figure 6.16. The network can be decomposed into the parallel connection of two simpler networks, as shown in Figure 6.16(b). The impedance and admittance matrices for these two sub networks can be written as

$$Z_a = \begin{bmatrix} Z_1 + Z_2 & Z_2 \\ Z_2 & Z_1 + Z_2 \end{bmatrix} \quad \text{and} \quad Y_b = \begin{bmatrix} \frac{1}{Z_3} & \frac{-1}{Z_3} \\ \frac{-1}{Z_3} & \frac{1}{Z_3} \end{bmatrix}$$

Inverting Z_a and applying the previous result for shunt-connected network gives the overall admittance matrix as

$$Y = Y_a + Y_b = \begin{bmatrix} \frac{1}{Z_3} + \frac{Z_1 + Z_2}{D} & -\left(\frac{1}{Z_3} + \frac{Z_2}{D}\right) \\ -\left(\frac{1}{Z_3} + \frac{Z_2}{D}\right) & \frac{Z_1 + Z_2}{D} + \frac{1}{Z_3} \end{bmatrix}, \qquad D = Z_1(Z_1 + 2Z_2)$$

The general definition of series and parallel interconnection of networks is shown in Figure 6.17.

Table 6.6
Conversions Between Two-Port Networks (from [1], by permission of the publisher)

	S	Z	Y	ABCD
S_{11}	S_{11}	$\dfrac{(Z_{11}-Z_o)(Z_{22}+Z_o)-Z_{12}Z_{21}}{\Delta Z}$	$\dfrac{(Y_o-Y_{11})(Y_o+Y_{11})+Y_{12}Y_{21}}{\Delta Y}$	$\dfrac{A+B/Z_o-CZ_o-D}{A+B/Z_o+CZ_o+D}$
S_{12}	S_{12}	$\dfrac{2Z_{12}Z_o}{\Delta Z}$	$\dfrac{-2Y_{12}Y_o}{\Delta Y}$	$\dfrac{2(AD-BC)}{A+B/Z_o+CZ_o+D}$
S_{21}	S_{21}	$\dfrac{2Z_{21}Z_o}{\Delta Z}$	$\dfrac{-2Y_{21}Y_o}{\Delta Y}$	$\dfrac{2}{A+B/Z_o+CZ_o+D}$
S_{22}	S_{22}	$\dfrac{(Z_{11}+Z_o)(Z_{22}-Z_o)-Z_{12}Z_{21}}{\Delta Z}$	$\dfrac{(Y_o+Y_{11})(Y_o-Y_{22})+Y_{12}Y_{21}}{\Delta Y}$	$\dfrac{-A+B/Z_o-CZ_o+D}{A+B/Z_o+CZ_o+D}$
Z_{11}	$Z_o\dfrac{(1+S_{11})(1-S_{22})+S_{12}S_{21}}{(1-S_{11})(1-S_{22})-S_{12}S_{21}}$	Z_{11}	$\dfrac{Y_{22}}{\lvert Y\rvert}$	$\dfrac{A}{C}$
Z_{12}	$Z_o\dfrac{2S_{12}}{(1-S_{11})(1-S_{22})-S_{12}S_{21}}$	Z_{12}	$\dfrac{-Y_{12}}{\lvert Y\rvert}$	$\dfrac{AD-BC}{C}$
Z_{21}	$Z_o\dfrac{2S_{21}}{(1-S_{11})(1-S_{22})-S_{12}S_{21}}$	Z_{21}	$\dfrac{-Y_{21}}{\lvert Y\rvert}$	$\dfrac{1}{C}$
Z_{22}	$Z_o\dfrac{(1-S_{11})(1+S_{22})+S_{12}S_{21}}{(1-S_{11})(1-S_{22})-S_{12}S_{21}}$	Z_{22}	$\dfrac{Y_{11}}{\lvert Y\rvert}$	$\dfrac{D}{C}$
Y_{11}	$Y_o\dfrac{(1-S_{11})(1+S_{22})+S_{12}S_{21}}{(1+S_{11})(1+S_{22})-S_{12}S_{21}}$	$\dfrac{Z_{22}}{\lvert Z\rvert}$	Y_{11}	$\dfrac{D}{B}$
Y_{12}	$Y_o\dfrac{-2S_{12}}{(1+S_{11})(1+S_{22})-S_{12}S_{21}}$	$\dfrac{-Z_{12}}{\lvert Z\rvert}$	Y_{12}	$\dfrac{BC-AD}{B}$
Y_{21}	$Y_o\dfrac{-2S_{21}}{(1+S_{11})(1+S_{22})-S_{12}S_{21}}$	$\dfrac{-Z_{21}}{\lvert Z\rvert}$	Y_{21}	$\dfrac{-1}{B}$
Y_{22}	$Y_o\dfrac{(1+S_{11})(1-S_{22})+S_{12}S_{21}}{(1+S_{11})(1+S_{22})-S_{12}S_{21}}$	$\dfrac{Z_{11}}{\lvert Z\rvert}$	Y_{22}	$\dfrac{A}{B}$
A	$\dfrac{(1+S_{11})(1-S_{22})+S_{12}S_{21}}{2S_{21}}$	$\dfrac{Z_{11}}{Z_{21}}$	$\dfrac{-Y_{22}}{Y_{21}}$	A
B	$Z_o\dfrac{(1+S_{11})(1+S_{22})-S_{12}S_{21}}{2S_{21}}$	$\dfrac{\lvert Z\rvert}{Z_{21}}$	$\dfrac{-1}{Y_{21}}$	B
C	$\dfrac{1}{Z_o}\dfrac{(1-S_{11})(1-S_{22})-S_{12}S_{21}}{2S_{21}}$	$\dfrac{1}{Z_{21}}$	$\dfrac{-\lvert Y\rvert}{Y_{21}}$	C
D	$\dfrac{(1-S_{11})(1+S_{22})-S_{12}S_{21}}{2S_{21}}$	$\dfrac{Z_{22}}{Z_{21}}$	$\dfrac{-Y_{11}}{Y_{21}}$	D

$\lvert Z\rvert = Z_{11}Z_{22}-Z_{12}Z_{21}$; $\lvert Y\rvert = Y_{11}Y_{22}-Y_{12}Y_{21}$; $\Delta Y = (Y_{11}+Y_o)(Y_{22}+Y_o)-Y_{12}Y_{21}$; $\Delta Z = (Z_{11}+Z_o)(Z_{22}+Z_o)-Z_{12}Z_{21}$; $Y_o = 1/Z_o$

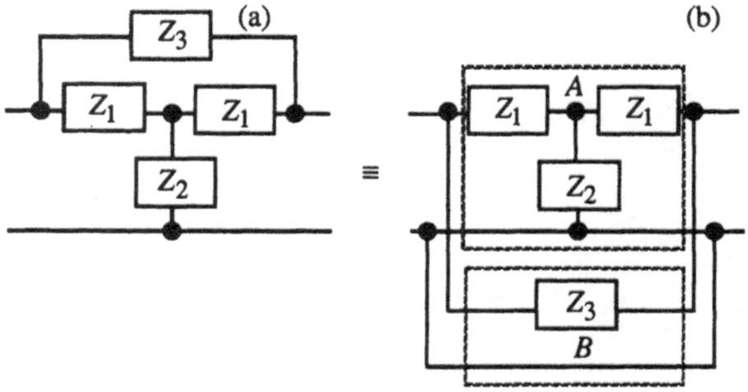

Figure 6.16 Decomposition of a bridged-T network into two simpler networks.

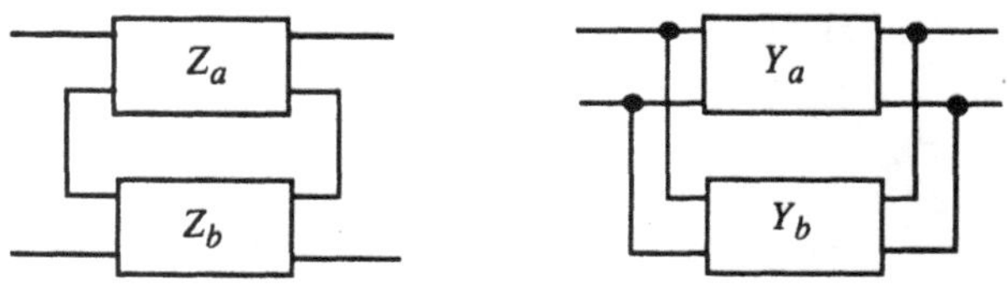

Figure 6.17 Series and parallel connections of two-port networks.

6.13 PASSIVE TOPOLOGIES FOR GROUP DELAY EQUALIZATION

It is known from the theory of four-terminal networks that, for every symmetrical ladder section, there corresponds an equivalent lattice schematic having equal transmission properties and the same input impedance [4]. According to Bartlett's method, to obtain the equivalent lattice, the ladder network has to be bisected and the short-circuit impedance of the half-section is used as the series arms of the lattice sections. The open-circuit impedances are used as diagonal arms of the lattice. This procedure is shown in Figure 6.18.

In Figure 6.19, the illustration of Bartlett's theorem for reactive networks is shown. Under certain conditions, the lattice cannot be transformed into a ladder because of the

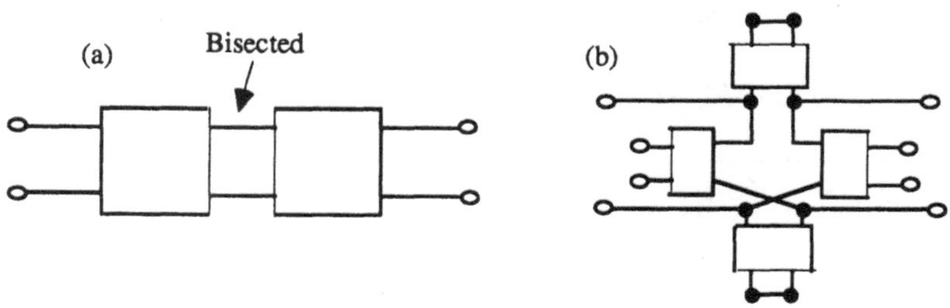

Figure 6.18 Symmetric ladder section and its equivalent lattice.

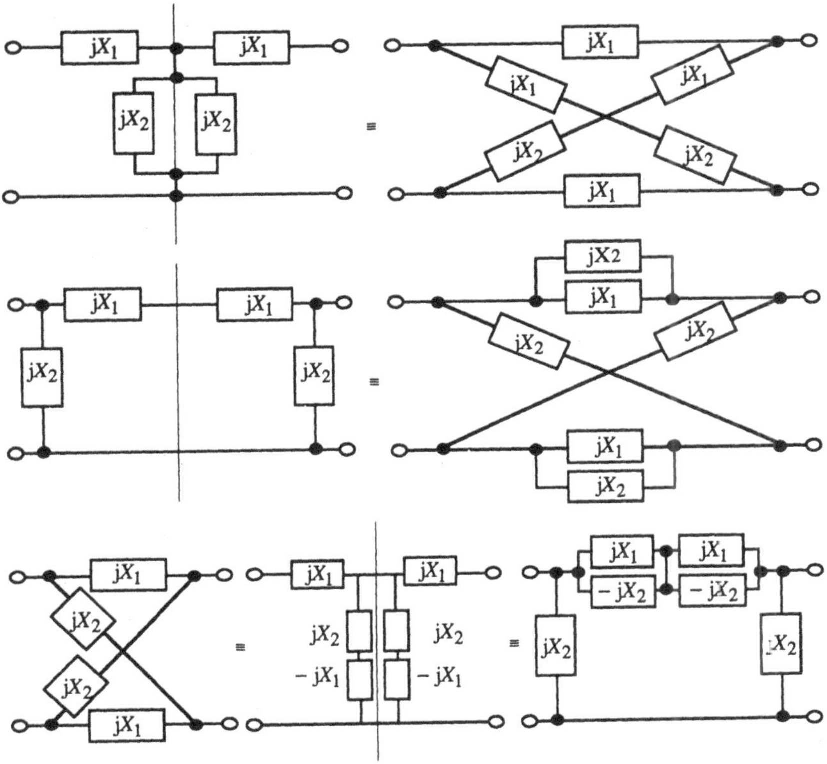

Figure 6.19 Network transformations based on Bartlett's theorem.

negative reactances involved in the transformation. On the other hand, every ladder has a lattice equivalent.

For the lattice configuration in Figure 6.20,

$$Z_{11} = Z_{22} = \frac{1}{2}[Z_b + Z_a] \tag{6.28}$$

$$Z_{12} = Z_{21} = \frac{1}{2}[Z_b - Z_a] \tag{6.29}$$

If a voltage is applied to the input, and the output is terminated in a 1 Ω load, then the input impedance, Z_{in}, becomes

$$Z_{in} = \frac{Z_{11} + Z_{11}Z_{22} - Z_{12}^2}{1 + Z_{22}} \tag{6.30}$$

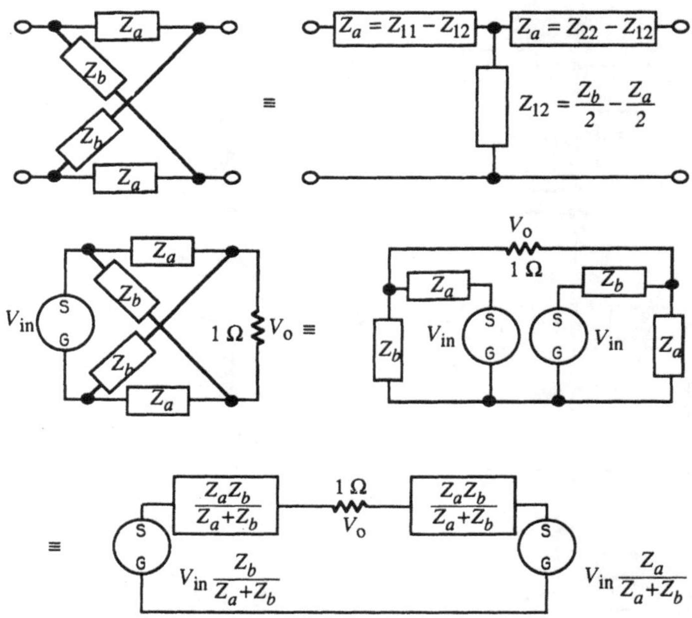

Figure 6.20 Lattice equivalent circuits.

The lattice network can be analyzed by applying Thevenin's theorem as shown in Figure 6.20. The voltage transfer function can be written by inspection of Figure 6.20 as

$$\frac{V_o}{V_{in}} = \frac{Z_b - Z_a}{2Z_aZ_b + Z_a + Z_b} \tag{6.31}$$

If

$$Z_a = \frac{1}{Z_b} \tag{6.32}$$

then there is a cancellation of terms reducing the transfer function to

$$\frac{V_o}{V_{in}} = \frac{1 - Z_a}{1 + Z_a} \tag{6.33}$$

which can produce an all-pass function if Z_a is purely reactive.

Using equations (6.28), (6.29), and (6.32), the input impedance given by (6.30) can be shown to be a constant 1 Ω resistance at all frequencies. This result makes it possible to cascade all-pass stages to a prototype filter design without affecting the amplitude response of the filter. The circuit in Figure 6.21 is capable of realizing any second-order all-pass function. The requirement that $Z_a = 1/Z_b$ can be expressed as

$$Z_a = \frac{\dfrac{1}{sC_1}}{\dfrac{1}{sC_1} + Ls} = \frac{L_1 s}{1 + L_1 C_1 s^2} \tag{6.34}$$

$$\frac{1}{Z_b} = \frac{sC_2}{1 + L_2 C_2 s} \tag{6.35}$$

which requires $L_1 = C_2$ and $L_2 = C_1$ for an all-pass lattice with a 1 Ω termination. Substituting these values into (6.33) yields a transfer function of

$$\frac{V_o}{V_{\text{in}}} = \frac{s^2 - \dfrac{1}{C_1}s + \dfrac{1}{L_1C_1}}{s^2 + \dfrac{1}{C_1}s + \dfrac{1}{L_1C_1}} \tag{6.36}$$

with $\omega_o^2 = 1/(L_1C_1)$ and $Q = \sqrt{C_1/L_1}$. If C_1 is allowed to become zero, then L_1 becomes infinity and the network defaults to a first-order all pass.

The remaining task is simply to convert the lattice to a ladder. The conversion usually results in a negative inductance, L_b, that can be effectively realized by a center-tapped coil as shown in Figure 6.22.

Figure 6.23 shows the conversion of a first-order all-pass lattice into a ladder network using Bartlett's bisection theorem. The transfer function is given by

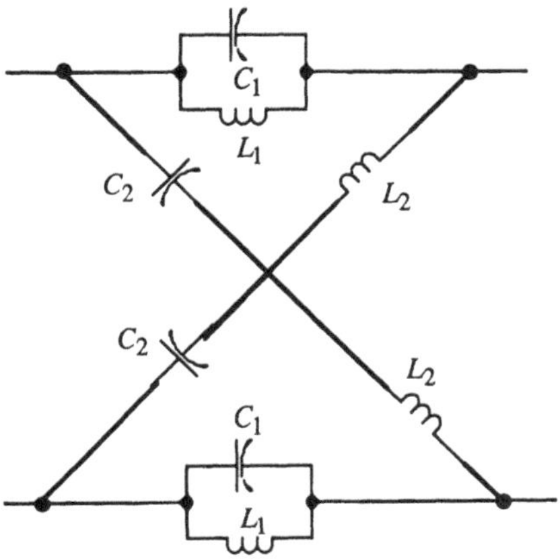

Figure 6.21 General all-pass lattice.

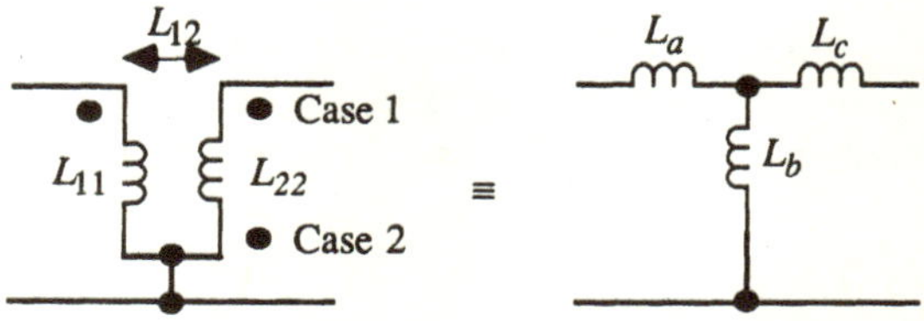

Two inductors with unity coefficient of coupling.
Total inductance is

$L_{12} = L_{11} + L_{22} \mp 2\sqrt{L_{11}L_{22}}$

$L_a = L_{11} \mp \sqrt{L_{11}L_{22}}$

$L_c = L_{22} \mp \sqrt{L_{11}L_{22}}$

$L_b = \pm\sqrt{L_{11}L_{22}}$

M = mutual inductance = $\sqrt{L_{11}L_{22}}$

Case 1: Opposite winding direction, L_a or L_c is negative. Use upper sign.

Case 2: Same winding direction, L_b is negative. Use lower sign

Figure 6.22 Realization of L_b as a negative inductance.

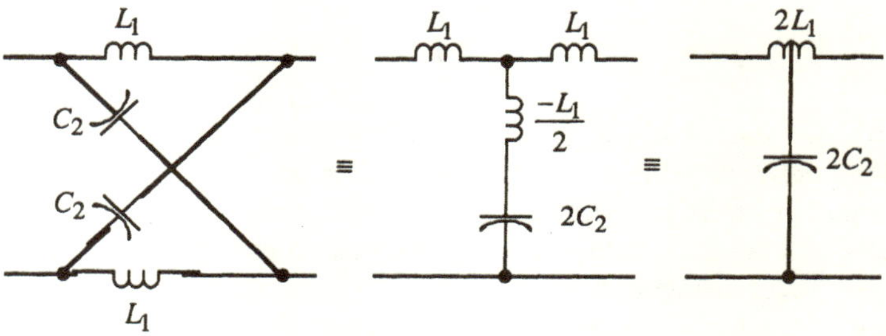

Figure 6.23 First-order all-pass lattice to ladder conversion.

$$\frac{V_o}{V_{in}} = \frac{1 - L_1 s}{1 + L_1 s} \tag{6.37}$$

and assumes that $C_2 = L_1$ and there is a 1 Ω termination resistance.

For the second-order all-pass lattice in Figure 6.21, Bartlett's bisection theorem is used to convert the lattice to a ladder as shown in Figure 6.24.

The final configurations of second-order all-pass networks in Figure 6.24 are derived from Figure 6.21 using a combination of Bartlett's bisection theorem, star-delta transformations, and Thevenin's equivalent circuits. If V_{in} and V_o are the input and output voltages of the all-pass network in Figure 6.24(b), then the voltage V_c can be shown to be $(V_o + V_{in})/2$ by application of Thevenin's equivalent. In 6.24(c), it can be shown that $V_a = V_b = (V_o + V_{in})/2$ regardless of the values of L_2 and C_2. All of the node voltages remain the same if either negative reactance is removed from the circuit. Removal of the negative

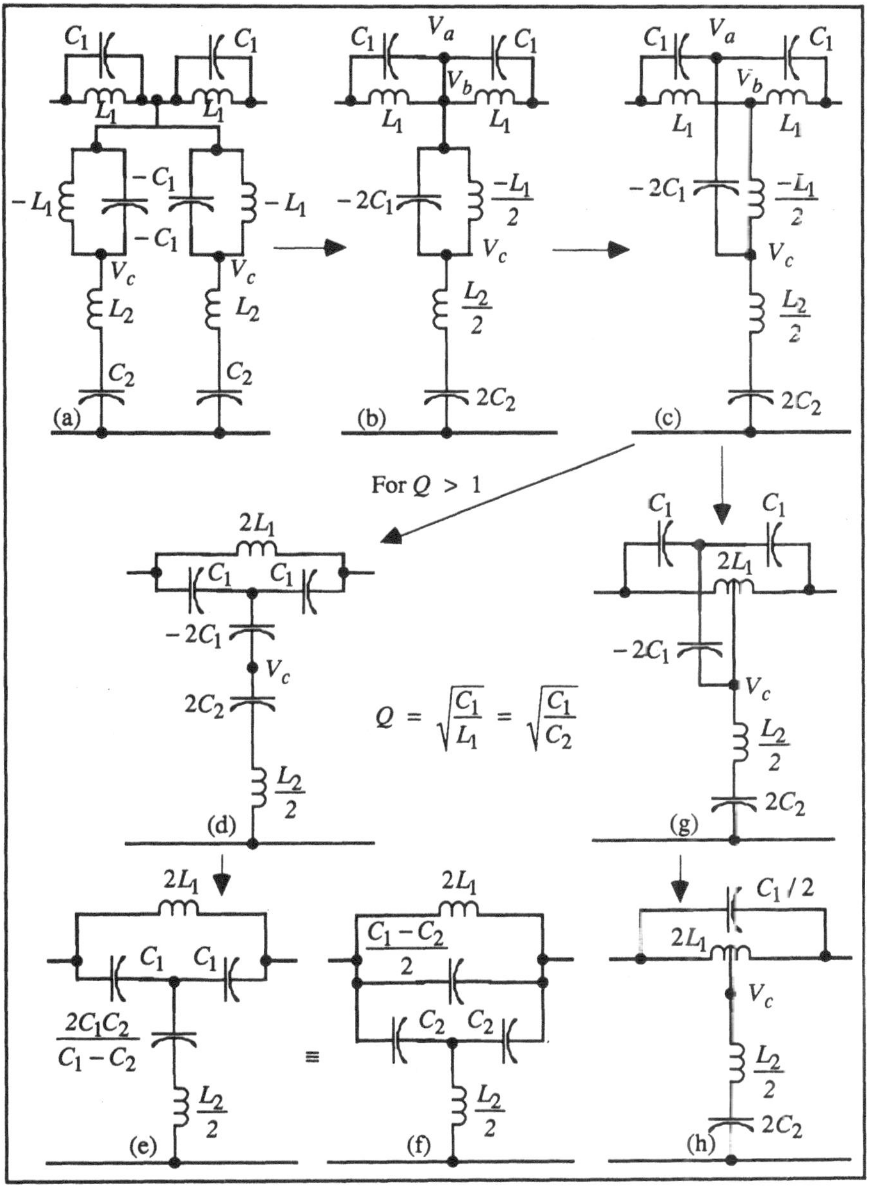

Figure 6.24 All-pass ladder equivalent networks derived from equation (6.19).

inductance results in a circuit that is realizable only for values of $Q > 1$. Figure 6.24(f) is produced from 6.24(e) by a star-delta transformation.

REFERENCES

[1] Pozar, David M., *Microwave Engineering,* Reading, MA: Addison-Wesley, 1990, pp. 220–231.
[2] Zverev, Anatol I., *Handbook of Filter Synthesis,* New York: John Wiley & Sons, 1967.
[3] Klappenberger, Albert J., ''Narrow Bandwidth Elliptic Bandpass Filters with Low Parts Value Spread,'' *RF Design Magazine,* December 1992, pp. 25–29.
[4] Sun, Hun H., *Synthesis of RC Networks,* Indianapolis: Hayden Book Company, 1967, pp. 35–40.

PROBLEMS

6.1 Find the dual of the network in Figure 6.11(d).

6.2 Design a Tchebyscheff Type I bandpass filter with 1 dB ripple between 1 MHz and 2 MHz and 60 dB attenuation below 500 KHz and above 4 MHz. Use 100 Ω source and load resistors.

6.3 Modify the lowpass in Figure 6.1 to insert a shunt L–C to absorb parasitic capacitance between the 15.19 uH inductor-834 pF capacitor to ground.

6.4 Use *ABCD* matrices to determine the frequency response of the circuit in Figure 6.8(a) with $L_A = 1$ H and $C_A = 2$ F.

6.5 Modify the circuit in Figure 6.1 for a load resistance of 200 Ω and a passband between 2 MHz and 4 MHz.

6.6 Find the component values for

(a) a high-pass filter with 0.1 dB ripple at a cutoff of 1 MHz and 60 dB attenuation below 100 KHz,

(b) a bandstop filter with 0.1 dB ripple at 500 KHz and 4 MHz and 60 dB attenuation between 1 MHz and 2 MHz.

In each case, start with a Tchebyscheff Type I low-pass prototype and 100 Ω source resistors. Use the minimum necessary filter order.

6.7 Find the component values for a second-order group delay equalizer in a ladder configuration with $Q = 2$, $\omega_o = 1$, and a load resistor of 1 Ω.
Repeat with a $Q = 0.5$.

6.8 Show that $V_c = V_b$ in Figure 6.24(c).

6.9 Show that V_c is the same in Figures 6.24(c), 6.24(d), and 6.24(h).

6.10 Show that equation (6.30) yields $Z_{\text{in}} = 1\ \Omega$ for a load resistance of 1 Ω.

Chapter 7
Synthesis of Unequally Terminated Passive Filters

7.1 DERIVATION OF THE INPUT IMPEDANCE

An unevenly terminated passive filter has unequal source and termination resistances. A general expression for the input impedance will be derived in this chapter, first for the synthesis of equally terminated passive filters, as shown in Figure 7.1, and then for the general case of unequal termination resistances. A general expression for the input impedance is required for the synthesis of passive filters. This expression can be derived from the voltage transfer function based on power transmitted to the load. In Figure 7.1, the power transmitted to the load resistor is given by

$$\text{Transmitted power} = \left[\frac{V_{\text{out}}}{V_{\text{in}}}\right]^2 = \left[\frac{\frac{1}{4}}{1 + \frac{F(s)F(-s)}{P(s)P(-s)}}\right] \tag{7.1}$$

If $V_{\text{in}} = 1$ volt, then the maximum possible output power = 1/4; therefore,

$$\text{Transmitted power} = \left[\frac{1}{1 + \frac{F(s)F(-s)}{P(s)P(-s)}}\right] \times 100\% \text{ of max} \tag{7.2}$$

In Figure 7.2, the input impedance is modeled by $Z_{\text{in}} = a + jb$. For $V_{\text{in}} = 1$ V, the power transmitted to the load is given by I^2a, because the network is assumed lossless, or

$$I^2a = \frac{a}{(a + 1)^2 + b^2} \tag{7.3}$$

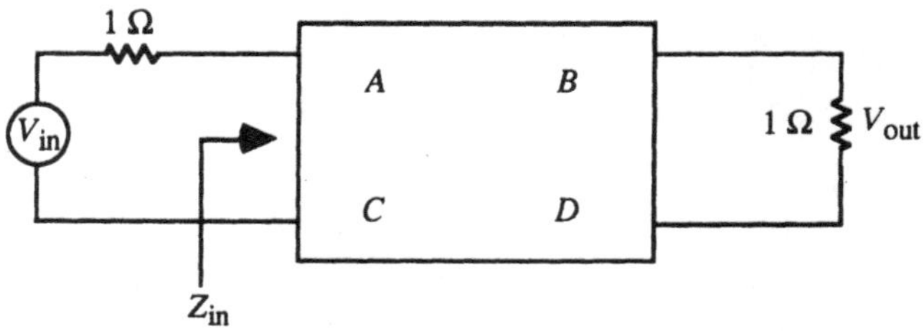

Figure 7.1 Network model for derivation of the input impedance, Z_{in}.

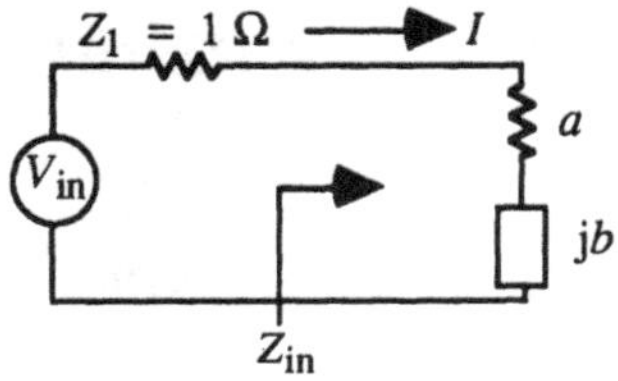

Figure 7.2 Network model for the input impedance, $Z_{in} = a + jb$.

If equal source and load terminations are assumed, the maximum possible transmitted power is 1/4; therefore,

$$\text{Transmitted power} = \frac{4a}{(a+1)^2 + b^2} \times 100\% \text{ of max.} \tag{7.4}$$

Rewriting equation (7.4) using the definitions in Figure 7.2 yields

$$\frac{4a}{(a+1)^2 + b^2} = \frac{(a+1)^2 + b^2 - [(a-1)^2 + b^2]}{(a+1)^2 + b^2} = 1 - \frac{(a-1)^2 + b^2}{(a+1)^2 + b^2}$$

$$= 1 - \left|\frac{a + jb - 1}{a + jb + 1}\right|^2 = 1 - \left|\frac{Z_{in} - Z_1}{Z_{in} + Z_1}\right|^2 = 1 - |\Gamma|^2, \quad \text{with } Z_1 = 1\ \Omega \tag{7.5}$$

The term Γ is the voltage reflection coefficient at the input of the terminated passive network, and its equivalence is shown here because Γ is often used as the starting point in other texts for the derivation of input impedance expressions. The return loss, *RL*, is often used in practice as a measure of how well Z_1 matches Z_{in} with *RL* defined by

$$\mathbf{RL = -20 \log\left[\left|\frac{Z_{in} - Z_1}{Z_{in} + Z_1}\right|\right]} \tag{7.6}$$

Equating (7.2) and (7.5) produces

$$\left[\frac{1}{1+\dfrac{F(s)F(-s)}{P(s)P(-s)}}\right] = G(s)G(-s) = 1 - |\Gamma|^2 = 1 - \left|\frac{Z_{in}-1}{Z_{in}+1}\right|^2 \tag{7.7}$$

or

$$\left|\frac{Z_{in}-1}{Z_{in}+1}\right|^2 = \left[\frac{\dfrac{F(s)F(-s)}{P(s)P(-s)}}{1+\dfrac{F(s)F(-s)}{P(s)P(-s)}}\right] = \frac{F(s)F(-s)}{P(s)P(-s)+F(s)F(-s)} = \frac{F(s)F(-s)}{E(s)E(-s)} \tag{7.8}$$

with

$$\mathbf{E(s)E(-s) = P(s)P(-s) + F(s)F(-s)} \tag{7.9}$$

Solving for Z_{in},

$$\frac{Z_{in}-1}{Z_{in}+1} = \frac{F(s)}{E(s)}$$

or

$$\mathbf{Z_{in} = \frac{E(s)+F(s)}{E(s)-F(s)}} \tag{7.10}$$

In (7.10), the first element must be a series impedance because the order of the numerator is higher than the order of the denominator; however, the theory of duality can be used to obtain a shunt component for the first reactive element, if required. Both $E(s)$ and $F(s)$ contain even and odd terms. Filters synthesized in this chapter using equation (7.10) will be low-pass prototypes with a 1 Ω source resistance, arbitrary load resistance, series lead inductor, and a cutoff frequency of 1 rad/sec. The techniques of the preceding chapter can apply an appropriate bandpass transformation to each element in the low-pass prototype, including impedance and frequency scaling.

Example 7.1

Find $E(s)$, $P(s)$, and $F(s)$ for (a) a third-order lowpass Butterworth filter and (b) a fourth-order Tchebyscheff Type I low-pass filter with 1 dB ripple.

Solution

(a) For a third-order Butterworth filter,

$$G(s)G(-s) = \frac{1}{1 + \omega^6} = \frac{1}{1 - s^6} = \frac{1}{1 + \dfrac{(s^3)(-s^3)}{(1)(1)}} = \frac{1}{1 + \dfrac{F(s)F(-s)}{P(s)P(-s)}}$$

therefore, $P(s) = 1$, $F(s) = s^3$, and $E(s)$ is constructed from the lefthand poles of $1 - s^6$, or $E(s) = s^3 + 2s^2 + 2s + 1$. The input impedance is

$$Z_{\text{in}} = \frac{2s^3 + 2s^2 + 2s + 1}{2s^2 + 2s + 1}$$

(b) For the fourth-order Tchebyscheff filter with $\epsilon^2 = 0.2589254$,

$$G(s)G(-s) = \frac{1}{1 + \dfrac{(s^4 + s^2 + 1/8)(s^4 + s^2 + 1/8)}{[1/(8\epsilon)][1/(8\epsilon)]}} = \frac{1}{1 + \dfrac{F(s)F(-s)}{P(s)P(-s)}}$$

or $P(s) = [1/(8\epsilon)]$, $F(s) = s^4 + s^2 + 1/8$, and $E(s) = s^4 + 0.9528114\ s^3 + 1.453925\ s^2 + 0.742619s + 0.2756276$. The input impedance is

$$Z_{\text{in}} = \frac{2s^4 + 0.9528114s^3 + 2.453925s^2 + 0.74261945s + 0.4006275}{0.9528114s^3 + 0.453925s^2 + 0.74261945s + 0.1506276}$$

7.2 INPUT IMPEDANCE DERIVATION FOR ARBITRARY LOAD RESISTANCE

With reference to Figure 7.2, an arbitrary load of R_2 will be used. The maximum load power for a 1 V source is

$$\text{Maximum load power} = \frac{R_2}{(R_2 + 1)^2} \tag{7.11}$$

By analogy to (7.7),

$$\left[\frac{1}{1 + \dfrac{F(s)F(-s)}{P(s)P(-s)}}\right] \times 100\% \text{ of max.} = \frac{4a\dfrac{1}{4}\dfrac{(R_2 + 1)^2}{R_2}}{(a + 1)^2 + b^2} \times$$

$$100\% \text{ of max.} = \frac{1}{4}\frac{(R_2 + 1)^2}{R_2}(1 - |\Gamma|^2)$$

which reduces to

$$\frac{P(s)P(-s)\dfrac{4R_2}{(1 + R_2)^2}}{P(s)P(-s) + F(s)F(-s)} = 1 - |\Gamma|^2 \tag{7.12}$$

or

$$\frac{P(s)P(-s) + F(s)F(-s) - P(s)P(-s)\dfrac{4R_2}{(1 + R_2)^2}}{P(s)P(-s) + F(s)F(-s)} = |\Gamma|^2 = \frac{M(s)M(-s)}{E(s)E(-s)} \tag{7.13}$$

Solving for Z_{in}, the input impedance becomes

$$Z_{in} = \frac{E(s) + M(s)}{E(s) - M(s)} \tag{7.14}$$

7.3 INPUT IMPEDANCE FOR ARBITRARY SOURCE AND LOAD RESISTANCES

By applying the same logic that led to (7.13), an expression for Z_{in} can be obtained from

$$\frac{E(s)E(-s) - P(s)P(-s)\dfrac{4R_1R_2}{(R_1 + R_2)^2}}{P(s)P(-s) + F(s)F(-s)} = \left|\frac{Z_{in} - R_1}{Z_{in} + R_1}\right|^2 \tag{7.15}$$

The input impedance is

$$Z_{in} = \frac{R_1E(s) + R_1M(s)}{E(s) - M(s)} \tag{7.16}$$

where

$$M(s)M(-s) = E(s)E(-s) - \frac{4R_1R_2}{(R_1 + R_2)^2}P(s)P(-s) \tag{7.17}$$

7.4 THE EVEN-ORDER TCHEBYSCHEFF TYPE I CASE

By far, evenly terminated filters are the most commonly designed. Equation (7.10) can be used to synthesize any evenly terminated passive filter; however, even-order Tchebyscheff Type I filters do not exist with even terminations. In this case, use of (7.10) will automatically produce the nearest theoretical resistance to a 1 Ω load.

The transfer function for a Tchebyscheff Type I filter is given by

$$\frac{1}{1 + \epsilon^2[T_N(\omega)]^2} = \frac{1}{1 + \dfrac{F(s)F(-s)}{P(s)P(-s)}} \tag{7.18}$$

The effect on the load resistor can be determined with a dc analysis of (7.18). If N is even, then at dc ($\omega = 0$), $T_N(0) = 1$. From (7.18),

$$\frac{1}{1 + \epsilon^2} = \frac{P(s)P(-s)}{P(s)P(-s) + F(s)F(-s)} = \frac{P(s)P(-s)}{E(s)E(-s)} \tag{7.19}$$

Identifying $P(s) = 1$, $E(s) = \sqrt{1 + \epsilon^2}$, and $F(s) = \epsilon$, the input impedance at dc becomes

$$R_2 = Z_{\text{in}} = \frac{E(s) + F(s)}{E(s) - F(s)} = \frac{\sqrt{1 + \epsilon^2} + \epsilon}{\sqrt{1 + \epsilon^2} - \epsilon} \tag{7.20}$$

where ϵ is related to the passband ripple by (3.12). Therefore R_2 is constrained for even-order Tchebyscheff Type I filters by

$$R_2 \geq \frac{\sqrt{1 + \epsilon^2} + \epsilon}{\sqrt{1 + \epsilon^2} - \epsilon} \tag{7.21}$$

Equation (7.13) can be modified for Tchebyscheff Type I filters such that

$$\frac{P(s)P(-s) + F(s)F(-s) - P(s)P(-s)(1 + \epsilon^2)\dfrac{4R_2}{(1 + R_2)^2}}{P(s)P(-s) + F(s)F(-s)} = \frac{M(s)M(-s)}{E(s)E(-s)} \tag{7.22}$$

then the expression for the input impedance,

$$Z_{\text{in}} = \frac{E(s) + M(s)}{E(s) - M(s)}$$

can be used to find a passive realization of an even-order Tchebyscheff Type I filter terminated in R_2.

7.5 REALIZATION PROCEDURE FOR ALL-POLE FILTERS

Butterworth and Tchebyscheff Type I filters can be synthesized from the input impedance by alternatively extracting impedances and admittances. The input impedance for the third-order Butterworth filter in example 7.1 was given by

$$Z_{in} = \frac{2s^3 + 2s^2 + 2s + 1}{2s^2 + 2s + 1} \tag{7.23}$$

The input impedance fits the topology in Figure 7.3. Dividing the denominator into the numerator provides a value of 1 H for L_1 from the following equation,

$$Z_{in} = s + \frac{s + 1}{2s^2 + 2s + 1} = L_1\, s + \frac{s + 1}{2s^2 + 2s + 1} \tag{7.24}$$

The capacitor, C_1, can be obtained by inverting the remainder term and extracting C_1 as an admittance. This C_1 is given as 2 farads by

$$Y_{in} = \frac{2s^2 + 2s + 1}{s + 1} = 2s + \frac{1}{s + 1} \tag{7.25}$$

Again, inverting the remainder term yields

$$Z_{in} = s + 1 \tag{7.26}$$

which gives the final two elements as a 1 H inductor in series with a 1 Ω resistor.

7.6 RECURSIVE FORMULAS FOR BUTTERWORTH-TCHEBYSCHEFF FILTERS

Simple recursion formulas [1] for the prototype element values of singly and doubly terminated filters are well-known and can be used as an alternate method of finding

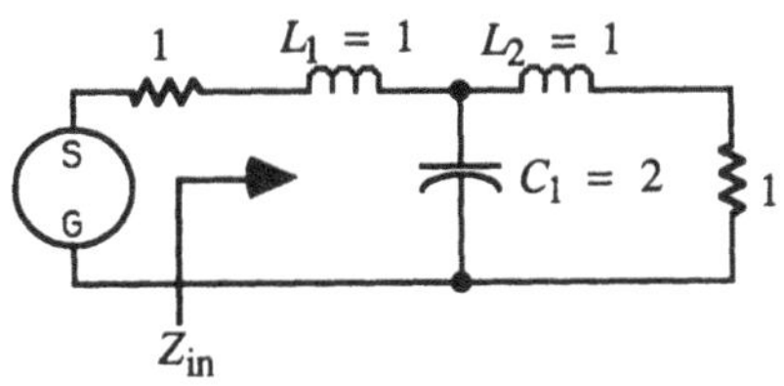

Figure 7.3 Third-order Butterworth low pass with series lead inductance.

element values. Table 7.1 contains the definitions of the filter parameters for Figures 7.4 and 7.5. In Figure 7.4, recursion formulas are given for doubly terminated filters. The value of A_t calculated by these formulas is the insertion loss expressed as a power ratio for an ideal Butterworth or odd-order Tchebyscheff filter. For an even-order Tchebyscheff filter, A_t is the insertion loss in excess of the required minimum. For odd-order filters with $R_L \leq 1$, there are shunt capacitors at both ends; whereas in the dual filter with $R_L \geq 1$, there are series inductors at both ends. Because the filter is reciprocal (the source and load can be interchanged), there are two possible configurations that exist for any odd-order filter. Even-order filters have a series inductor at the low-resistance end and a shunt capacitor at the high-resistance end in both cases. Figure 7.5 shows the singly terminated prototype filters. The final element is always either a capacitor with an open circuit filter termination or an inductor with a short circuit termination.

Table 7.1

Filter Parameters for Figures 7.4 and 7.5

n: number of poles or filter elements

Am: passband ripple in dB

$$Ap = 10^{(Am/10)} \qquad \beta = 2\sinh^{-1}\left(\frac{1}{\epsilon}\right)$$

$$\epsilon = \sqrt{Ap - 1}$$

$$\sinh(x) = 0.5 \cdot (e^{x} - e^{-x}) \qquad \sinh^{-1}(x) = \log_e\left(x + \sqrt{x^2 + 1}\right)$$

$$\tanh(x) = \frac{e^{x} - e^{-x}}{e^{x} + e^{-x}} \qquad \cosh(x) = 0.5 \cdot (e^{x} + e^{-x})$$

$$\coth(x) = \frac{e^{x} + e^{-x}}{e^{x} - e^{-x}}$$

Example 7.2

Find the element values for a third-order Butterworth filter with a 1 Ω source resistance and a 2 Ω load resistance.

Solution

The voltage transfer function, $G(s)$, is given by

$$G(s)G(-s) = \frac{1}{1 - s^6} = \frac{1}{1 + \dfrac{(s^3)(-s^3)}{(1)(1)}} = \frac{1}{1 + \dfrac{F(s)F(-s)}{P(s)P(-s)}}, \quad \text{or}$$

$$G(s) = \frac{1}{s^3 + 2s^2 + 2s + 1} = \frac{P(s)}{E(s)}$$

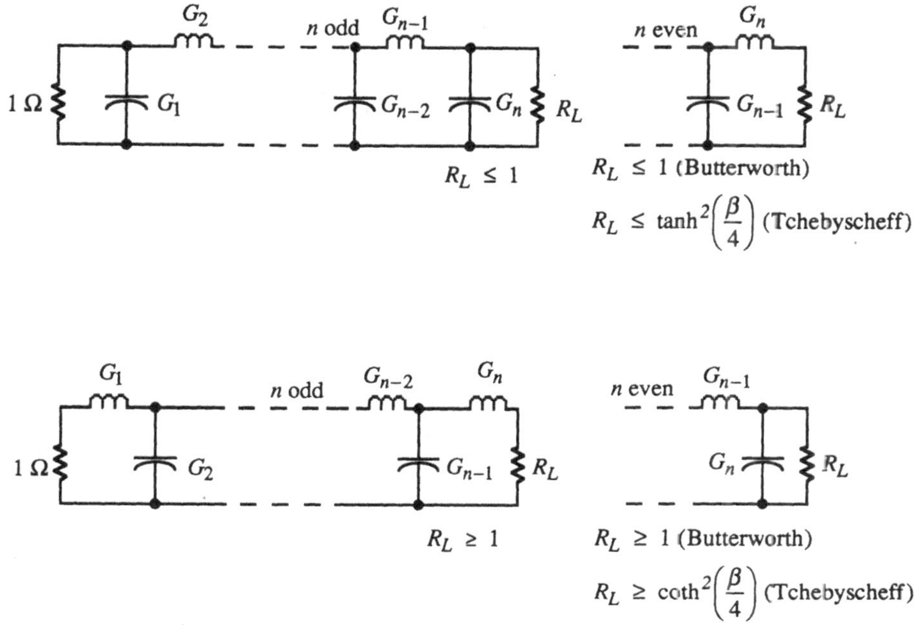

Tchebyscheff:

$$A_t = \frac{4R_L}{(R_L+1)^2} \quad (n \text{ odd}) \qquad A_t = \frac{4R_L A_p}{(R_L+1)^2} \quad (n \text{ even})$$

$$\gamma = \sinh\left(\frac{\beta}{2n}\right) \qquad d = \sinh\left[\left(\frac{1}{n}\right)\sinh^{-1}\left(\sqrt{\frac{1-A_t}{\varepsilon^2}}\right)\right]$$

$$b_k = \gamma^2 + d^2 - 2\,\gamma d \cos\left(\frac{k\pi}{n}\right) + \sin^2\left(\frac{k\pi}{n}\right) \qquad k = 1,2,\ldots,n$$

Butterworth:

$$A_t = \frac{4R_L}{(R_L+1)^2}$$

$$\gamma = 1 \qquad d = (1-A_t)^{1/(2n)}$$

$$b_k = \gamma^2 + d^2 - 2\,\gamma d \cos\left(\frac{k\pi}{n}\right) \qquad k = 1,2,\ldots,n$$

$$a_k = \sin\left[\frac{(2k-1)\pi}{2n}\right] \quad k = 1,2,\ldots,n \qquad G_1 = \frac{2a_1}{\gamma - d} \qquad G_k = \frac{4a_k\, a_{k-1}}{b_{k-1}\, G_{k-1}} \quad k = 2,3,\ldots,n$$

Figure 7.4 Doubly terminated Butterworth and Tchebyscheff Type I filters. (from [1], p. 63, by permission of the publisher).

From equation (7.13),

$$\frac{E(s)E(-s) - P(s)P(-s)\dfrac{4R_2}{(1+R_2)^2}}{E(s)E(-s)} = \frac{1 - s^6 - \dfrac{8}{9}}{1 - s^6} = \frac{M(s)M(-s)}{E(s)E(-s)}$$

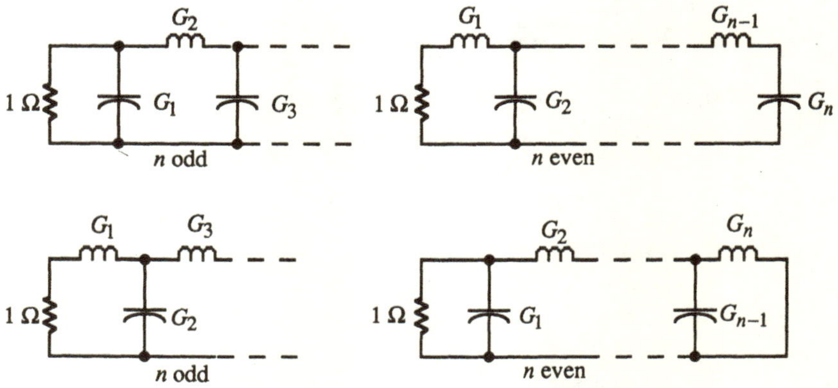

Tchebyscheff:

$$\gamma = \sinh\left(\frac{\beta}{2n}\right)$$

$$b_k = \left[\gamma^2 + \sin^2\left(\frac{k\pi}{2n}\right)\right]\cos^2\left(\frac{k\pi}{2n}\right) \quad k = 1,2,\ldots,n$$

Butterworth:

$$\gamma = 1$$

$$b_k = \cos^2\left(\frac{k\pi}{n}\right) \quad k = 1,2,\ldots,n$$

$$a_k = \sin\left[\frac{(2k-1)\pi}{2n}\right] \quad k = 1,2,\ldots,n \qquad G_1 = \frac{a_1}{\gamma}$$

$$G_k = \frac{a_k\, a_{k-1}}{b_{k-1}\, G_{k-1}} \quad k = 2,3,\ldots,n$$

Figure 7.5 Singly terminated Butterworth and Tchebyscheff Type I filters (from [1], p. 63, by permission of the publisher).

Because $M(s)M(-s) = 1/9 - s^6$, $M(s)$ can be found from the six roots of $M(s)M(-s)$, construct a polynomial in s using the three roots in the left half-plane (with negative real components). Then $M(s) = s^3 + 1.3867225s^2 + 0.9614991s + 0.333333$. The input impedance becomes

$$Z_{in} = \frac{2s^3 + 3.3867225s^2 + 2.9614991s + 1.33333}{0.6132775s^2 + 1.0385009s + 0.6666666}$$

Alternatively extracting impedances and admittances yields $L_1 = 3.2611644$, $C_1 = 0.7788755$, $L_2 = 1.1810826$, and $R_2 = 2\ \Omega$, with reference to the topology in Figure 7.3.

7.7 THE FOUR TYPES OF ELLIPTIC FILTERS

Filter literature classifies elliptic filters types a, b, c, and s. Type s refers to odd-order elliptic low-pass prototypes, which have a symmetric topology. In a type-s elliptic low-pass prototype of order N, the numerator always has order $N - 1$. The number of traps, or transmission zeroes, is $(N - 1)/2$.

The three antimetric filters in Figure 7.6 have only even-degree N. Antimetric filters have $N/2$ traps. The type-*a* filter has the exact elliptic response defined by equation (5.1) with $R_N(\omega)$ given in equation (5.3). However, the realization for the type-*a* must include the ideal transformers shown in Figure 7.7 or have at least one negative element for the topology shown in Figure 7.6(a). Also type-*a* filters have unequal termination resistances.

Type *b* eliminates the need for a negative element or ideal transformers by moving the highest two finite-frequency stopband attenuation poles to infinity. This causes ω_1 to increase slightly so that the type-*b* response is not quite as fast as the ideal elliptic response.

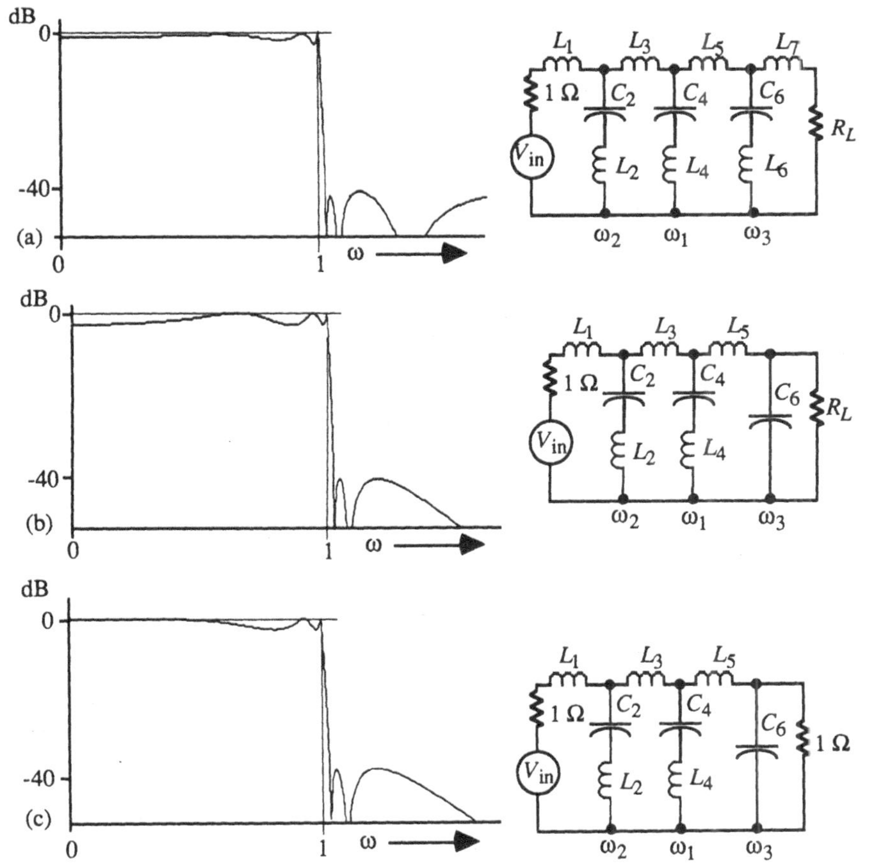

Figure 7.6 Antimetric elliptic filters (types *a*, *b*, and *c*) for sixth-order realizations with 3 dB passband ripple, and 40 dB stopband attenuation.

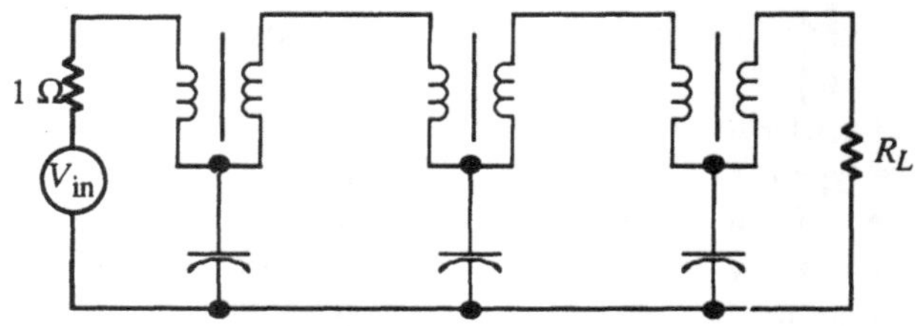

Figure 7.7 Alternate topology for a sixth-order type-*a* filter showing mutual inductances.

The type-*c* elliptic filter moves the lowest two passband attenuation zeros to the origin, so that the termination resistances can be made equal. The filter cutoff is further degraded but is usually accurate enough because the next higher integer filter degree is required in any case.

Type-*s* filters have a symmetric topology with odd-degree *N*. The number of traps is $(N - 1)/2$ and termination resistances can always be made equal. Programs B4 and B5 are given in Appendix B for the design of antimetric and symmetric elliptic filters (with equal terminating resistances when possible) and includes the examples presented in Figures 7.6 and 7.8, whose component values are shown in Table 7.2.

7.8 DERIVATIONS OF TYPE-*B* AND TYPE-*C* ELLIPTIC FILTERS

The type-*a* functions are not conducive to implementation as a passive filter because negative elements, or mutual inductances, are required. Daniels [2] gives a simple transformation for moving the highest transmission zero to infinity, which results in the type-*b* transfer function. Both the type-*a* and the type-*b* circuits have a loss at dc equal to the amount of ripple in the passband. Synthesis of these circuits in a passive realization always results in topology with unequal source and load resistances. By also moving the lowest

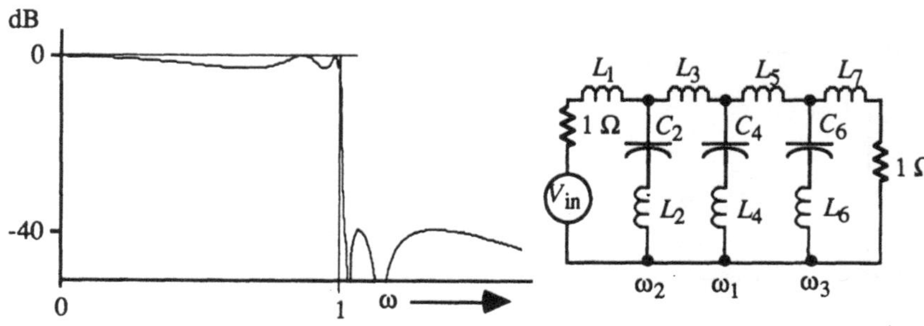

Figure 7.8 Symmetric elliptic filters (type *s*) for seventh-order realizations with 3 dB passband ripple and 40 dB stopband attenuation.

Table 7.2
Component Values for Figures 7.6 and 7.8 in Henries, Farads, Ohms, and Rad/Sec.

	Type a	*Type b*	*Type c*	*Type s*
R_L	5.808899	5.808899	1.0	1.0
C_2	0.412129	0.4508446	0.823417	0.5605527
C_4	0.254237	0.2742942	0.7382455	0.1729089
C_6	0.5442446	0.6095749	2.329889	0.2683593
L_1	2.4129814	2.536746	1.836813	2.88658
L_2	1.628364	1.366332	0.7037314	0.7969948
L_3	2.112437	2.278798	1.143136	2.219765
L_4	3.43785	3.108146	1.127527	5.450775
L_5	2.755555	2.9376737	0.9116018	1.485185
L_6	0.263616	0.0	0.0	3.128971
L_7	−0.24865358	0.0	0.0	1.93277
ω_1	1.0696388	1.0830312	1.0960645	1.4961117
ω_2	1.2206966	1.2741145	1.313671	1.0300592
ω_3	2.64008	∞	∞	1.09129

transmission pole to the origin, the type-*b* transfer can be transformed into a type *c* and implemented as an evenly terminated elliptic filter.

The transfer functions for the type-*a* antimetric filter shown in Figure 7.6 is given as

$$G(s)G(-s) = \frac{1}{1 + \left[\dfrac{\left(s^2 + \dfrac{\omega_x^2}{(2.64008339)^2}\right)\left(s^2 + \dfrac{\omega_x^2}{(1.2206966)^2}\right)\left(s^2 + \dfrac{\omega_x^2}{(1.0698388)^2}\right)}{0.01(s^2 + (2.64008339)^2)(s^2 + (1.2206966)^2)(s^2 + (1.0698388)^2)}\right]^2} \tag{7.27}$$

with a passband edge of 1 rad/sec, stopband edge, ω_x, of 1.0589145 rad/sec, and transmission zeroes at $0 \pm j2.64008339$, $0 \pm j1.2206966$, and $0 \pm j1.0698388$. The denominator constant, 0.01, can be determined from the requirement that $|G(s)|^2 = -40$ dB as ω approaches ∞. From [2] the transformation

$$\omega'^2 = (\omega_c^2 - \omega_b^2)\frac{\omega^2}{\omega_c^2 - \omega^2} \tag{7.28}$$

1. Transforms $\omega = 0$ to $\omega = 0$.
2. Transforms $\omega = \omega_b$ to $\omega = \omega_b$.
3. Transforms $\omega = \omega_c$ to $\omega = \infty$.

Using values of ω_c = 2.64008339, ω_b = 1, and $\omega^2 = (1.2206966)^2$, $(1.0698388)^2$, and $(1.0589145)^2$ for equation (7.28), the corresponding type-b transfer function becomes

$$G'(s)G'(-s) = \frac{1}{1 + \left[\dfrac{(s^2 + 0.141084)(s^2 + 0.7226148)(s^2 + 0.9763833)}{0.0523193(s^2 + 1.6234462)(s^2 + 1.1729564)}\right]^2} \quad (7.29)$$

The denominator constant, 0.0523193, is chosen for consistency with the specifications that the filter response be −3 dB at dc. The frequency at which the stopband first meets the prescribed loss of −40 dB is $\omega_x' = 1.06938852$ rad/sec.

The transformation

$$\omega''^2 = \frac{1}{(1 - \omega_o^2)}(\omega'^2 - \omega_o^2) \quad (7.30)$$

converts the type-b transfer function in equation (7.29) into a type-c function by moving the lowest transmission pole, ω_o, to dc. With $\omega_o^2 = 0.141084$ and $\omega'^2 = 0.7226148$, 0.9763833, 1.6234462, 1.1729564, the corresponding type-c transfer function becomes

$$G''(s)G''(-s) = \frac{1}{1 + \left[\dfrac{s^2\left(s^2 + \dfrac{\omega_x''^2}{(1.313671)^2}\right)\left(s^2 + \dfrac{\omega_x''^2}{(1.096065)^2}\right)}{0.06098294(s^2 + (1.313671)^2)(s^2 + (1.096065)^2)}\right]^2} \quad (7.31)$$

with $\omega_x'' = 1.080875$ rad/sec.

7.9 A TRANSFORM VARIABLE FOR FINDING $E(s)$

The transfer function, $G(s)G(-s)$, can be written as

$$\mathbf{G(s)G(-s)} = \frac{1}{1 + \left[\dfrac{F(s)F(-s)}{P(s)P(-s)}\right]} = \frac{P(s)P(-s)}{P(s)P(-s) + F(s)F(-s)} = \frac{P(s)P(-s)}{E(s)E(-s)} \quad \mathbf{(7.32)}$$

The polynomial $E(s)$ must be determined before synthesis of the passive elliptical filter can be started. In example 5.1, polynomial addition was used to add $P(s)P(-s)$ and $F(s)F(-s)$. $E(s)$ was then determined from the lefthand poles of $E(s)E(-s)$. For elliptic filter of order 9 or higher, the roots of $E(s)E(-s)$ tend to cluster near the passband edge, and they cannot be determined accurately. The transformation

$$Z^2 = 1 + \frac{\omega_b^2}{s^2} \tag{7.33}$$

is often used to spread the passband $0 \le \omega \le \omega_b$ over the frequency range $0 \le \Omega \le \infty$. If a normalized low pass is used, then $\omega_b = 1$ and the process can be summarized by the following steps:

1. Replace s^2 with $1/(Z^2 - 1)$ in both $F(s)F(-s)$ and $P(s)P(-s)$.
2. Solve for $E(Z)$ by finding the roots of $E(Z)E(-Z) = F(Z)F(-Z) + P(Z)P(-Z)$.
3. Write $E(Z)E(-Z)$ as a product of quadratic factors, replace Z^2 with $1 + 1/s^2$ and construct $E(s)$ from the lefthand poles of $E(s)E(-s)$.

Although any root finding technique can be used to find the roots of $E(Z)E(-Z)$, computer program B3 reconstructs $E(s)$ using Bairstow's root finding algorithm. A useful formula for reduction of a fourth-order even polynomial is

$$s^4 + 2bs^2 + a^2 = \left[s^2 - 2\left(\frac{a-b}{2}\right)^{1/2} s + a\right]\left[s^2 + 2\left(\frac{a-b}{2}\right)^{1/2} s + a\right] \tag{7.34}$$

Example 7.3

Find $P(Z)$ and $F(Z)$ for a normalized low-pass filter that is given by

$$\frac{1}{1 + \dfrac{F(s)F(-s)}{P(s)P(-s)}} = \frac{1}{1 + \left[\dfrac{s(s^2+0.8)(s^2+0.95)}{(s^2+1.1)(s^2+1.5)}\right]\left[\dfrac{-s(s^2+0.8)(s^2+0.95)}{(s^2+1.1)(s^2+1.5)}\right]}$$

Solution

Substituting $1/(Z^2 - 1)$ for s^2 yields

$$\frac{F(Z)}{P(Z)} = \frac{(0.8)(0.95)(Z^2+0.25)(Z^2+0.0526)}{1.65\sqrt{Z^2-1}(Z^2-0.090909)(Z^2-0.33333)}$$

Example 7.4

A normalized low-pass filter is given by

$$\frac{1}{1+\dfrac{F(s)F(-s)}{P(s)P(-s)}}=\frac{1}{1+\left[\dfrac{s^2+0.9}{s^2+1.1}\right]\left[\dfrac{s^2+0.9}{s^2+1.1}\right]}$$

Use the transform variable to find $E(s)$.

Solution

Substituting $1/(Z^2-1)$ for s^2 yields

$$\frac{1}{1+\dfrac{F(Z)F(-Z)}{P(Z)P(-Z)}}=\frac{1}{1+\left[\dfrac{0.9(Z^2+0.11111)}{1.1(Z^2-0.090909)}\right]\left[\dfrac{0.9(Z^2+0.11111)}{1.1(Z^2-0.090909)}\right]}$$

$E(Z)E(-Z) = F(Z)F(-Z) + P(Z)P(-Z)$, or $E(Z)E(-Z) = 2.02(Z^4 - 0.0198Z^2 + 0.0099)$. Substituting $1/s^2 + 1$ for Z^2, $E(s)E(-s) = K(s^4 + 2s^2 + 1.01)$, or

$$E(s)=K'\left[s^2+2\left[\frac{\sqrt{1.01}-1}{2}\right]^{1/2}s+\sqrt{1.01}\right]=K'(s^2+0.0998755s+1.0049876)$$

Example 7.5

Find the voltage transfer function, $G(s)$, for the filters in Figure 7.6 and 7.8 in terms of $P(s)$ and $E(s)$.

Solution

From equations (7.27), (7.29), and (7.31), the transfer functions, $G(s)$, are

$$G(s)=\frac{P(s)}{E(s)}=$$

for type a,

$$\frac{0.009990864(s^2 + 1.1446646)(s^2 + 1.49036)(s^2 + 6.9722455)}{(s^2 + 0.4075898s + 0.2177597)(s^2 + 0.1331146s + 0.7804224)(s^2 + 0.023372s + 0.9877281)}$$

for type b,

$$\frac{0.0523193(s^2 + 1.1729564)(s^2 + 1.6234462)}{(s^2 + 0.02719s + 0.98566)(s^2 + 0.1469468s + 0.7515987)(s^2 + 0.390682s + 0.1901794)}$$

for type c,

$$\frac{0.06098294(s^2 + 1.7257315)(s^2 + 1.2013585)}{(s^2 + 0.0316928s + 0.9833745)(s^2 + 0.1758838s + 0.7137071)(s^2 + 0.6508336s + 0.1801424)}$$

for type s,

$$\frac{0.03866234(s^2 + 2.2383511)(s^2 + 1.1909182)(s^2 + 1.0610215)}{(s + 0.2386)(s^2 + 0.25208s + 0.513646)(s^2 + 0.0623952s + 0.8967)(s^2 + 0.010285s + 0.9945986)}$$

7.10 MODELS FOR ELLIPTIC FILTER SYNTHESIS

The roots of $P(s)$ are the zeroes of $G(s)$, and they are often referred to as the *trap frequencies*, or the frequencies at which no energy is transmitted from the input to the output. Not every permutation of trap frequencies, ω_z, leads to positive element values. In Figure 7.9, the location of these zeroes alternates from end to end. By alternating the trap frequencies for any order elliptic filter as shown in Figure 7.9, positive elements values can always be obtained.

The expression for input impedance given by (7.10), (7.14), and (7.16) do not explicitly contain information about the zeroes of transmission (roots of $P(s)$). During the synthesis procedure, these traps must be extracted in the order shown in Figure 7.9. The classical synthesis procedure for filters that contain both poles and zeroes of transmission will be considered next.

7.11 CLASSICAL FILTER SYNTHESIS

The synthesis techniques will be presented using the basic normalized structures shown in Figure 7.9(a) without loss of generality. Transformations to more elaborate structures can easily be done using the methods derived in the previous chapter. Using the expression for input impedance from equation (7.10) or (7.16) and the values of ω_z, the following steps will yield the proper element values.

1. At frequency ω_5, the input impedance is $Z_{in} = jL_A\omega_5$, because L_5–C_5 presents a short to ground, thus $L_A = Z_{in}|_{s=j\omega_5}/j\omega_5$. The impedance, sL_A, is extracted from Z_{in} to form

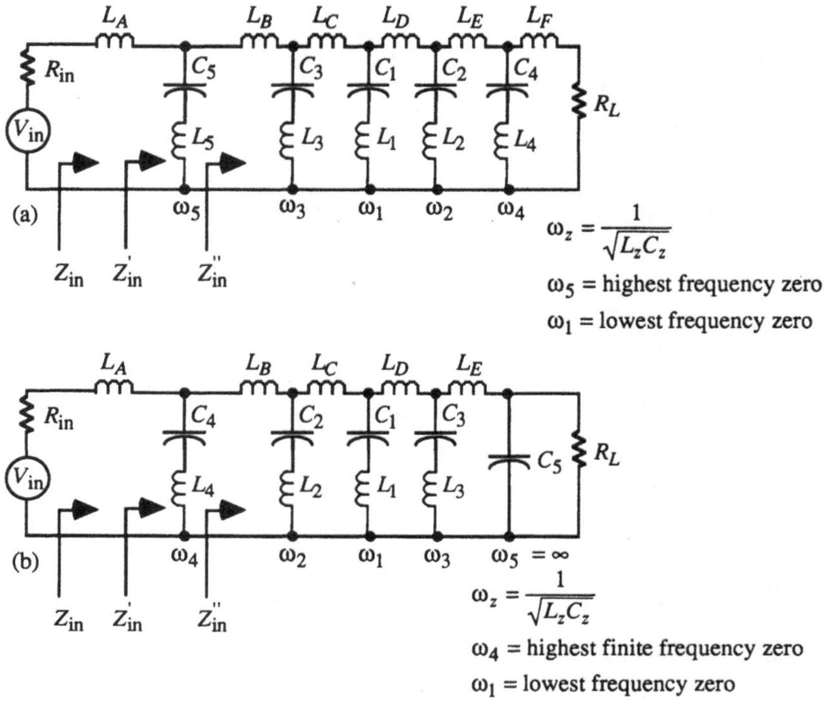

Figure 7.9 Trap frequency locations for realization of positive element values.

the input impedance, Z'_{in}, for the remainder of the network behind L_A with $Z'_{in} = Z_{in} - sL_A$.

2. The inductance L_5 is determined from $L_5 = [sZ'_{in}/(s^2 + \omega_5^2)]|_{s^2=-\omega_5^2}$. The capacitance C_5 is obtained from the resonant frequency, ω_5, by the relation, $C_5 = 1/(L_5\omega_5^2)$. The input impedance, Z''_{in}, for the remainder of the network beyond $C_5 - L_5$ is obtained from $1/Z''_{in} = (1/Z'_{in}) - (s/L_5)/(s^2 + \omega_5^2)$.
3. The process is repeated until all the element values are identified.

Table 7.3 summarizes the removal process for elliptic filters.

Example 7.6

Determine the element values for a sixth-order type-*c* elliptic filter with 1 dB passband ripple and a minimum attenuation of 50 dB in the stopband. Design for a source resistance of 1 Ω and the load resistance of 2 Ω.

Table 7.3
Removal Process for Synthesis for Elliptic Filters

	Pole	*Removal*	*Remainder Function*
1	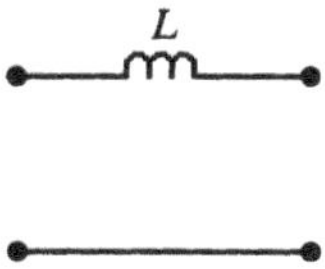	$L = \frac{Z_{in}}{s}\Big\vert_{s=\infty}$	$Z'_{in} = Z_{in} - sL$
2	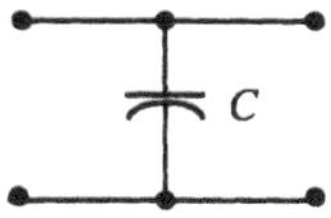	$C = \frac{1/Z_{in}}{s}\Big\vert_{s=\infty}$	$\frac{1}{Z'_{in}} = \frac{1}{Z_{in}} - sC$
3	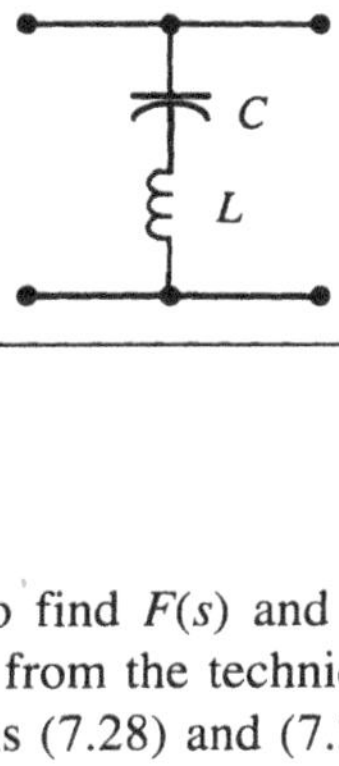	$L = \frac{sZ_{in}}{s^2 + \omega_z^2}\Big\vert_{s^2 = -\omega_z^2}$ $C = \frac{1}{L\omega_z^2}$	$\frac{1}{Z'_{in}} = \frac{1}{Z_{in}} - \frac{s/L}{s^2 + \omega_z^2}$

Solution

Step 1 is to find $F(s)$ and $P(s)$. The polynomials $F(s)$ and $P(s)$ for the type-*c* can be determined from the techniques presented in Chapter 5, using the transformations given by equations (7.28) and (7.30). These polynomials are

$$F(s) = s^2\left[s^2 + \left(\frac{\omega_x}{1.273134}\right)^2\right]\left[s^2 + \left(\frac{\omega_x}{1.630707}\right)^2\right], \qquad \text{with } \omega_x = 1.244 \text{ rad/sec}$$

$$P(s) = 0.0356362[s^2 + (1.273134)^2][s^2 + (1.630707)^2]$$

Step 2 is to find $E(s)$ from $F(s)$ and $P(s)$. Here $E(s)$ can be determined from $E(s)E(-s) = P(s)P(-s) + F(s)F(-s)$. It is not necessary to use the change of variable given by equation (7.33) for finding $E(s)$ because the polynomial is of relatively low order. Hence, $E(s) = s^6 + 1.179969s^5 + 2.234428s^4 + 1.781004s^3 + 1.345695s^2 + 0.602656s + 0.1519941$.

Step 3 is to determine $M(s)$. Because the source and load terminations are unequal, $M(s)$ must be found using equation (7.17) with $R_1 = 1$ and $R_2 = 2$ such that

$$M(s)M(-s) = E(s)E(-s) - \frac{4R_1R_2}{(R_1 + R_2)^2}P(s)P(-s), \qquad \text{or}$$

$$M(s) = s^6 + 0.6276991s^5 + 1.735267s^4 + 0.8786084s^3 + 0.7858717s^2 + 0.273009s + 0.0506646$$

Step 4 is to find the input impedance, Z_{in}. The input impedance, Z_{in}, given by $Z_{in} = [R_1E(s) + R_1M(s)]/[E(s) - M(s)]$, becomes

$$Z_{in} = \frac{2s^6 + 1.80767s^5 + 3.9697s^4 + 2.65961s^3 + 2.13157s^2 + 0.8757s + 0.20266}{0.552270s^5 + 0.4991s^4 + 0.90240s^3 + 0.55982s^2 + 0.3296s + 0.10133}$$

Step 5 is to extract the first series inductor. Evaluating Z_{in} at $\omega = \omega_2 = 1.630707$ to find L_1 yields

$$Z_{in}|_{\omega=1.630707} = 5.0129312 \angle 90°, \qquad \text{or } L_1 = \frac{5.0129312}{1.630707} = 3.0740846 \text{ H}$$

The remainder function, Z'_{in}, is generated from $Z'_{in} = Z_{in} - 3.0720846s$, or

$$Z'_{in} = \frac{0.30118s^6 + 0.2732s^5 + 1.1956s^4 + 0.93867s^3 + 1.1182s^2 + 0.5642s + 0.20266}{0.552270s^5 + 0.4991s^4 + 0.90240s^3 + 0.55982s^2 + 0.3296s + 0.10133}$$

Step 6 is to extract the first trap at ω_2. From Table 7.3, the trap elements can be obtained from $L_2 = [sZ'_{in}/(s^2 + \omega_z^2)]|_{s^2=-\omega_z^2}$ and $C_2 = 1/(L_2\,\omega_z^2)$ with $\omega_z = 1.63070$, yielding $L_2 = 0.6384241$ H and $C_2 = 0.589032$ F. The remainder function is given by $1/Z''_{in} = (1/Z'_{in}) - [(s/L_2)/(s^2 + \omega_z^2)]$ or

$$Z''_{in} = \frac{0.3022756s^4 + 0.273205s^3 + 0.3918417s^2 + 0.2121596s + 0.0762164}{0.0787984s^3 + 0.0712245s^2 + 0.0790906s + 0.038105}$$

Step 7 is to extract the second series inductor. Evaluating Z''_{in} at $\omega = \omega_1 = 1.273134$ to find L_3 yields $Z''_{in}|_{\omega=1.273134} = 3.7980235 \angle 90°$, or $L_3 = 3.7980235/1.273134 = 2.983208$ H. The remainder function Z'''_{in} is generated from $Z'''_{in} = Z''_{in} - 2.983208s$, or

$$Z'''_{in} = \frac{0.0672036s^4 + 0.0607275s^3 + 0.155898s^2 + 0.098485s + 0.076264}{0.0787984s^3 + 0.0712245s^2 + 0.0790906s + 0.038105}$$

Step 8 is to extract the second trap at ω_1. From Table 7.3, the elements for the second trap can be obtained from $L_4 = [sZ'_{in}/(s^2 + \omega_z^2)]|_{s^2=-\omega_z^2}$ and $C_4 = 1/(L_4\omega_z^2)$ with $\omega_z =$

1.273134, yielding $L_4 = 1.2732238$ H and $C_4 = 0.4845594$ F. The remainder function is given by $1/Z''''_{in} = (1/Z'''_{in}) - [s/L_4/(s^2 + \omega_z^2)]$ or

$$Z''''_{in} = \frac{0.0672036s^2 + 0.607275s + 0.0469697}{0.0260162s + 0.0235286}$$

Step 9 is to extract the third series inductor. The extraction of L_5 occurs at $\omega = \omega_3 = \infty$ and is done using process 1 in Table 7.3. The value of L_5 is 2.5831443 H with an input impedance remainder function of $Z'''''_{in} = 0.0469697/(0.0260162s + 0.0235286)$.

Step 10 is to extract the shunt capacitor and load resistor. The corresponding admittance is $Y''''''_{in} = 0.5538933s + 0.5$, which is recognized as a capacitor, C_6, of 0.5538933 F in parallel with a 2 Ω resistor. The final circuit is given in Figure 7.10.

7.12 THE PERMUTATION METHOD TO CALCULATE TRAP-SECTION ELEMENTS*

Amstutz [3] reported on a procedure for calculating the component values of doubly terminated elliptic filters. His method, used in programs B4 and B5, will be described next because it can be used to determine element values without the loss of accuracy that plagues every other known method.

Amstutz's process depends on the equivalence of any two networks having the same topology but permuted trap positions. The key to his method is to permute each of the traps into the input position and apply a recursive algorithm that yields all of the element values. This analysis is based on the topology in Figure 7.6(c). As noted previously, the impedance of a series LC branch resonant at the trap frequency, ω_z, is zero. Then the input impedance for Figure 7.6(c) at $\omega = \omega_2$ is $j\omega_2 L_1$. In the neighborhood of ω_2, the input impedance can be written as

$$Z_{in}(s) = sL_1 + L_2\left(s - \frac{s_2^2}{s}\right) \tag{7.35}$$

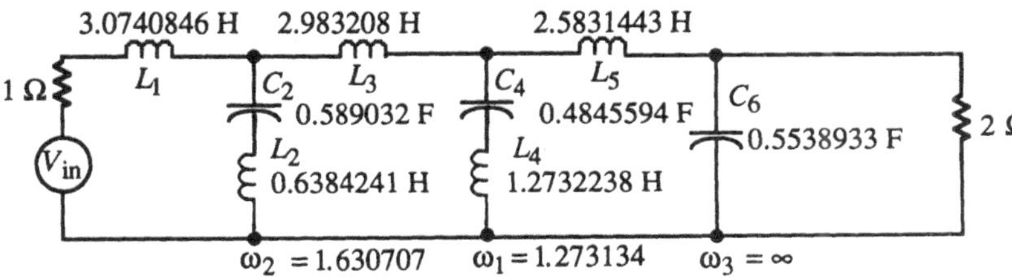

Figure 7.10 Final design for sixth-order type-c elliptic filter with 2 Ω termination.

*Much of the remaining material is taken from P. Amstutz, "Elliptic Approximations and Elliptic Filter Design on Small Computers," *IEEE Trans. Circuits Syst.*, © copyright 1978, by permission of the IEEE.

with $s_2 = j\omega_2$. Note that the impedance, Z, seen to the right of the (L_2, C_2) branch in Figure 7.6(c) is not zero. Differentiating (7.35) with respect to the complex frequency, s, yields

$$Z'_{in}(s) = L_1 + L_2\left(1 + \frac{s_2^2}{s^2}\right) \tag{7.36}$$

A Taylor series approximation valid in the neighborhood of trap frequency, ω_2, for equation (7.35) is

$$Z_{in}(s) = L_1 s_2 + (L_1 + 2L_2)(s - s_2) + \ldots \tag{7.37}$$

Now it is possible to identify element values of the first three elements in Figure 7.6(c) that cause the notch at ω_2. Setting $s = j\omega_2$ in (7.37) yields

$$L_1 = \frac{Z_{in}(s_2)}{s_2} \tag{7.38}$$

Setting $s = s_2$ in equation (7.36) yields

$$L_2 = \frac{Z'_{in}(s_2) - L_1}{2} \tag{7.39}$$

L_2 and C_2 are resonant at ω_2; therefore,

$$C_2 = \frac{1}{L_2\omega_2^2} \tag{7.40}$$

Continued fraction expansion of impedance functions can be applied to realize the element values of passive networks. Expressing (7.23) in continued fraction form yields

$$Z_{in} = \frac{2s^3 + 2s^2 + 2s + 1}{2s^2 + 2s + 1} = s + \cfrac{1}{2s + \cfrac{1}{s + 1/1}} = Z_1 + \cfrac{1}{Y_2 + \cfrac{1}{Z_3 + 1/Y_4}} \tag{7.41}$$

which produces the element values given in Figure 7.3.

To develop the equivalence used by Amstutz, the continued fraction expansion for the network input impedance in Figure 7.11(a) can be written as

$$Z_1 = L_{11}s + \cfrac{1}{\cfrac{1}{M_{11}s + \cfrac{1}{K_{11}s}} + \cfrac{1}{L_{22}s + \cfrac{1}{\cfrac{1}{M_{22}s + \cfrac{1}{K_{22}s} + \dots}}}} \tag{7.42}$$

For convenience, form Z_1/s by dividing every impedance phrase in (7.42) by s and define the inverse squared-frequency variables

$$h = \frac{-1}{s^2} \quad \text{and} \quad h_i = \frac{1}{\omega_i^2} = M_{ii}K_{ii} \tag{7.43}$$

for the ith trap. Then equation (7.42) becomes

$$\frac{Z_1}{s} = L_{11} + \cfrac{1}{\cfrac{K_{11}}{h_1 - h} + \cfrac{1}{L_{22} + \cfrac{1}{\cfrac{K_{22}}{h_2 - h} + \dots}}} \tag{7.44}$$

By analogy,

$$\frac{Z_2}{s} = L_{21} + \cfrac{1}{\cfrac{K_{21}}{h_2 - h} + \dots} \tag{7.45}$$

It should be noted that $Z_2 = Z_1$ for all s. Choose $s = s_2$ and equate (7.44) and (7.45):

$$L_{21} = L_{11} + \cfrac{1}{\cfrac{K_{11}}{h_1 - h_2} + \cfrac{1}{L_{22}}} \tag{7.46}$$

Amstutz defined the constants

$$U = L_{11} - L_{21}, \qquad \text{and} \qquad \frac{1}{V} = \frac{U}{(h_2 - h_1)B_{11}} - 1 \tag{7.47}$$

The solution for L_{22} is then given by

$$L_{22} = UV \tag{7.48}$$

One more relation is required. The derivatives of (7.44) and (7.45) with respect to h can be equated. The resulting expressions will be evaluated at $h = h_2$. In this context, the network in Figure 7.11(b) does not involve L_{21} and the elements to its right, so that (7.45) is truncated to yield the derivative

$$\left.\frac{d(Z_2/s)}{dh}\right|_{h=h_2} = \frac{-1}{K_{21}} \tag{7.49}$$

Also, the derivative of Z_1/s with respect to h does not involve anything to the right of trap 2 in Figure 7.11(a). Therefore, (7.44) is truncated, and its derivative is equated to (7.49), which yields an expression for K_{22} given by

$$K_{22} = \frac{[(h_2 - h_1) + (L_{21} - L_{11})K_{11}]^2}{(h_2 - h_1)/K_{21} - (L_{21} - L_{11})^2 K_{11}} \tag{7.50}$$

Amstutz rearranges (7.50) into the remaining expression

$$B_{22} = V^2 B_{21} - (V + 1)^2 B_{11} \tag{7.51}$$

The complete permutation algorithm is summarized in Table 7.4.

Example 7.7

Repeat example 7.6 using the permutation method given the input impedance

$$Z_{\text{in}} = \frac{2s^6 + 1.80767s^5 + 3.9697s^4 + 2.65961s^3 + 2.13157s^2 + 0.8757s + 0.20266}{0.552270s^5 + 0.4991s^4 + 0.90240s^3 + 0.55982s^2 + 0.3296s + 0.10133}$$

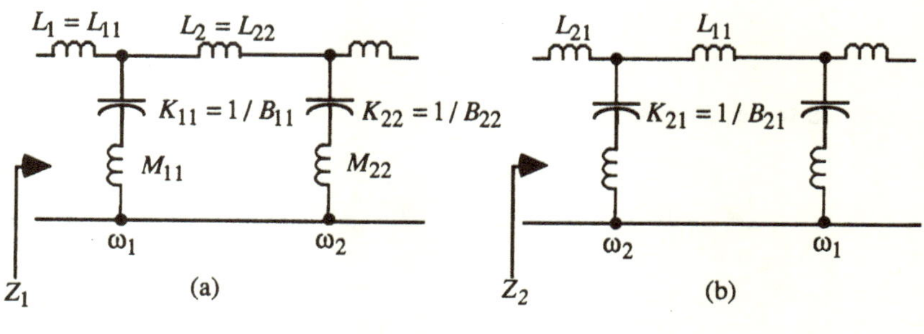

Figure 7.11 Equivalent two-port networks creating the same attenuation poles.

Table 7.4
Array of Permutation Coefficients for $M = 4$ Traps (from [4], by permission of publisher)

Trap k	$i = 1$	$i = 2$	$i = 3$	$i = 4$
ω_1	L_{11}, B_{11}			
ω_2	L_{21}, B_{21}	L_{22}, B_{22}		
ω_3	L_{31}, B_{31}	L_{32}, B_{32}	L_{33}, B_{33}	
ω_4	L_{41}, B_{41}	L_{42}, B_{42}	L_{43}, B_{43}	L_{44}, B_{44}

For $i = 1$ (first column):

For $k = 1,2,\ldots,M$:

$$L_{k1} = X_1(\omega_k)/\omega_k, \qquad \text{where } Z_1 = R_1 + jX_1$$

$$M_{k1} = [Z'(\omega_k) - L_{k1}]/2$$

$$B_{k1} = M_{k1}\,\omega_k^2$$

For $i = 2,3,\ldots,M$:

For $j = i, i+1,\ldots,M$:

$$U = L_{i-1,i-1} - L_{j,k-1}$$

$$1/V = U/\left[(\omega_j^{-2} - \omega_{i-1}^{-2})B_{i-1,i-1}\right] - 1$$

$$L_{ji} = UV$$

$$B_{ji} = V^2 B_{j,i-1} - (V+1)^2 B_{i-1,i-1}$$

Solution

See Figure 7.12. For step 1, the initial set of element values are determined by extraction at $\omega = 1.630707$, $\omega = 1.273134$, and $\omega = \infty$, using equations (7.38) through (7.40). For step 2, part (d), the constants are $h_1 = 0.3760522$, $h_2 = 0.616956$, $U = 0.4739852$, $V = 6.2927157$, and $L = UV = 2.9827$.

Also $B_{22} = 2.065639$ yielding $C = 0.4841118$ and $L = 1.274401$. In step 2, part (e), the constants are $h_1 = 0.3760522$, $h_2 = 0$, $U = -0.5473331$, $V = -0.142681$, and $L = UV = 3.8360626$. Therefore $B_{22} = 1.1224314$ yielding $C = 0.89039$.

Finally, for step 3, the constants are $h_1 = 0.6169526$, $h_2 = 0$, $V = -3.0234828$, $L = UV = 2.5787$, $B_{22} = 1.80907$ yielding $C = 0.55277$.

7.13 SUMMARY

Simple equations for the design of evenly terminated Butterworth and Tchebyscheff Type I filters were given previously by equation (2.11) and Table 3.1, respectively. Either

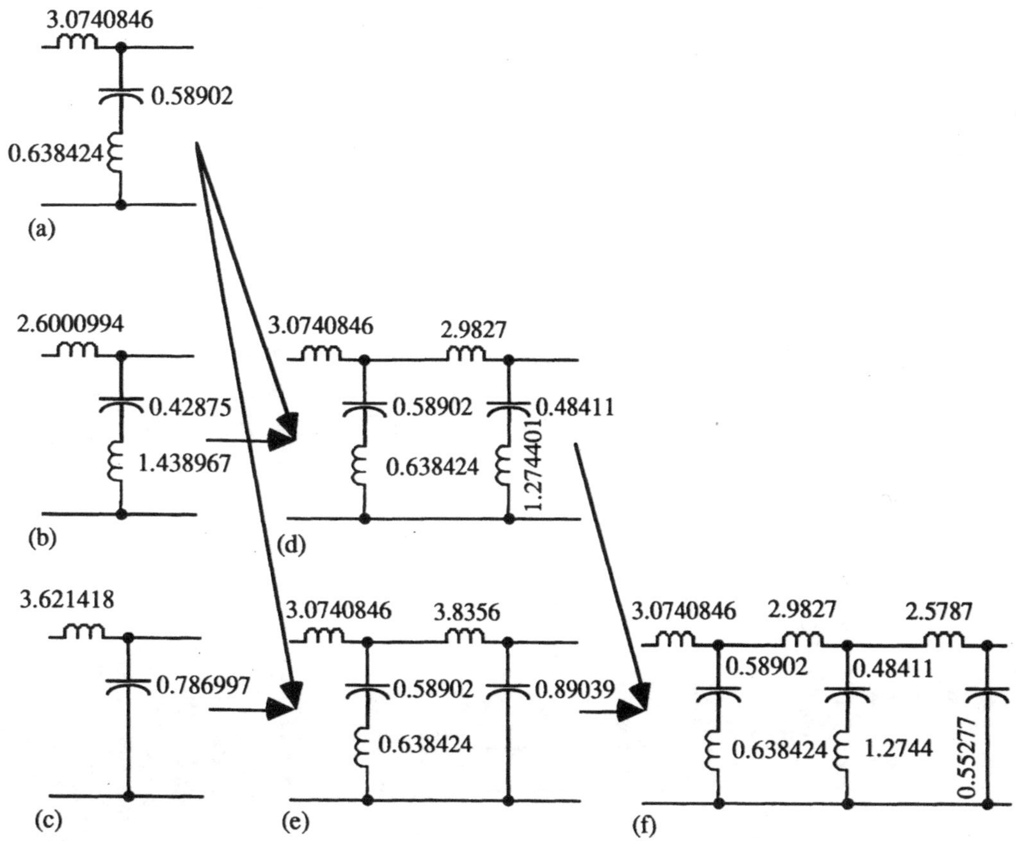

Figure 7.12 Extraction of element values using the permutation method. (a,b,c) Step 1; (d,e) step 2; (f) step 3.

continued fraction expansion or the recursive formulas in Figures 7.5 and 7.6 can be used to synthesize Butterworth and Tchebyscheff Type I filters with arbitrary terminations. An expression for the input impedance can easily be derived for all polynomial filters as a preliminary step for element extraction. A low-pass prototype with a lead series inductor was used for the examples in this chapter since the bandpass transformations discussed previously can apply any assortment of frequency or impedance scaling, and passband transformations to the low-pass prototype.

The design of arbitrarily terminated filters was extended to elliptic filters using both classical synthesis and the permutation method. Amstutz's permutation method is not susceptible to the exponential accumulation of round-off errors that plague all other known methods. The order of extraction of the traps was shown to be important in avoiding negative element values for elliptic filters. Programs B4 and B5 in Appendix B are given as a starting point for programmers interested in computer-aided filter design, however, these programs design only evenly terminated elliptic filters. Both classical synthesis techniques and the permutation method can be modified to design unevenly terminated elliptic filters as shown in examples 7.6 and 7.7. Optimization routines that modify component values on evenly terminated filters in an attempt to generate a corresponding set of element values suitable for arbitrary terminations almost always converge to a suboptimal solution. Synthesis based on an expression for input impedance guarantees an exact filter response for arbitrarily terminated filters.

Tchebyscheff Type II element values for passive filters will rarely be seen tabulated in filter tables because they often contain negative elements values that cannot be absorbed by network transformations. The design of even-order type-*a* elliptic filters also requires one, or more, negative element values. Simple transformations for the elliptic type-*a* were provided for moving the highest attenuation zero to infinity and the lowest transmission pole to zero in order to realize topologies with nonnegative component values.

REFERENCES

[1] Porter, Jack, "Chebyshev Filters with Arbitrary Source and Load Resistances," *RF Design Magazine*, August 1989, pp. 63–68.

[2] Daniels, Richard W., *Approximation Methods for Electronic Filter Design*, New York: McGraw-Hill Book Company, 1975, pp. 100–103.

[3] Amstutz, P., "Elliptic Approximation and Elliptic Filter Design on Small Computers," *IEEE Trans. Circuits Syst.*, December 1978, pp. 1001–1011.

[4] Cuthbert, Thomas R., Jr., *Circuit Design Using Personal Computers*, New York: John Wiley & Sons, 1983.

PROBLEMS

7.1 Find an expression as a function of ω for the return loss, *RL*, of a normalized tenth-order Tchebyscheff Type I low pass with 0.5 dB passband ripple.

7.2 Derive equation (7.15) using the circuit models in Figures 7.1 and 7.2.

7.3 Derive an expression similar to equation (7.10) for the input impedance of a passive filter such that synthesis yields the first element as a shunt, instead of a series.

7.4 Derive equation (7.22).

7.5 Repeat example 7.2 using Bartlett's bisection theorem. Are the resulting component values the same as in example 7.2?

7.6 For a normalized tenth-order Tchebyscheff Type I low pass, find the closest resistance possible to a 1 Ω load if the passband ripple is 0.01 dB.

7.7 Use the equations given in Figure 7.4 to find the component values of a sixth-order normalized Butterworth filter with 1 Ω source resistance and a 10 Ω load. Repeat for a Tchebyscheff Type I filter with 1 dB passband ripple.

7.8 Use the equation given in Figure 7.5 to find the component values of a sixth-order normalized Butterworth filter with a 1 Ω source resistance and an open circuit load.

7.9 Find the element values for a normalized low-pass filter with 1 dB passband ripple and 80 dB stopband attenuation for

(a) a sixth-order elliptic type-*a* configuration,
(b) a sixth-order elliptic type-*b* configuration,
(c) a sixth-order elliptic type-*c* configuration,
(d) a seventh-order elliptic type-*s* configuration.

7.10 Find the bandpass equivalent of the circuit in Figure 7.10, if a bandpass region from 1 MHz to 2 MHz is required.

Chapter 8
Active Filters

8.1 VOLTAGE OPERATIONAL AMPLIFIER CHARACTERISTICS

Active filters are usually designed with operational amplifiers to take advantage of their high input impedance, low output impedance, large open-loop gain, and low cost. These filters can be realized using just resistors and capacitors; the large inductors, necessary in low-frequency passive filters, are not required in active filters.

Most designs based on operational amplifiers assume ideal op-amp characteristics in the frequency range of interest. The following two rules are useful in the analysis of any voltage op-amp filter circuit.

1. If a feedback circuit exists between the output of the op-amp and the inverting input, then the voltage at the inverting input is the same as the voltage at the noninverting input.
2. The inverting input and noninverting input draw no current.

Figure 8.1 shows a basic inverting op-amp configuration. Because the impedance Z_f connects the output to the inverting input, the voltage at the inverting input must be the same as the voltage at the noninverting input. Voltage V^- is at virtual ground and can be assumed to be zero volts because the voltage at the noninverting input, V^+, is zero. Therefore the current through the impedance Z_{in} is

$$I_{in} = \frac{V_{in} - 0}{Z_{in}} \tag{8.1}$$

The output voltage, V_o, is the voltage between the output of the op-amp and ground. Because Z_f is connected between the output of the op-amp and virtual ground, V_o is also the voltage across the impedance Z_f. The op-amp input terminals draw essentially no

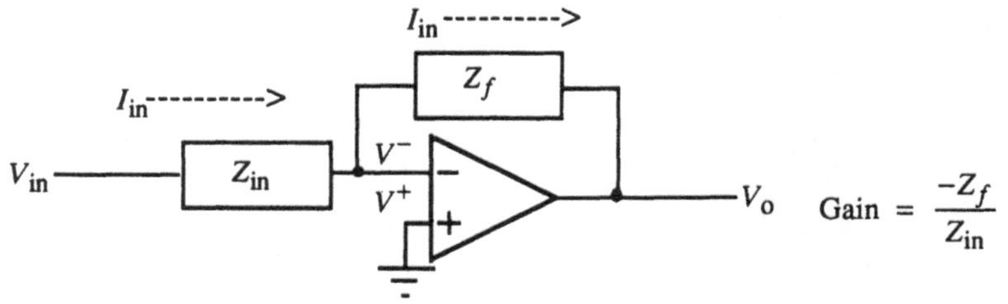

Figure 8.1 Inverting op-amp configuration.

current and the entire input current, I_{in}, flows through Z_f. The voltage across Z_f is the output voltage, V_o, given by

$$V_o = -I_{in} Z_f \tag{8.2}$$

Substituting equation (8.2) into (8.1), the voltage transfer function is

$$\text{Gain} = \frac{V_o}{V_{in}} = \frac{-Z_f}{Z_{in}} \tag{8.3}$$

In Figure 8.2, a noninverting op-amp configuration is shown. Because of the presence of negative feedback, the voltage, V^-, at the inverting op-amp input is the same as the input voltage, V_{in}. The current through Z_i is given by

$$I_i = \frac{V_{in}}{Z_i} \tag{8.4}$$

Because I_i also flows through Z_f, the voltage across Z_f is given by

$$V_{Zf} = I_i Z_f \tag{8.5}$$

Using Kirchhoff's voltage law, the output voltage, V_o, is the sum of V_{in} and the voltage across Z_f, or

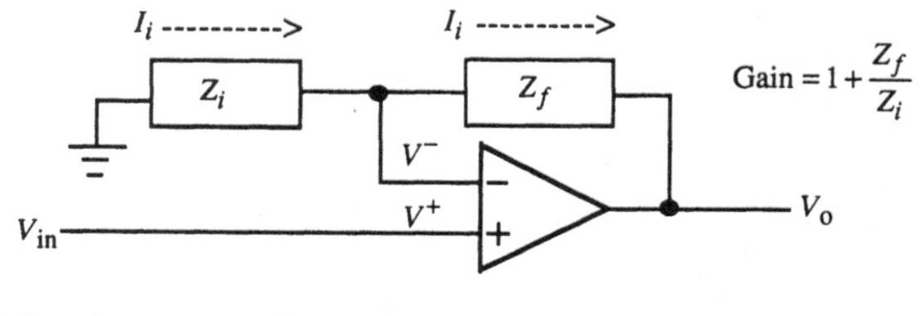

Figure 8.2 Noninverting op-amp configuration.

$$V_o = V_{\text{in}} + V_{\text{in}}\frac{Z_f}{Z_i} = V_{\text{in}}\left(1 + \frac{Z_f}{Z_i}\right) \tag{8.6}$$

and the voltage gain is

$$\text{Gain} = \left(1 + \frac{Z_f}{Z_i}\right) \tag{8.7}$$

8.2 DECOMPOSITION OF THE TRANSFER FUNCTION

The filter voltage transfer function, $G(s)$, of order N has the general form

$$G(s) = \frac{A_N s^N + A_{N-1}s^{N-1} + \ldots + A_1 s^1 + A_0}{B_N s^N + B_{N-1}s^{N-1} + \ldots + B_1 s^1 + B_0} \tag{8.8}$$

and can be decomposed three different ways for active synthesis. If the poles and zeroes of $G(s)$ are not known, then a root finder should be employed to determine both the real and the complex conjugate roots of the numerator and denominator of (8.8). The poles of $G(s)$ must lie in the left half $j\omega$ plane; however, the zeros can lie in either the left or right half-planes or on the $j\omega$ axis. The order of the numerator should be less than or equal to the order of the denominator.

The function $G(s)$ can be separated into quadratic factors and rewritten for N even and $m = N/2$ as

$$G(s) = \frac{A_{12}s^2 + A_{11}s + A_{10}}{B_{12}s^2 + B_{11}s + B_{10}}\,\frac{A_{22}s^2 + A_{21}s + A_{20}}{B_{22}s^2 + B_{21}s + B_{20}}\cdots\frac{A_{m2}s^2 + A_{m1}s + A_{m0}}{B_{m2}s^2 + B_{m1}s + B_{m0}} \tag{8.9}$$

or for N odd and $m = (N + 1)/2$ as

$$G(s) = \frac{A_{11}s + A_{10}}{B_{11}s + B_{10}}\,\frac{A_{22}s^2 + A_{21}s + A_{20}}{B_{22}s^2 + B_{21}s + B_{20}}\cdots\frac{A_{m2}s^2 + A_{m1}s + A_{m0}}{B_{m2}s^2 + B_{m1}s + B_{m0}} \tag{8.10}$$

or for N odd and $m = (N - 1)/2$ as

$$G(s) = \frac{A_{13}s^3 + A_{12}s^2 + A_{11}s + A_{10}}{B_{13}s^3 + B_{12}s^2 + B_{11}s + B_{10}}\,\frac{A_{22}s^2 + A_{21}s + A_{20}}{B_{22}s^2 + B_{21}s + B_{20}}\cdots\frac{A_{m2}s^2 + A_{m1}s + A_{m0}}{B_{m2}s^2 + B_{m1}s + B_{m0}} \tag{8.11}$$

Each quadratic term has a Q and a natural resonant frequency, ω_o, related by the expression

$$\frac{A_{x2}s^2 + A_{x1}s + A_{x0}}{B_{x2}s^2 + B_{x1}s + B_{x0}} = \frac{\frac{A_{x2}}{B_{x2}}s^2 + \frac{A_{x1}}{B_{x2}}s + \frac{A_{x0}}{B_{x2}}}{s^2 + \frac{B_{x1}}{B_{x2}}s + \frac{B_{x0}}{B_{x2}}} = \frac{\frac{A_{x2}}{B_{x2}}s^2 + \frac{A_{x1}}{B_{x2}}s + \frac{A_{x0}}{B_{x2}}}{s^2 + \frac{\omega_{ox}}{Q_x}s + \frac{B_{x0}}{B_{x2}}} \tag{8.12}$$

where Q_x and ω_{ox} are the Q and the natural resonant frequency of the xth stage, respectively.

Example 8.1

Calculate the output voltage, V_o, as a function of V_1, V_2, V_3, and V_4 in Figure 8.3.

Solution

If the resistors values are integer multiples of each other, then the answer can usually be written down by inspection. By superposition, the contribution that each input voltage makes to the output can be determined independently by shorting, in turn, the other input voltages to ground. The output voltage, V_o, can then be written in terms of the individual contributing terms. Using the voltage divider to calculate V^+,

$$V^+ = \frac{V_3}{3} + \frac{V_4}{3}$$

From equation (8.7), V^+ is multiplied by a gain factor of 3 so that

$$V_o' = 3V^+ = V_3 + V_4, \qquad \text{with } Z_f = 10K \text{ and } Z_i = 5K$$

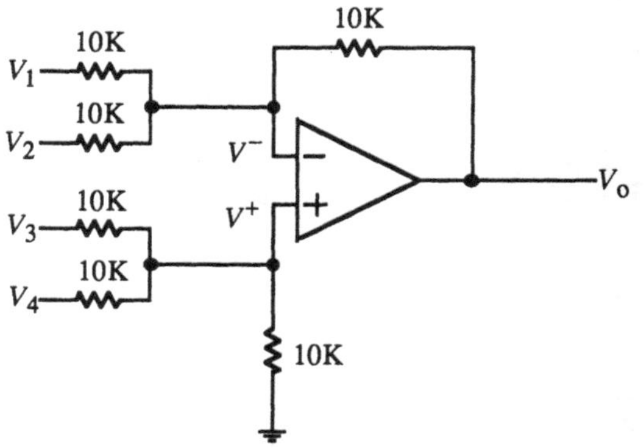

Figure 8.3 Op-amp summing amplifier.

The contribution of V_1 and V_2 can be determined using (8.3) such that

$$V_o'' = -V_1 - V_2$$

The total output voltage is

$$V_o = V_o' + V_o'' = -V_1 - V_2 + V_3 + V_4$$

Example 8.2

$G(s)$ is the transfer function for a low-pass Tchebyscheff Type I filter with 1 dB passband ripple and a cutoff frequency of 5000 Hz. Prepare $G(s)$ for synthesis by decomposing it into the product of quadratic factors, each with unity gain at dc, and find the Q of each stage.

$$G(s) = \frac{3.758738 \times 10^{21}}{s^5 + 29431s^4 + (1.666795 \times 10^9)s^3 + (3.02124 \times 10^{13})s^2 + (5.654391 \times 10^{17})s + (3.758738 \times 10^{21})}$$

Solution

The roots of the denominator are the poles of $G(s)$. Using a root finder, the five poles are -9094.701; $-7{,}357.768 \pm j19{,}224.03$; $-2{,}810.417 \pm j31{,}105.13$. Then $G(s)$ can be rewritten as

$$\underset{\text{Stage 1}}{\frac{9094.701}{s + 9094.701}} \quad \underset{\text{Stage 2}}{\frac{4.237 \times 10^8}{s^2 + 14{,}715.54s + (4.237 \times 10^8)}} \quad \underset{\text{Stage 3}}{\frac{9.754272 \times 10^8}{s^2 + 5{,}620.835s + (9.754272 \times 10^8)}}$$

Using the definition in equation (8.12), the values of Q of the second and third stages are

$$\text{Stage 2: } Q = \frac{\sqrt{4.237 \times 10^8}}{14{,}715.54} = 1.398792$$

$$\text{Stage 3: } Q = \frac{\sqrt{9.754 \times 10^8}}{5{,}620.835} = 5.556441$$

8.3 INFINITE-GAIN MULTIPLE-FEEDBACK REALIZATIONS

The infinite-gain multiple-feedback circuit [1] has the advantage that the output impedance is low; therefore networks may be cascaded with negligible interaction. A disadvantage is that it is not possible to obtain high Q without resorting to large spreads of element values and also incurring large Q sensitivities. Trimming the component values of these filters to adjust the Q and the resonant frequency can be formidable. Generally these circuits should be restricted to implementations requiring a Q less than 10, or a passband gain of less than 10.

The circuit shown in Figure 8.4 is a lowpass capable of synthesizing quadratic factors of the form

$$G(s) = \frac{-H\omega_o^2}{s^2 + \alpha\omega_o s + \omega_o^2} \tag{8.13}$$

The voltage transfer function, $G(s)$, can be found by writing the nodal equations for the currents. Because the voltage at the inverting input must be 0 volts, the current, I_3, is the output voltage, V_o, divided by the impedance of C_2, or

$$I_3 = sC_2V_o \tag{8.14}$$

The inverting input draws no current, thus I_3 also flows through R_3 and the voltage, V_1, can be calculated as

$$V_1 = -I_3R_3 = -sC_2R_3V_o. \tag{8.15}$$

Using ohm's law to write down expressions for the other currents,

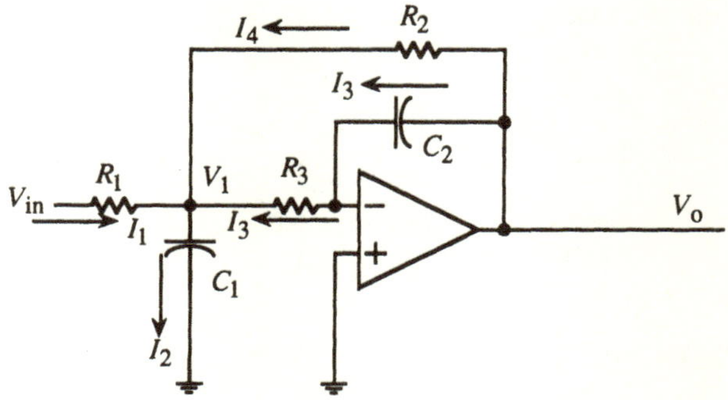

Figure 8.4 Multiple feedback low pass.

$$I_1 = \frac{V_{\text{in}} - V_1}{R_1} = \frac{V_{\text{in}} + sC_2R_3V_o}{R_1} \tag{8.16}$$

$$I_2 = \frac{V_1}{\dfrac{1}{sC_1}} = sC_1V_1 = -s^2C_1C_2R_3V_o \tag{8.17}$$

$$I_4 = \frac{V_o - V_1}{R_2} = \frac{V_o(1 + sC_2R_3)}{R_2} \tag{8.18}$$

Because the currents entering a node must equal the currents leaving,

$$I_2 = I_1 + I_3 + I_4 \tag{8.19}$$

Substituting for the currents and rearranging terms, the transfer function becomes

$$G(s) = \frac{\dfrac{-1}{C_1C_2R_1R_3}}{s^2 + s\left(\dfrac{1}{C_1}\right)\left(\dfrac{1}{R_1} + \dfrac{1}{R_2} + \dfrac{1}{R_3}\right) + \dfrac{1}{C_1C_2R_2R_3}} \tag{8.20}$$

Comparing equations (8.13) to (8.20) and solving for R_1,

$$R_1 = \frac{\alpha\omega_oC_1H \pm \sqrt{(\alpha\omega_oC_1H)^2 - 4(H + 1)H^2\omega_o^2C_1C_2}}{2H^2\omega_o^2C_1C_2} \tag{8.21}$$

To avoid taking the square root of a negative number

$$C_1 \geq \frac{4(H + 1)C_2}{\alpha^2} \tag{8.22}$$

The derivations for the bandpass and high-pass multiple-feedback filters are similar to the low-pass case. These three filters, based on the multiple-feedback topology, are summarized in Figure 8.5.

8.4 VOLTAGE-CONTROLLED VOLTAGE-SOURCE (VCVS) REALIZATIONS

The voltage-controlled voltage-source (VCVS) realizations have an advantage over the multiple-feedback infinite-gain configurations in that it is possible to obtain higher values of Q without resorting to large spreads of element values and also without incurring large

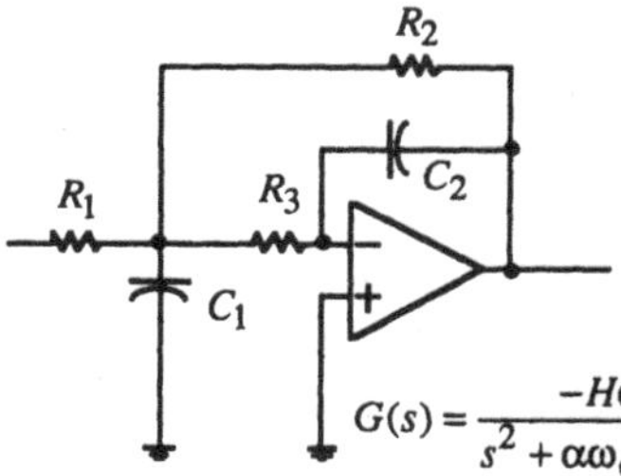

$$G(s) = \frac{-H\omega_o^2}{s^2 + \alpha\omega_o s + \omega_o^2}$$

$$G(s) = \frac{\dfrac{-1}{C_1 C_2 R_1 R_3}}{s^2 + s\left(\dfrac{1}{C_1}\right)\left(\dfrac{1}{R_1} + \dfrac{1}{R_2} + \dfrac{1}{R_3}\right) + \dfrac{1}{C_1 C_2 R_2 R_3}}$$

Choose C_2 = any convenient value:

$$C_1 \geq \frac{4(H+1)C_2}{\alpha^2}$$

$$R_1 = \frac{\alpha}{2H\ \omega_o\ C_2}\left(1 + \sqrt{1 - \frac{4(H+1)C_2}{\alpha^2 C_1}}\right)$$

$$R_2 = HR_1 \qquad R_3 = \frac{1}{\omega_o^2 C_1 C_2 R_2}$$

MFB low - pass section

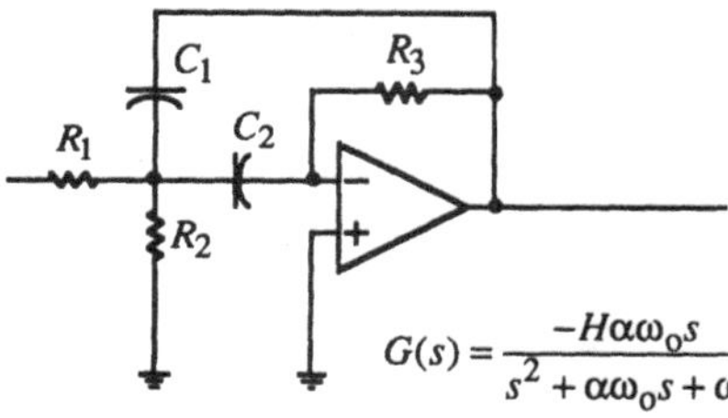

$$G(s) = \frac{-H\alpha\omega_o s}{s^2 + \alpha\omega_o s + \omega_o^2}$$

$$G(s) = \frac{\dfrac{-s}{C_1 R_1}}{s^2 + s\left(\dfrac{C_1 + C_2}{R_3 C_1 C_2}\right) + \dfrac{R_1 + R_2}{C_1 C_2 R_1 R_2 R_3}}$$

Choose $C_1 = C_2$ = any convenient value.

$$R_1 = \frac{1}{\alpha\omega_o H C_1}$$

$$R_2 = \frac{\alpha}{(2 - H\alpha^2)\omega_o C_1}$$

$$R_3 = \frac{2}{\alpha\omega_o C_1}$$

MFB bandpass section

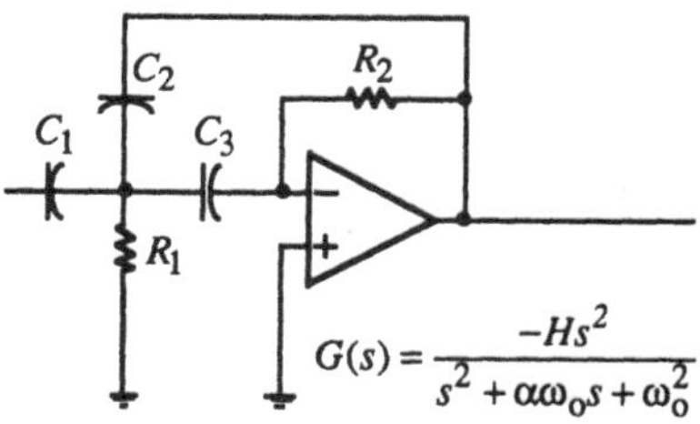

$$G(s) = \frac{-Hs^2}{s^2 + \alpha\omega_o s + \omega_o^2}$$

$$G(s) = \frac{\dfrac{-sC_1}{C_2}}{s^2 + s\left(\dfrac{1}{R_2}\right)\left(\dfrac{C_1}{C_2 C_3} + \dfrac{1}{C_2} + \dfrac{1}{C_3}\right) + \dfrac{1}{C_2 C_3 R_1 R_2}}$$

Choose $C_1 = C_3$ = any convenient value.

$$C_2 = \frac{C_1}{H}$$

$$R_1 = \frac{\alpha H}{(2H+1)\omega_o C_1} \qquad R_2 = \frac{2H+1}{\alpha\omega_o C_1}$$

Note: If C_1 and C_3 are chosen from standard capacitor values, then it may be necessary to modify H if C_2 is desired to be a standard value.

MFB high - pass section

Figure 8.5 Multiple feedback (MFB) topologies and design equations.

Q sensitivities. The VCVS low-pass and high-pass circuits allow reasonable sensitivities up to a Q of 50. The controlled source circuits are easier to tune than multiple-feedback circuits, and component values can be adjusted over wide ranges without interaction of the network parameters.

The circuit shown in Figure 8.6 is a low-pass capable of synthesizing quadratic factors of the form

$$G(s) = \frac{H\omega_o^2}{s^2 + \alpha\omega_o s + \omega_o^2} \tag{8.23}$$

The voltage transfer function, in terms of the resistors and capacitors, can be derived by nodal analysis similar to the MFB circuits. The resistor R_5 is optional and required only if the dc gain, H, is less than unity. Resistors R_3 and R_4 are also optional and are required only for dc gains greater than unity. From the previous chapter on frequency scaling, it should be obvious that increasing, or decreasing, R_1, R_2, and R_5 by the same percentage will decrease, or increase, ω_o without affecting α or H. Capacitors C_1 and C_2 can be adjusted in the same way for the same result. A complete summary of the design equations is given in Figure 8.7 for the low-pass, bandpass, and high-pass cases.

All of the VCVS design equations for the low-pass, bandpass, and high-pass circuits can be derived from the VCVS low pass shown in Figure 8.8. Using Kirchhoff's current law, the currents I_1, I_2, and I_3 must satisfy the equation

$$I_1 + I_3 = I_2 \tag{8.24}$$

The currents are related to the node voltages by the equations

$$I_1 = \frac{V_{in} - V_1}{R_x} \tag{8.25}$$

$$I_2 = \frac{sV_oC_2}{K} \tag{8.26}$$

$$I_3 = sC_1(V_o - V_1) \tag{8.27}$$

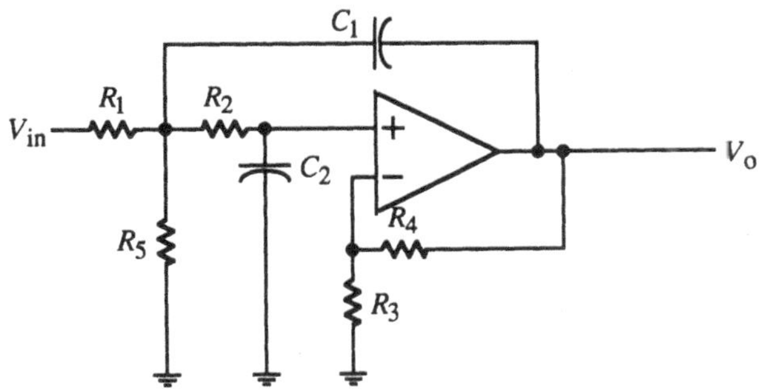

Figure 8.6 VCVS low-pass filter.

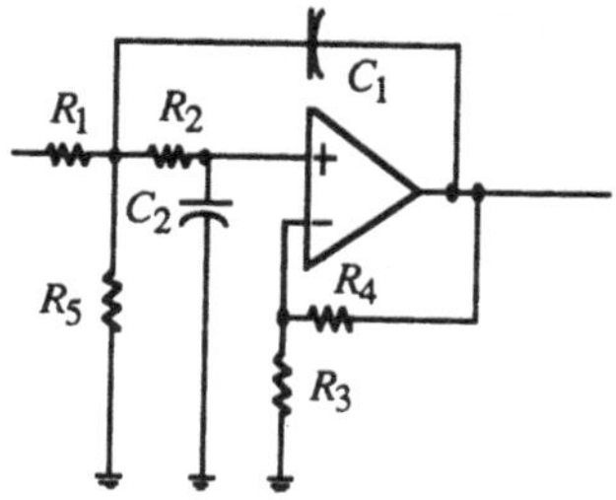

$$G(s) = \frac{\dfrac{K}{C_1C_2R_1R_2}}{s^2 + s\left(\dfrac{1}{R_1C_1} + \dfrac{1}{R_2C_1} + \dfrac{1-K}{R2C_2}\right) + \dfrac{1}{C_1C_2R_1R_2}}$$

If $H < 1$ then $K = 1.$ $\quad C_1 \geq \dfrac{4C_2}{\alpha^2 - 4(1-K)}$

If $H \geq 1$ then $K = H.$

Choose $C_2 = 1.$ $\quad R_2 = 1 / C_1R_1\omega_o^2$

$$R_1 = \frac{\alpha\omega_o C_1 \pm \sqrt{(\alpha\omega_o C_1)^2 - 4(C_1 + (1-K)C_1{}^2)\omega_o^2}}{2\omega_o^2(C_1 + (1-K)C_1{}^2)}$$

$$G(s) = \frac{H\omega_o^2}{s^2 + \alpha\omega_o s + \omega_o^2}$$

VCVS low - pass section

Impedance scale R_1, R_2, C_1, and C_2 for reasonable values.

If $K = 1$ then $R_3 = \infty$ and $R_4 = 0$ else $R_3 = 10{,}000$ and $R_4 = R_3(K-1)$

If $H < 1$ then $R_5 = R_1 / (1-H)$ & then set $R_1 = R_1 / H$

If $H \geq 1$ then $R_5 = \infty$

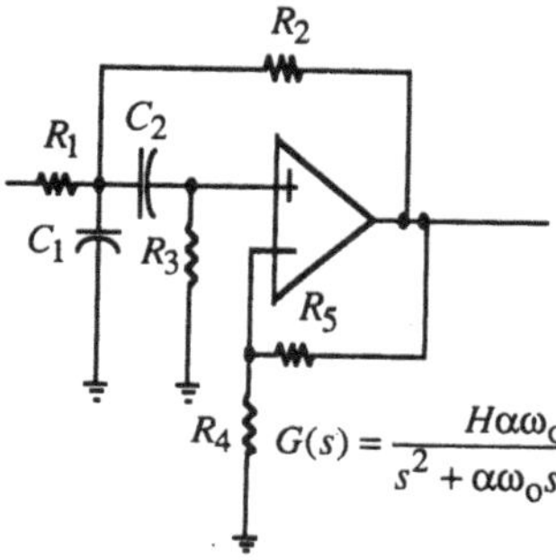

$$G(s) = \frac{H\alpha\omega_o s}{s^2 + \alpha\omega_o s + \omega_o^2}$$

$$G(s) = \frac{\dfrac{Ks}{C_1R_1}}{s^2 + s\left(\dfrac{1}{R_1C_1} + \dfrac{1}{R_3C_1} + \dfrac{1}{R_3C_2} + \dfrac{1-K}{R_2C_1}\right) + \dfrac{R_1 + R_2}{C_1C_2R_1R_2R_3}}$$

Choose $C_1 = C_2 = R_3 = 1$

$G_1 = \omega_o^2 - G_2$ $\quad R_1 = \dfrac{1}{G_1}$

$G_2 = \dfrac{\omega_o^2(2 - \alpha\omega_o + \omega_o^2)}{2 + \omega_o^2 + \alpha\omega_o(H-1)}$ $\quad R_2 = \dfrac{1}{G_2}$

VCVS bandpass section

Impedance scale R_1, R_2, R_3, and C_2 for reasonable values.

$K = H\alpha\omega_o R_1C_1 \qquad R_4 = 10{,}000 \qquad R_5 = R_4(K-1)$

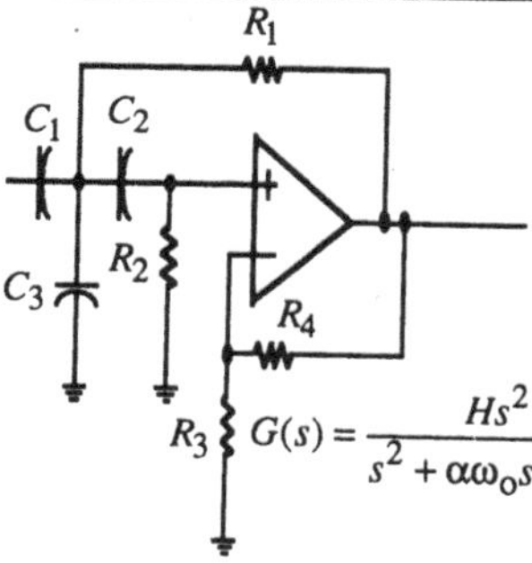

$$G(s) = \frac{Hs^2}{s^2 + \alpha\omega_o s + \omega_o^2}$$

$$G(s) = \frac{Ks^2}{s^2 + s\left(\dfrac{1}{R_2C_1} + \dfrac{1}{R_2C_2} + \dfrac{1-K}{R_1C_1}\right) + \dfrac{1}{C_1C_2R_1R_2}}$$

Choose $C_1 = C_2 =$ any convenient value. $\quad H = K.$

If $H < 1$ then $H_1 = 1$ else $H_1 = H$

$R_1 = \dfrac{\alpha \pm \sqrt{\alpha^2 + 8(H_1 - 1)}}{4\omega_o C_1}$ $\quad R_2 = \dfrac{1}{\omega_o^2 R_1C_1C_2}$

If $H < 1$ then $C_3 = C_1(1-H)$ and set $C_1 = HC_1$ else $C_3 = 0$

If $H = 1$ then $R_3 = \infty$ and $R_4 = 0$ else $R_3 = 10{,}000$ and $R_4 = R_3 / (H-1)$

VCVS high - pass section

Figure 8.7 VCVS topologies and design equations.

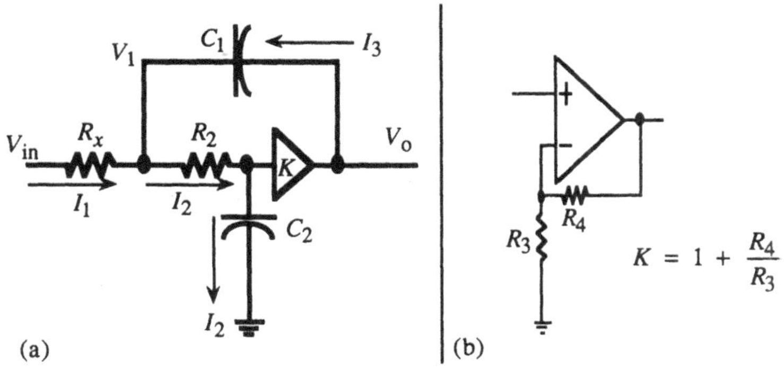

Figure 8.8 (a) VCVS low-pass circuit, and (b) amplifier with a gain of K.

The constant K represents the gain of the amplifier circuit, where

$$K \geq 1 \quad \text{and} \quad K = 1 + \frac{R_4}{R_3} \tag{8.28}$$

and R_3 and R_4 are defined in Figure 8.8(b). Because the voltage at the input of the amplifier must be V_o/K, the voltage across R_2 can be written as

$$I_2R_2 = V_1 - \frac{V_o}{K} \tag{8.29}$$

Substituting equations (8.25), (8.27), and (8.29) into (8.24) yields

$$\frac{V_{in} - V_1}{R_x} + sC_1(V_o - V_1) = \frac{V_1 - \dfrac{V_o}{K}}{R_2} \tag{8.30}$$

The voltage V_o/K can be written in terms of V_1, using the voltage divider relationship between R_2 and C_2. The voltage V_o/K is

$$\frac{V_o}{K} = \frac{V_1}{R_2sC_2 + 1} \tag{8.31}$$

Using (8.31) to eliminate V_1 from (8.30) leads directly to the transfer function

$$\frac{V_o}{V_{in}} = \frac{\dfrac{K}{C_1C_2R_xR_2}}{s^2 + s\left(\dfrac{1}{R_xC_1} + \dfrac{1}{R_2C_1} + \dfrac{1 - K}{R_2C_2}\right) + \dfrac{1}{C_1C_2R_xR_2}} \tag{8.32}$$

Equation (8.32) does not allow gains less than unity because K is restricted to values greater than or equal to unity. Using Thevenin's theorem, R_x, may be replaced by a voltage divider consisting of R_1 and R_5 such that

$$R_1 \parallel R_5 = R_x, \quad \text{Gain} = \frac{R_5}{R_1 + R_5}, \quad \text{and} \quad K = 1 + \frac{R_4}{R_3} \tag{8.33}$$

The exact transfer function becomes

$$\frac{V_o}{V_{in}} = \frac{\dfrac{K(R_1 + R_5)\,\text{Gain}}{C_1C_2R_1R_2R_5}}{s^2 + s\left(\dfrac{R_1 + R_5}{R_1C_1R_5} + \dfrac{1}{R_2C_1} + \dfrac{1 - K\,\text{Gain}}{R_2C_2}\right) + \dfrac{R_1 + R_5}{C_1C_2R_1R_2R_5}} \tag{8.34}$$

To simplify the selection of components, if Gain ≤ 1, then K is taken as unity and $R_3 = \infty$ and $R_4 = 0$. If $K \geq 1$ then Gain is taken as unity and $R_x = R_1$ with $R_5 = \infty$. The complete schematic, including R_5, is shown in Figure 8.7 for the VCVS low pass.

The use of (8.32) to select component values can be simplified by letting $C_2 = 1$. Then by direct comparison to (8.23)

$$R_2 = \frac{1}{C_1R_x\omega_o^2} \tag{8.35}$$

$$\alpha\omega_o = \left(\frac{1}{R_xC_1} + \frac{1}{R_2C_1} + \frac{1 - K}{R_2C_2}\right) \tag{8.36}$$

Substituting (8.35) into (8.36) and solving for R_x yields

$$R_x = \frac{\alpha\omega_oC_1 \pm \sqrt{(\alpha\omega_oC_1)^2 - 4(C_1 + (1 - K)C_1^2)\omega_o^2}}{2\omega_o^2(C_1 + (1 - K)C_1^2)} \qquad C_2 = 1 \tag{8.37}$$

The quantity under the radical cannot be negative; therefore,

$$C_1 \geq \frac{4C_2}{\alpha^2 - 4(1 - K)}, \qquad \text{with } C_2 = 1 \tag{8.38}$$

If $H \geq 1$ then R_3 and R_4 can be determined from

$$H = 1 + \frac{R_4}{R_3} \quad \text{with } R_5 = \infty \quad \text{and} \quad R_1 = R_x \tag{8.39}$$

For values of the dc gain, H, less than 1, R_1 and R_5 can be obtained from

$$H = \frac{R_5}{R_1 + R_5} \quad \text{with } R_3 = \infty \quad \text{and} \quad R_4 = 0 \tag{8.40}$$

The voltage transfer function for the VCVS high pass can be derived directly from the low-pass transfer function, given by equation (8.32), by the substitutions $R_x \to 1/(sC_x)$, $C_1 \to 1/(sR_1)$, $R_2 \to 1/(sC_2)$, $C_2 \to 1/(sR_2)$. The arrows are interpreted to read "is replaced by" and the components refer to Figure 8.9. Making these substitutions and rearranging terms yields

$$\frac{V_o}{V_{in}} = \frac{Ks^2}{s^2 + s\left(\frac{1}{R_2'C_x} + \frac{1}{R_2'C_2'} + \frac{1-K}{R_1'C_x}\right) + \frac{1}{C_xC_2'R_1'R_2'}} \tag{8.41}$$

The derivation of component values for the high-pass VCVS is completely analogous to the low-pass case. The capacitor, C_3, shown for the VCVS high pass in Figure 8.7 is optional and required only if the high-frequency gain is less than unity.

8.5 STATE VARIABLE REALIZATIONS

The state variable active filter stage [2] can be used to realize transfer functions of the form

$$G(s) = \frac{As^2 - Bs + C}{s^2 + \alpha\omega_o s + \omega_o^2} \tag{8.42}$$

This can generally give good response for values of Q up to 100. The state variable implementation can be used to realize all of the quadratic stages and gives a highly

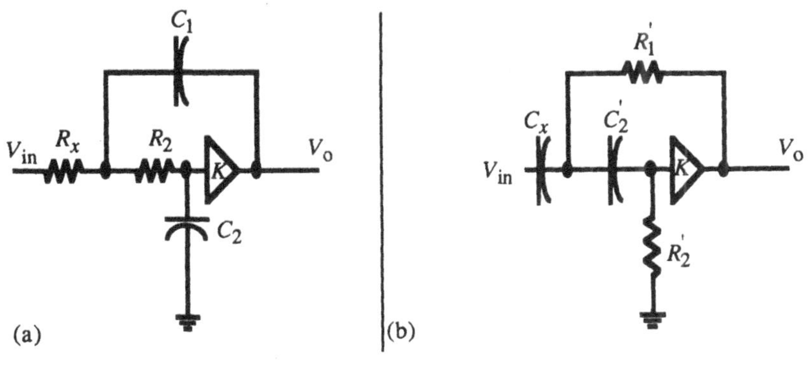

Figure 8.9 (a) VCVS low-pass circuit and (b) VCVS high-pass circuit.

repeatable circuit design with minimal sensitivity to component tolerance. The transfer function of the circuit in Figure 8.10 can be written down by inspection. Expressions for both $\alpha\omega_o$ and ω_o^2 are more easily derived for the state variable filter than for the previous circuits because each op-amp is buffered from the adjacent one. The term ω_o^2 is the gain around the major feedback loop such that

$$\omega_o^2 = \frac{R_{10}}{R_5 R_2 C_1 R_3 C_2} \tag{8.43}$$

Likewise the term $\alpha\omega_o$ is the gain around the minor feedback loop given by

$$\alpha\omega_o = \left(1 + \frac{R_{10}}{R_1 \| R_5}\right)\left(\frac{1}{R_2 C_1}\right)\left(\frac{R_9}{R_9 + R_4}\right) \tag{8.44}$$

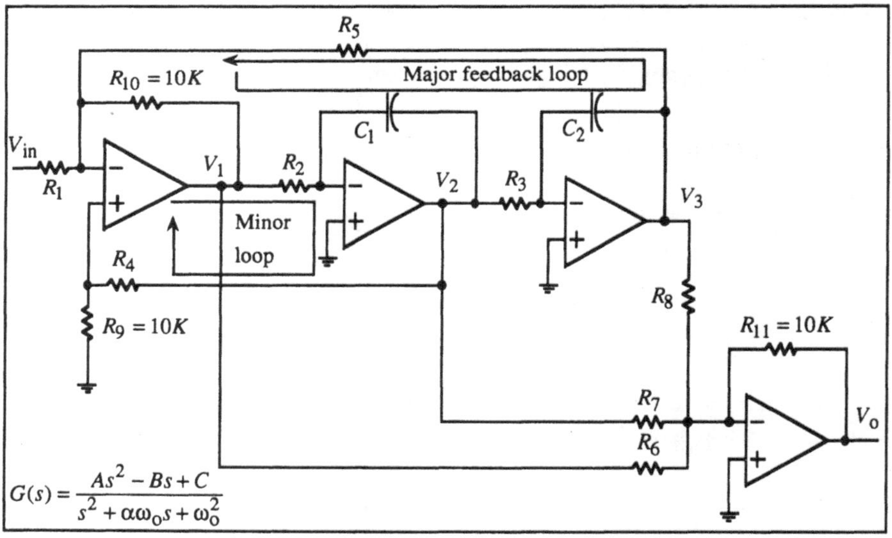

1. Choose $R_1 = R_5 = R_9 = R_{10} = R_{11}$ = any convenient value (10K in the example). Note that setting $R_{10}/R_1 = 1$ guarantees that V_1/V_{in} will not be greater than unity at high frequencies. Also setting $R_5/R_1 = 1$ guarantees that V_3/V_{in} will not be greater than unity at low frequencies.
2. Choose $C_1 = C_2$ = any convenient value.
3. Set $R_6 = R_{10}R_{11}/(R_1 A)$, $R_7 = R_{10}R_{11}/(BC_1R_2R_1)$, $R_8 = (R_{11}R_{10}/(R_1C))\ (1/(R_2C_1))\ (1/(R_3C_2))$.
4. If $A = 0$, then $R_6 = \infty$. If $B = 0$, then $R_7 = \infty$. If $C = 0$, then $R_8 = \infty$.
5. $R_2 = 1/(\omega_o C_1)$, $R_3 = 1/(\omega_o C_2)$, set $R_4 = R_9(3 - \alpha)/\alpha$. Note that $\alpha = 1/Q$, which implies that Q must be greater than 1/3. Step 3 splits the gain of the main feedback loop evenly between the second and third op-amps. The following sets of components can be impedance scaled independently: (R_1, R_5, R_{10}), (C_1, R_2), (C_2, R_3), (R_4, R_9), (R_{11}, R_8, R_7, R_6).

Figure 8.10 Simplified state-variable design equations.

The coefficients A, B, and C are related to the component values by

$$A = \frac{R_{10}R_{11}}{R_1R_6}, \qquad B = \frac{R_{10}R_{11}}{R_1R_2C_1R_7}, \qquad C = \frac{R_{10}R_{11}}{R_1R_2C_1C_2R_3R_8} \tag{8.45}$$

By choosing $R_2 = R_3$ and $C_1 = C_2$ these equations can easily be used to obtain component values from the transfer function coefficients; however, the performance of the state variable circuit, based on these equations, is not optimal for either noise performance or dynamic range. In the next section a modified procedure is shown for improved signal-to-noise ratio and dynamic range.

8.6 GAIN OPTIMIZATION IN ACTIVE FILTERS

Active filters have a dynamic range that is limited for weak signals by the noise floor and limited for strong signals by the power supply voltage. To maximize the dynamic range, the output of each op-amp should also come close to the maximum possible undistorted voltage level that each op-amp can provide. For operation from a ±15 V power supply, a voltage swing of ±10 V peak to peak is usually possible at the op-amps output. In filter circuits, this peak voltage level may occur at only one frequency or over a narrow range of frequencies.

The poles and zeros of a transfer function specify the filter characteristics completely, except for those of the H_i, which are arbitrary gain constants. Cascade designs should be done by placing the lowest Q filter sections at the input with the other sections ordered in ascending values of Q so that the accumulation of any significant deviation from theoretical will not occur until the signal reaches the last sections of the cascade. As an example, H_i is optimized according to the following procedure.

For Figure 8.11, assuming that the input voltage, V_{in}, ranges over ±10 V, the output of each stage can supply up to ±10 V peak-to-peak. Therefore the cumulative gain from the input to any intermediate output should never exceed unity, for this example. The optimization steps for each stage are as follows:

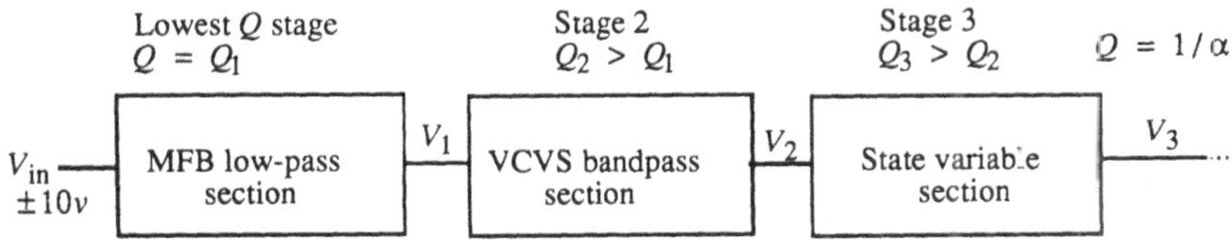

$$G_1(s) = \frac{H_1\,\omega_{01}^2}{s^2 + \alpha_1\,\omega_{01}\,s + \omega_{01}^2} \qquad G_2(s) = \frac{H_2\,\alpha_2\,\omega_{02}}{s^2 + \alpha_2\,\omega_{02}\,s + \omega_{02}^2} \qquad G_3(s) = \frac{H_3\,(As^2 - Bs + C)}{s^2 + \alpha_3\,\omega_{03}\,s + \omega_{03}^2}$$

Figure 8.11 Gain optimization model.

1. Choose H_1 so that $|V_1/V_{in}| \leq 1$ for all frequencies and $|V_1/V_{in}| = 1$ for at least one frequency. This will require computing the magnitude, M_1, of the transfer function given by

$$M_1 = \max\left|\frac{\omega_{o1}^2}{s^2 + \alpha_1\omega_{o1}s + \omega_{o1}^2}\right| = \max\left[\frac{\omega_{o1}^2}{\sqrt{(\omega_{o1}^2 - \omega^2)^2 + (\alpha_1\omega_{o1}\omega)^2}}\right]$$

at all frequencies in the passband and choosing H_1 so that the maximum voltage gain, max $[V_1/V_{in}]$, is unity. Usually evaluation of M_1 at 100 different values of ω in the passband is sufficient. Then

$$H_1 = \frac{1}{M_1}$$

2. Choose H_2 so that $|V_2/V_{in}| \leq 1$ for all frequencies and is exactly unity for some frequency in the passband. The magnitude of the cumulative transfer function, M_2, is computed by

$$M_2 = H_1 \cdot \max\left|\frac{\omega_{o1}^2}{s^2 + \alpha_1\omega_{o1}^2 s + \omega_{o1}^2}\right| \cdot \left|\frac{\alpha_2\omega_{o2}s}{s^2 + \alpha_2\omega_{o2}s + \omega_{o2}^2}\right|$$

at all frequencies in the passband and selecting H_2 so that the maximum voltage gain, $\max[V_2/V_{in}]$, is unity. Note that all the values computed in step 1 should be saved because these values are needed in step 2. All the cumulative values computed in step 2 will be used in step 3, and so on. H_2 is given by

$$H_2 = \frac{1}{M_2}$$

3. The third stage has two zeros in addition to the two poles that determine α and ω. The value of H_3 is chosen by first computing the magnitude, M_3, of the cumulative transfer function

$$M_3 = \max H_1\left|\frac{\omega_{o1}^2}{s^2 + \alpha_1\omega_{o1}s + \omega_{o1}^2}\right| \cdot H_2\left|\frac{\alpha_2\omega_{o2}s}{s^2 + \alpha_2\omega_{o2}s + \omega_{o2}^2}\right| \cdot \left|\frac{As^2 - Bs + C}{s^2 + \alpha_3\omega_{o3}s + \omega_{o3}^2}\right|$$

at all frequencies in the passband and then selecting H_3 so that the maximum voltage gain is unity. The value of H_3 needed to scale the cumulative transfer function to unity is given by

$$H_3 = \frac{1}{M_3}$$

If the third stage were a MFB or VCVS, there would be only one op-amp to optimize and the procedure would be complete. Stage 3 is a state variable and therefore the dynamic range has to be optimized for the output of *each* op-amp.

First, with reference to Figure 8.10, calculate

$$M_{3A} = \max H_1\left|\frac{\omega_{o1}^2}{s^2 + \alpha_1\omega_{o1}s + \omega_{o1}^2}\right| \cdot H_2\left|\frac{\alpha_2\omega_{o2}s}{s^2 + \alpha_2\omega_{o2}s + \omega_{o2}^2}\right| \cdot \left|\frac{s^2}{s^2 + \alpha_3\omega_{o3}s + \omega_{o3}^2}\right|$$

Choose R_{10} = any convenient value and $R_1 = R_{10}M_{3A}$.

Second, choose C_1 = any convenient value and calculate

$$M_{3B} = \max H_1\left|\frac{\omega_{o1}^2}{s^2 + \alpha_1\omega_{o1}s + \omega_{o1}^2}\right| \cdot H_2\left|\frac{\alpha_2\omega_{o2}s}{s^2 + \alpha_2\omega_{o2}s + \omega_{o2}^2}\right| \cdot \left|\frac{s}{s^2 + \alpha_3\omega_{o3}s + \omega_{o3}^2}\right|$$

then $R_2C_1 = 1/(M_{3A}M_{3B})$.

Finally, compute

$$M_{3C} = \max H_1\left|\frac{\omega_{o1}^2}{s^2 + \alpha_1\omega_{o1}s + \omega_{o1}^2}\right| \cdot H_2\left|\frac{\alpha_2\omega_{o2}s}{s^2 + \alpha_2\omega_{o2}s + \omega_{o2}^2}\right| \cdot \left|\frac{1}{s^2 + \alpha_3\omega_{o3}s + \omega_{o3}^2}\right|$$

Choose C_2 = any convenient value, then $R_3C_2 = 1/(M_{3A}M_{3B}M_{3C})$.

Select R_9 and R_{11} = any convenient values. The other components are

$$R_5 = \frac{R_{10}}{\omega_{o3}^2 R_2C_1R_3C_2}, \qquad R_4 = \frac{R_9(3 - \alpha_3)}{\alpha_3}, \qquad R_6 = \frac{R_{11}R_{10}}{AR_1H_3},$$

$$R_7 = \frac{R_{11}R_{10}}{BR_1R_2C_1H_3}, \qquad R_8 = \frac{R_{11}R_{10}}{CR_1R_2R_3C_1C_2H_3}.$$

If the procedures for gain optimization are properly understood, none of the formulas presented in this section will need to be memorized or referenced again. All the equations can be written by inspection directly from the block diagram in Figure 8.11 and the schematic in Figure 8.10.

8.7 SPECIAL ACTIVE FILTER CONFIGURATIONS

The state-variable topology is the only filter configuration presented so far that can synthesize a generalized quadratic transfer function. This section concludes the basic discussion by giving three other building blocks for realizing (1) real pole first-order transfer functions, (2) "elliptic" second-order sections, and (3) third-order low-pass and high-pass implementations.

Circuits for realizing a single real pole are shown in Figure 8.12 with the design equations. An odd-order low-pass or high-pass transfer function must contain at least one real pole, and the circuit for realizing the single pole is generally placed at the input of the cascade. The all pass in Figure 8.12 has the unusual feature of having a unity gain at all frequencies with a phase shift of 0–180° depending on the frequency. At dc, the gain is +1 while at very high frequencies the gain is −1. Interchanging the positions of R_2 and C_1 causes the phase characteristics to reverse but does not affect the magnitude of the gain. Cascades of all-pass filters can be used as group delay equalizers for other types of filters.

The "elliptic" filter sections [3] in Figure 8.13 are used to realize a transfer function of the form

$$G(s) = \frac{H(s^2 + A)}{s^2 + \alpha\omega_o s + \omega_o^2} \tag{8.46}$$

Both implementations given in Figure 8.13 can realize (8.46), however, the gain, K, will differ depending on which implementation is selected. The circuit with the lower value of K will usually give better stability, where K is as defined in Figure 8.8(b).

The component values for the third-order sections [4] in Figure 8.14 can be found by solving a third-degree polynomial. A nonnegative solution to these polynomials is guaranteed because the coefficients of the polynomials alternate in sign. The transfer functions for Figure 8.14(a) and 8.14(c) are easily derived by applying Thevenin's theorem to the VCVS low pass in Figure 8.7. Likewise the low-pass transfer function of Figure 8.14(b) is obtainable from the MFB low pass of Figure 8.5.

Gain optimization should always be used where active filter sections are cascaded for determining the values of H. The "elliptic" sections in Figure 8.13 are the only building blocks presented so far in which H is not a free variable. Conceptually, the gain optimization procedure can easily be modified to either decrease the gain of the previous stage, or to increase the gain of the following stage, depending on the value of H.

8.8 SENSITIVITY OF ACTIVE FILTERS

Sensitivity is a measure of the change in some performance characteristic of a network resulting from a change in value of one or more of the elements of the network. The symbol S is used to denote sensitivity. A superscript character is used to indicate the performance characteristic that is changing, and a subscript character is used to indicate the specific element or parameter that is causing the change. The sensitivity of ω_o with respect to R_1 is defined by

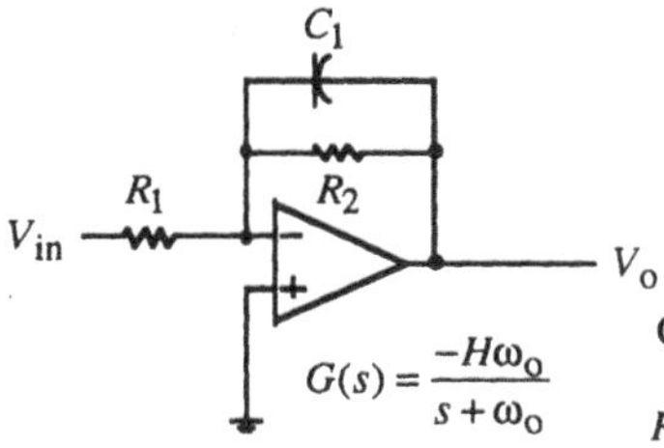

$$G(s) = \frac{-H\omega_o}{s+\omega_o}$$

$$G(s) = \frac{\dfrac{-1}{R_1C_1}}{s + \dfrac{1}{R_2C_1}}$$

Choose C_1 = any convenient value

$$R_1 = \frac{1}{HC_1\omega_o} \qquad R_2 = \frac{1}{\omega_o C_1}$$

First order low-pass

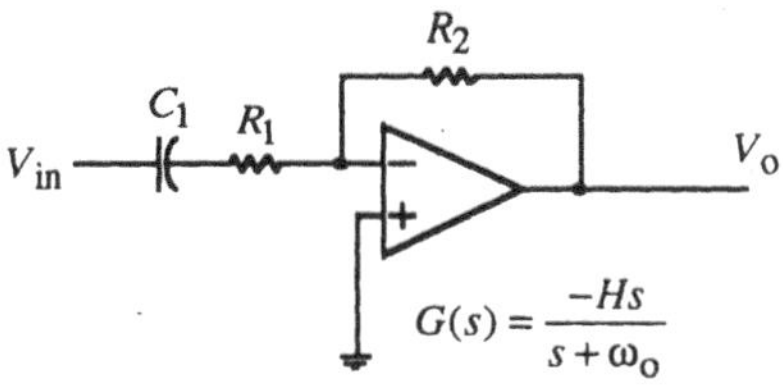

$$G(s) = \frac{-Hs}{s+\omega_o}$$

$$G(s) = \frac{\dfrac{-R_2\,s}{R_1}}{s + \dfrac{1}{R_1C_1}}$$

Choose C_1 = any convenient value

$$R_1 = \frac{1}{\omega_o\, C_1} \qquad R_2 = H\,R_1$$

First order high-pass

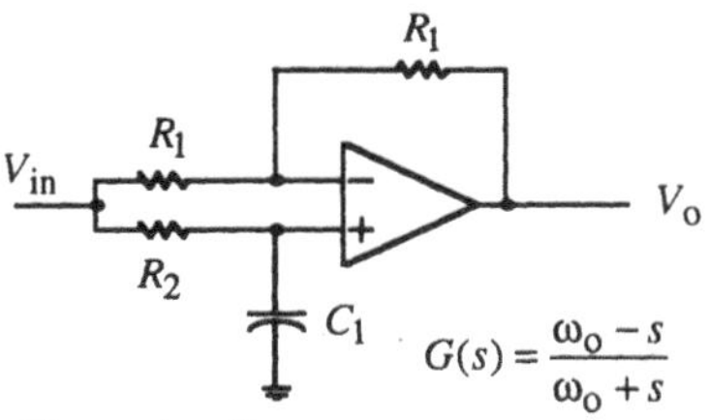

$$G(s) = \frac{\omega_o - s}{\omega_o + s}$$

$$G(s) = \frac{1 - R_2sC_1}{1 + R_2sC_1}$$

Choose C_1 = any convenient value

Choose R_1 = any convenient value

$$R_2 = \frac{1}{\omega_o\, C_1}$$

First order all-pass

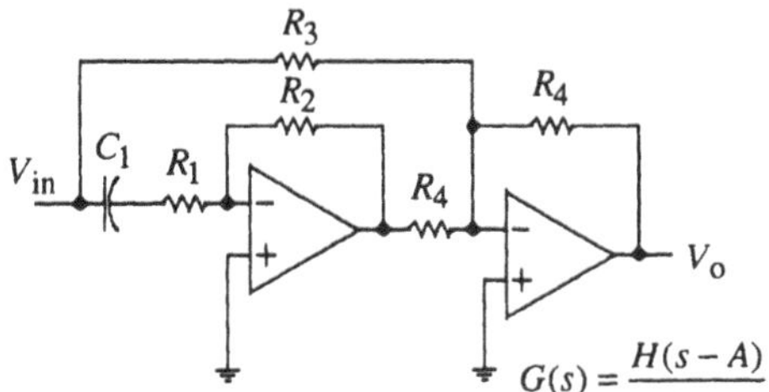

$$G(s) = \frac{H(s-A)}{s+\omega_o}$$

$$G(s) = \frac{\left(\dfrac{R_2}{R_1} - \dfrac{R_4}{R_3}\right)s - \dfrac{R_4}{R_3R_1C_1}}{s + \dfrac{1}{R_1C_1}}$$

Choose C_1 = any convenient value

Choose R_4 = any convenient value

$$R_1 = \frac{1}{\omega_o\, C_1}$$

$$R_3 = \frac{R_4}{R_1\, C_1\, H\, A} \qquad R_2 = \left[H + \frac{R_4}{R_3}\right]R_1$$

Generalized first order

Figure 8.12 First-order topologies and design equations.

Choose R = impedance scaling constant

$$G(s)=\frac{H(s^2+A)}{s^2+\alpha\omega_0 s+\omega_0^2}$$

$$R_1=\frac{A+\omega_0^2}{3A}R$$

$$R_2=2R_1$$

$$R_3=\frac{A+\omega_0^2}{4.5\omega_0^2}R$$

$$R_4=4.5R_3$$

$$C_1=\frac{4.5A}{(A+\omega_0^2)\,R\,\omega_0}$$

$$C_2=\frac{C_1}{4.5}$$

$$C_3=\frac{C_1\,\omega_0^2}{1.5A}$$

$$C_4=\frac{C_3}{2}$$

$$R_4=R_{4a}+R_{4b}$$

$$H_a=\frac{R_{4b}\ (1+R_6/R_5)\omega_0^2}{A(R_1+R_2+R_4)}$$

$$H_b=\frac{R_{4b}\ (1+R_8/R_7)\omega_0^2}{A(R_1+R_2+R_4)}$$

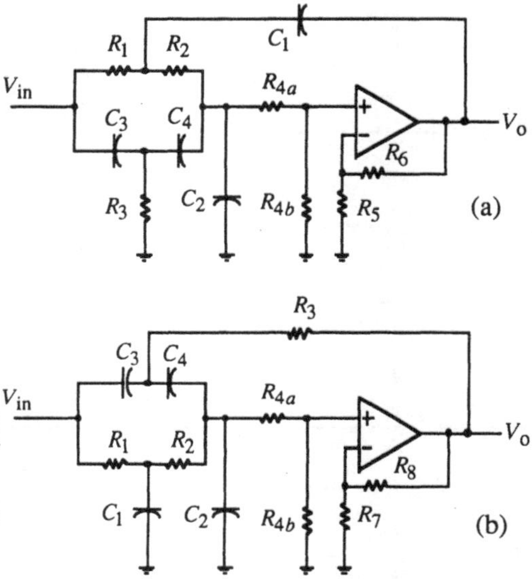

For the circuit in (a)

$$K_a=\frac{(2.5-\alpha)\left(A+\omega_0^2\right)}{1.5A}=1+\frac{R_6}{R_5}$$

For $K_a\le 1$, $R_{4a}=R_4(1-K_a)$, $R_{4b}=R_4K_a$, $R_6=0$, $R_5=\infty$

For $K_a>1$, $R_{4a}=0$, $R_{4b}=R_4$

For the circuit in (b)

$$K_b=\frac{(2.5-\alpha)\left(A+\omega_0^2\right)}{1.5\omega_0^2}=1+\frac{R_8}{R_7}$$

For $K_b\le 1$, $R_{4a}=R_4(1-K_b)$, $R_{4b}=R_4K_b$, $R_8=0$, $R_7=\infty$

For $K_b>1$, $R_{4a}=0$, $R_{4b}=R_4$

Figure 8.13 Elliptic topologies and design equations [4].

symbol S is used to denote sensitivity. A superscript character is used to indicate the performance characteristic that is changing, and a subscript character is used to indicate the specific element or parameter that is causing the change. The sensitivity of ω_o with respect to R_1 is defined by

$$S_{R_1}^{\omega_o}=\frac{d\omega_o/\omega_o}{dR_1/R_1}=\frac{R_1}{\omega_o}\frac{d\omega_o}{dR_1}\tag{8.47}$$

Third order low - pass with unity D.C. gain

$$G(s) = \frac{D_0}{s^3 + D_2 s^2 + D_1 s + D_0}$$

Let $R_1 = R_2 = R_3 = 1$

$$C_1 = \frac{D_1}{D_0} - 3C_3$$

$$C_2 = \frac{1}{C_1 C_3 D_0}$$

$$C_3 = \frac{D_2}{(2C_1 + 2C_2)D_0}$$

$$D_0 = \frac{1}{R_1 R_2 R_3 C_1 C_2 C_3}$$

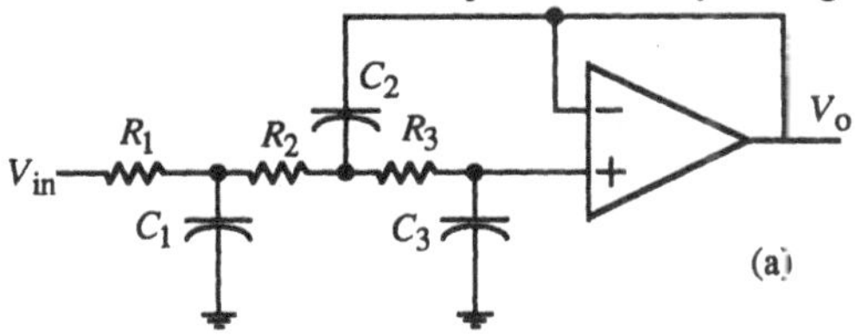

(a)

Find C_1 by solving $2C_1^3 - 2C_1^2 \frac{D_1}{D_0} + 3C_1 \frac{D_2}{D_0} - \frac{6}{D_0} = 0$

$$D_1 = \frac{R_1 C3 + R_2 C_3 + R_3 C_3 + R_1 C_1}{R_1 R_2 R_3 C_1 C_2 C_3}, \quad D_2 = \frac{1}{R_2 C_1} + \frac{1}{R_1 C_1} + \frac{1}{R_3 C_2} + \frac{1}{R_2 C_2}$$

Third order low - pass with adjustable gain

$$G(s) = \frac{-N_0}{s^3 + D_2 s^2 + D_1 s + D_0}$$

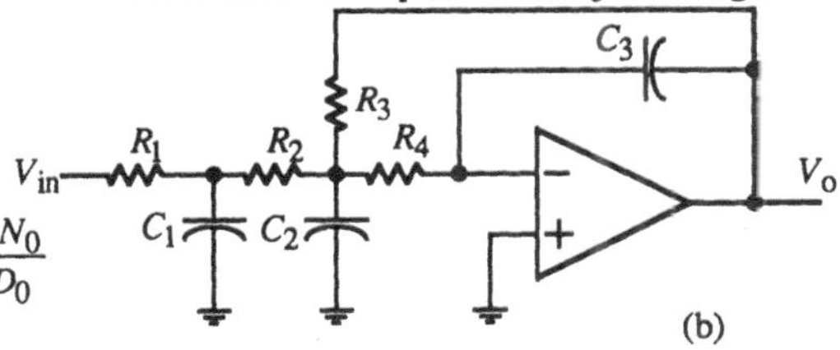

(b)

Let $R_1 = R_2 = R_4 = 1$, then $R_3 = \frac{2N_0}{D_0}$

Find C_1 from

$$C_1^3(D_0 + 2R_3 D_0) - C_1^2(4R_3 D_1 + 2D_1) + C_1(6D_2 R_3 + 4D_2) - (12R_3 + 8) = 0$$

$$D_1 = \frac{1}{R_1 R_2 C_1 C_2} + \frac{1}{R_2 R_3 C_1 C_2} + \frac{1}{R_2 R_4 C_1 C_2} + \frac{1}{R_1 R_3 C_1 C_2} + \frac{1}{R_1 R_4 C_1 C_2} + \frac{1}{R_3 R_4 C_2 C_3}$$

$$D_2 = \frac{1}{R_2 C_1} + \frac{1}{R_1 C_1} + \frac{1}{R_2 C_2} + \frac{1}{R_4 C_2} + \frac{1}{R_3 C_2}, \quad N_0 = \frac{1}{R_1 R_2 R_4 C_1 C_2 C_3}, \quad D_0 = \frac{R_1 + R_2}{R_1 R_2 R_3 R_4 C_1 C_2 C_3}$$

Third order high - pass

$$G(s) = \frac{s^3}{s^3 + D_2 s^2 + D_1 s + D_0}$$

Let $C_1 = C_2 = C_3 = 1$

$$D_0 = \frac{1}{R_1 R_2 R_3}$$

$$D_1 = \frac{2}{R_2 R_3} + \frac{2}{R_1 R_3}$$

$$D_2 = \frac{3}{R_3} + \frac{1}{R_1}$$

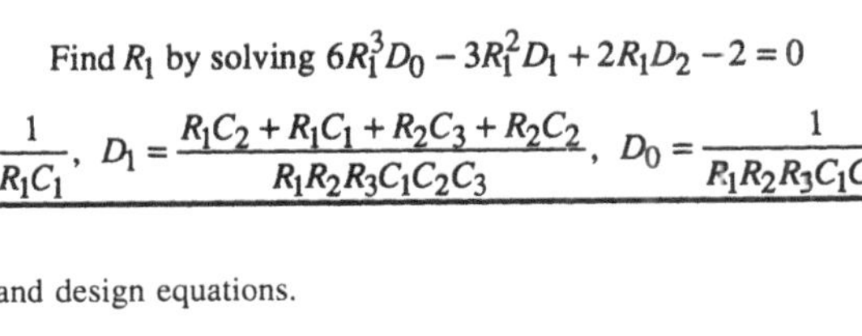

(c)

Find R_1 by solving $6R_1^3 D_0 - 3R_1^2 D_1 + 2R_1 D_2 - 2 = 0$

$$D_2 = \frac{1}{R_3 C_1} + \frac{1}{R_3 C_2} + \frac{1}{R_3 C_3} + \frac{1}{R_1 C_1}, \quad D_1 = \frac{R_1 C_2 + R_1 C_1 + R_2 C_3 + R_2 C_2}{R_1 R_2 R_3 C_1 C_2 C_3}, \quad D_0 = \frac{1}{R_1 R_2 R_3 C_1 C_2 C_3}$$

Figure 8.14 Third-order topologies and design equations.

is changed. A sensitivity of 1/2 implies that a 1% change in R_1 can be expected to produce a 1/2% change in ω_o.

Various methods exist for using the definition of sensitivity to produce circuits that are more tolerant to component variation. Capacitors are often the components that have the largest temperature coefficients. For the VCVS filters, it is possible to select K such that the overall α sensitivity is minimized, assuming that the capacitors drift equally. For the VCVS low pass in Figure 8.7 (with $R_5 = \infty$),

$$\alpha = \frac{1}{\omega_o}\left(\frac{1}{R_1C_1} + \frac{1}{R_2C_1} + \frac{1-K}{R_2C_2}\right) \tag{8.48}$$

$S^{\alpha}_{C_1}$ is found first taking $d\alpha/dC_1$. However, ω_o is also a function for C_1; therefore, equation (8.48) can be expanded and rewritten as

$$\alpha = \left(\frac{R_2C_2}{R_1C_1}\right)^{1/2} + \left(\frac{R_1C_2}{R_2C_1}\right)^{1/2} + \left(\frac{R_1C_1}{R_2C_2}\right)^{1/2} - K\left(\frac{R_1C_1}{R_2C_2}\right)^{1/2} \tag{8.49}$$

Finally,

$$S^{\alpha}_{C_1} = \frac{d\alpha}{dC_1}\frac{C_1}{\alpha} = \frac{1}{2} - \frac{1}{\alpha\omega_o C_1}\left(\frac{1}{R_1} + \frac{1}{R_2}\right) \tag{8.50}$$

In general, the sensitivities of the VCVS low pass are

$$S^{\omega_o}_{R_1} = S^{\omega_o}_{R_2} = S^{\omega_o}_{C_1} = S^{\omega_o}_{C_2} = \frac{-1}{2} \tag{8.51a}$$

$$S^{H}_{K} = 1 \tag{8.51b}$$

$$S^{\alpha}_{R_1} = \frac{1}{2} - \frac{1}{\alpha\omega_o R_1C_1} \tag{8.51c}$$

$$S^{\alpha}_{C_1} = \frac{1}{2} - \frac{1}{\alpha\omega_o C_1}\left(\frac{1}{R_1} + \frac{1}{R_2}\right) \tag{8.51d}$$

$$S^{\alpha}_{C_2} = \frac{1}{2} - \frac{1-K}{\alpha\omega_o R_2C_2} \tag{8.51e}$$

$$S^{\alpha}_{R_2} = \frac{1}{2} - \frac{1}{\alpha\omega_o R_2}\left(\frac{1}{C_1} + \frac{1-K}{C_2}\right) \tag{8.51f}$$

To cancel the variation of α with respect to equal changes in C_1 and C_2, we set

$$S_{C_1}^{\alpha} = -S_{C_2}^{\alpha} \tag{8.52}$$

Choose $C = C_1 = C_2$ and let $R_1 = R_2 = R$, then $K = 3 - \alpha$ and $R = 1/(\omega_o C)$.

A complete set of VCVS filters based on the alternate design of equation (8.52) is given in Figure 8.15. However, (8.52) cannot be used to optimize the tolerance of capacitor values in the multiple feedback filters of Figure 8.5 for reasons that would become apparent once the sensitivities of the MFB filters were derived. The penalty for designing for minimum sensitivity is typically to lose control over the gain of the quadratic section. For a long cascade this almost always results in a poor dynamic range and excessive output noise. Sensitivity design should be restricted to filter cascades that contain only one, or two, quadratic sections. However, sensitivity analysis should be one of the factors used in selecting a circuit topology. In the next section, a new topology is presented in which the sensitivity is low enough to permit stable values of Q in excess of 1000.

8.9 VERY HIGH-Q INSENSITIVE ACTIVE FILTERS

The circuit topology shown in Figure 8.16 can realize stable values of Q up to 5000 [5]. This new configuration overcomes one of the limitations of the state-variable filter in that the frequency response is much less dependent on the high-frequency limitations of the operational amplifiers. The voltage transfer functions for the network in Figure 8.16 are

$$E_1 = -K_1 V_{in} + K_2 V_o \tag{8.53}$$

$$\frac{V_1}{E_1} = \frac{1 - sC_1R_1}{1 + sC_1R_1} \tag{8.54}$$

$$\frac{V_2}{V_1} = \frac{-K_3(1 - sC_2R_2)}{1 + sC_2R_2} \tag{8.55}$$

$$\frac{V_o}{V_2} = K_4 \tag{8.56}$$

Collecting terms, the transfer function becomes

$$\frac{V_o}{V_{in}} = \frac{K_1K_3K_4(1 - sC_1R_1)(1 - sC_2R_2)}{(1 + K_2K_3K_4)R_1C_1R_2C_2s^2 + (R_1C_1 + R_2C_2)(1 - K_2K_3K_4)s + (1 + K_2K_3K_4)} \tag{8.57}$$

If $R_1C_1 = R_2C_2$ is temporarily set equal to 1 then

$$\frac{2(1 - K_2K_3K_4)}{1 + K_2K_3K_4} = \frac{1}{Q} \tag{8.58}$$

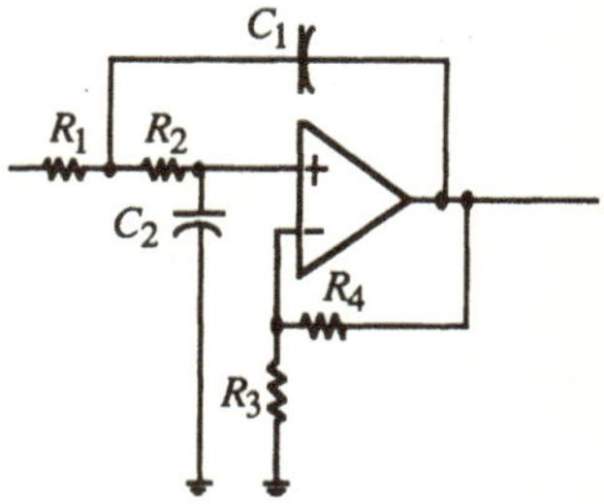

$$G(s) = \frac{\dfrac{K}{C_1C_2R_1R_2}}{s^2 + s\left(\dfrac{1}{R_1C_1} + \dfrac{1}{R_2C_1} + \dfrac{1-K}{R_2C_2}\right) + \dfrac{1}{C_1C_2R_1R_2}}$$

$$G(s) = \frac{H\omega_o^2}{s^2 + \alpha\omega_o s + \omega_o^2}$$

Choose $C_1 = C_2 =$ any convenient value

$$R_1 = R_2 = \frac{1}{\omega_o C_1}$$

Choose $R_3 =$ any convenient value

$$R_4 = (2-\alpha)R_3$$

$$K = 3-\alpha$$

VCVS low - pass section

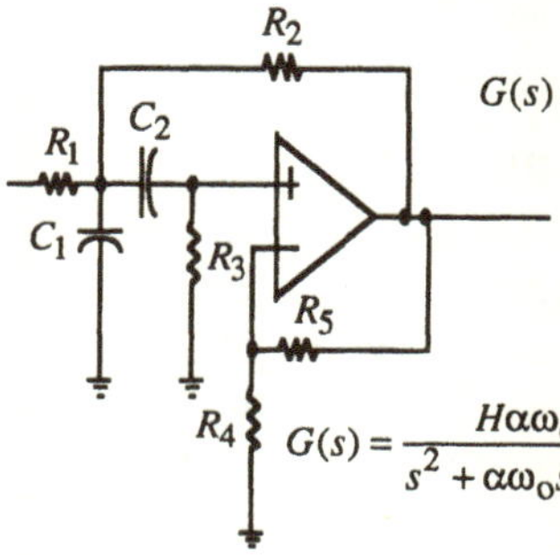

$$G(s) = \frac{\dfrac{Ks}{C_1R_1}}{s^2 + s\left(\dfrac{1}{R_1C_1} + \dfrac{1}{R_3C_1} + \dfrac{1}{R_3C_2} + \dfrac{1-K}{R_2C_1}\right) + \dfrac{R_1+R_2}{C_1C_2R_1R_2R_3}}$$

$$G(s) = \frac{H\alpha\omega_o s}{s^2 + \alpha\omega_o s + \omega_o^2}$$

Choose $C_1 = C_2 =$ any convenient value

$$R_1 = R_2 = R_3 = \frac{1}{\omega_o C_1}$$

Choose $R_4 =$ any convenient value

$$R_5 = (3-\alpha\sqrt{2})R_4$$

$$K = 4-\alpha\sqrt{2}$$

VCVS bandpass section

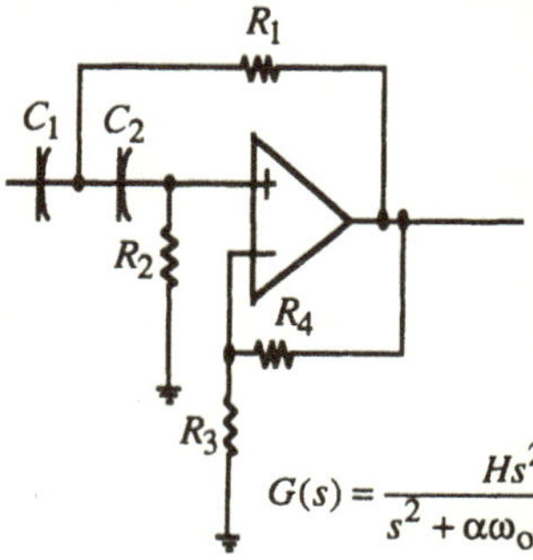

$$G(s) = \frac{Ks^2}{s^2 + s\left(\dfrac{1}{R_2C_1} + \dfrac{1}{R_2C_2} + \dfrac{1-K}{R_1C_1}\right) + \dfrac{1}{C_1C_2R_1R_2}}$$

$$G(s) = \frac{Hs^2}{s^2 + \alpha\omega_o s + \omega_o^2}$$

Choose $C_1 = C_2 =$ any convenient value

$$R_1 = R_2 = \frac{1}{\omega_o C_1}$$

Choose $R_3 =$ any convenient value

$$R_4 = (2-\alpha)R_3$$

$$K = 3-\alpha$$

VCVS high - pass section

Figure 8.15 VCVS design equations based on reduced sensitivity to capacitor variations.

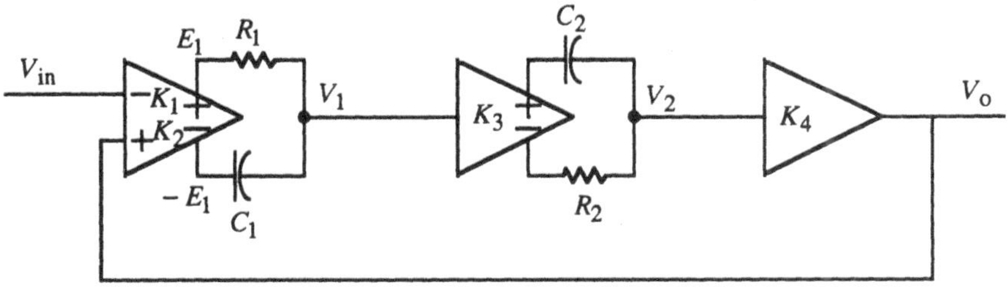

Figure 8.16 High-Q circuit model.

Setting $K_3 = K_4 = 1$ and solving for K_2 yields

$$K_2 = \frac{2Q - 1}{2Q + 1} = \frac{X}{X + 1} \qquad \text{where } X = \frac{2Q - 1}{2} \tag{8.59}$$

The term K_2 is slightly less than unity and must be maintained with a very tight tolerance. This is easily done as follows. The K_3 and K_4 gains are unity, as shown in Figure 8.17(a). The configuration for realizing $-K_1$ and K_2 is shown in Figure 8.17(b). Ideally the gain of an operational amplifier is infinite so that the desired gains are

$$-K_1 = \frac{(R_d/R_c)[1 + R_b/R_a]}{1 + 2(R_b/R_a) + (R_bR_d)/(R_aR_c)} = \frac{V_{out}}{V_{in1}} \tag{8.60}$$

$$K_2 = \frac{1 + R_d/R_c}{1 + 2(R_b/R_a) + (R_bR_d)/(R_aR_c)} = \frac{-V_{out}}{V_{in2}} \tag{8.61}$$

One solution for solving for the resistor ratios is to define

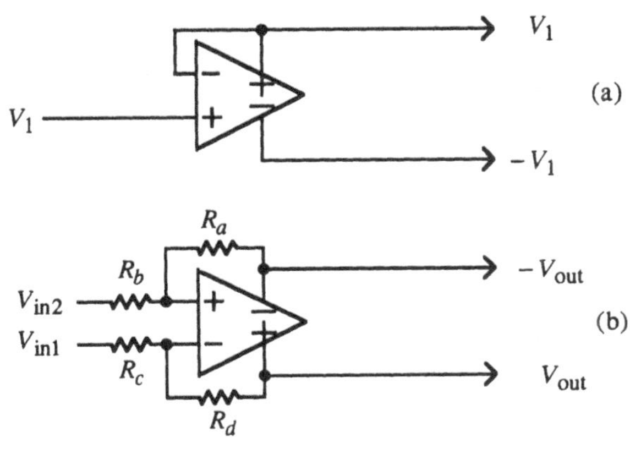

Figure 8.17 (a) Unity gain amplifier, (b) differential output finite gain amplifier.

$$\frac{R_d}{R_c} = \frac{1}{X} \quad \text{and} \quad \frac{R_b}{R_a} = \frac{1}{X} \tag{8.62}$$

Using equation (8.62), the following values are obtained

$$-K_1 = \frac{1/X}{1 + (1/X)} \quad \text{and} \quad K_2 = \frac{1}{1 + (1/X)} \tag{8.63}$$

The complete design equations are given in Table 8.1.

Example 8.3

Use the Motorola 1520 differential output op-amp to design a bandpass filter with a center frequency of 10,000 Hz and a $Q = 725$.

Table 8.1
High-Q Topology Design Equations

Step 1. Given $G(s) = \dfrac{As^2 + Bs + C}{s^2 + \alpha\omega_o s + \omega_o^2}$, choose $R_a = R_c = 100\ \Omega$

Step 2. Calculate $Q = 1/\alpha$; $X = (2Q - 1)/2$

Step 3. Then $R_b = R_d = 100\ X\Omega$

Step 4. Choose $C = C_1 = C_2$ = any convenient value

Step 5. Calculate $R_1 = R_2 = 1/(\omega_o C)$

Step 6. Choose R_6 = any convenient value

Step 7. $K_1 = \dfrac{2}{2Q + 1}$, $\quad K_2 = \dfrac{2Q - 1}{2Q + 1}$, $\quad K_3 = \dfrac{1 + K_2}{4K_1}\left(A - \dfrac{B}{\omega_o} + \dfrac{C}{\omega_o^2}\right)$

$K_4 = \dfrac{1 + K_2}{2K_1}\left(A - \dfrac{C}{\omega_o^2}\right)$, $\quad K_5 = -\dfrac{1 + K_2}{4K_1}\left(A + \dfrac{B}{\omega_o} + \dfrac{C}{\omega_o^2}\right)$

Step 7. $R_3 = \dfrac{R_6}{K_3}$, $\quad R_4 = \dfrac{R_6}{K_4}$, $\quad R_5 = \dfrac{R_6}{K_5}$

Step 8. If $R_5 < 0$, then $R_5 = -R_5$, $R_4 = -R_4$, $R_3 = -R_3$

If $R_4 > 0$, then connect R_4 to E_2

If $R_4 < 0$, then $R_4 = -R_4$, connect R_4 to $-E_2$

If $R_3 > 0$, then connect R_3 to $-E_1$

If $R_3 < 0$, then $R_3 = -R_3$, connect R_3 to E_1

Solution

Calculate $X = 749.5$. Let $R_a = R_c = 100\ \Omega$. Then $R_b = R_d = 100\ X\ \Omega$, and $\omega_o = 2\pi 10{,}000$ rad/sec. Pick $C_1 = C_2 = 1500$ pF. Then $R_1C_1 = R_2C_2 = 1/\omega_o$, so $R_1 = R_2 = 10{,}600\ \Omega$. Based on a prototype model, the error in Q was 3.3% due only to a 0.5 Hz error in the bandwidth. The measured center frequency was in error by only 0.1 Hz.

8.10 GAIN OPTIMIZATION OF THE HIGH-Q INSENSITIVE NETWORK

With reference to Figure 8.18, the output of U_1, U_2, and U_3 must peak at exactly the same frequency and output level. The input signal (from the previous stage) can therefore be padded, or boosted, so that these three op-amps are driven as close as possible to their full-scale output. The feedback resistor, R_6, can be adjusted to drive U_4 to full-scale output or, if another high-Q stage follows, to provide the proper output level to drive the next stage to its full dynamic range.

8.11 SIMULATION PITFALLS

Circuit simulators have become popular in recent years as a convenient way to test the theoretical performance of a circuit design. These computer-aided design tools are particularly useful as a supplemental check for breadboard prototype designs, especially for switched capacitor circuits where the expense and time required for each additional integrated circuit iteration can be prohibitive.

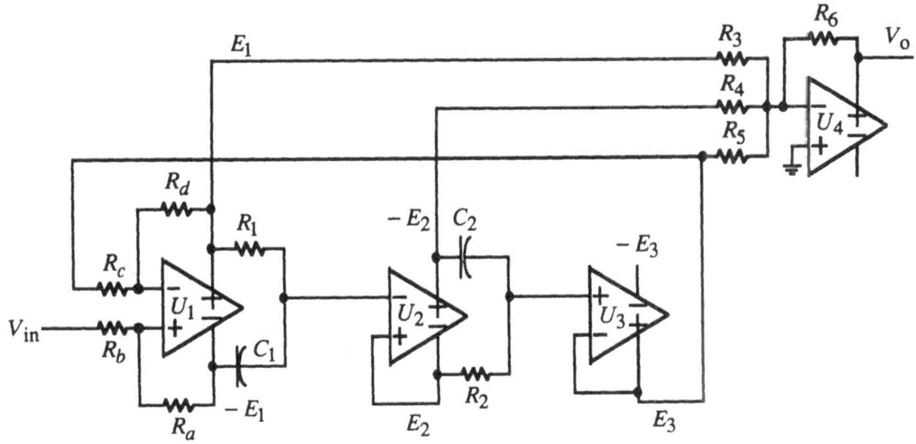

Figure 8.18 High-Q topology.

It is a mistake to blindly trust the results of any computer simulation program for active and switched capacitor filters. A common and dangerous error occurs if the inverting and noninverting op-amp inputs are swapped on the MFB, VCVS, state variable, and so forth, designs given in this chapter, because the resulting circuit will become nonlinear and latch to either the positive supply rail or the negative supply rail, for a constant dc voltage applied to the input. However, simulation programs show that the filter behaves the same regardless of how the op-amp inverting and noninverting inputs are swapped. The only exception applies to linear circuits using op-amps with both a differential input and a differential output. In this case, the simulation tools will correctly identify errors caused by swapping the op-amp inputs.

To understand this phenomena, the circuit in Figure 8.19 will be analyzed. The nonlinear circuit in Figure 8.19 represents a hysteresis comparator and differs from a linear summer only because the op-amp inputs are swapped. The open loop gain of the op-amp can be approximated by

$$\frac{G}{s + \omega_a} \tag{8.64}$$

where G is the open loop gain at dc and ω_a is the frequency of the dominant op-amp pole. From Figure 8.19, the mathematical expression for the output voltage can be written as

$$V_o = \frac{G}{s + \omega_a}\left(V_o\frac{10K}{10K + 100K} + V_{in}\frac{100K}{10K + 100K}\right) \tag{8.65}$$

Note that equation (8.65), written this way, actually has three solutions. Either V_o can be $+\infty$ (positive rail lock-up), $-\infty$ (negative rail lock-up), or $-10V_{in}$. The $-10V_{in}$ solution represents a metastable solution that cannot possibly exist in an actual circuit; however, it is the solution that all simulators find.

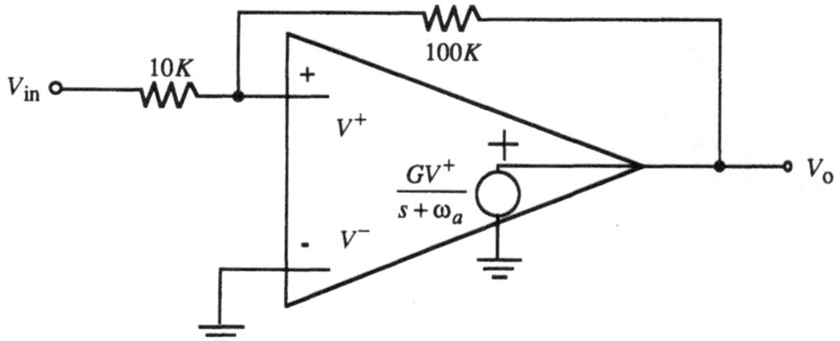

Figure 8.19 Nonlinear op-amp circuit.

If the op-amp inputs are placed correctly, as in Figure 8.1, then the mathematical expression for the output voltage becomes

$$V_o = -\frac{G}{s + \omega_a}\left(V_o\frac{10K}{10K + 100K} + V_{\text{in}}\frac{100K}{10K + 100K}\right) \tag{8.66}$$

which has only one solution ($V_o = -10V_{\text{in}}$) and can therefore be uniquely resolved by a circuit simulator.

8.12 CURRENT OPERATIONAL AMPLIFIER CHARACTERISTICS

The previous sections pertained exclusively to voltage operational amplifiers. With the introduction of current feedback op-amps in the early 1980s, previously unattainable gains and bandwidths became easily available. The basic differences in the voltage feedback op-amp and the current feedback op-amp follow:

	Voltage Feedback Op-Amp	*Current Feedback Op-Amp*
Noninverting input	Very high input impedance	Very high input impedance
Inverting input	Very high input impedance	Very low input impedance
Max. frequency response	Approximately 10 MHz	Approximately 400 MHz

The equivalent amplifier circuit [6] of Figure 8.20 will be used to develop the noninverting transfer function for the current feedback topology because the inverting transfer function development is a subset of the noninverting.

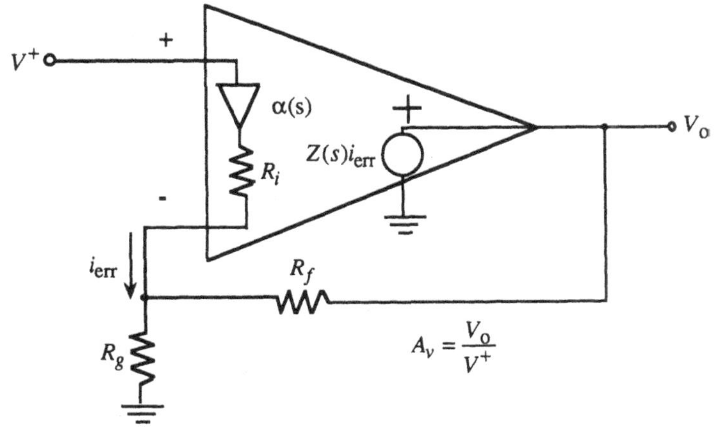

Figure 8.20 Current feedback amplifier internal equivalent.

The amplifier's noninverting input presents a high impedance to the input voltage, V^+, so as to not load the driving source. Any voltage appearing at the input node is passed through an open loop, unity gain, buffer that has a frequency dependent gain, $\alpha(s)$. Here $\alpha(s)$ is very nearly equal to 1 at dc (typically 0.996 or higher, but always less than 1.00) and typically has a −3 dB point beyond 500 MHz. The output of the buffer ideally presents a zero output impedance at the inverting input, V^-. It actually shows a frequency-dependent impedance, Z_i, that is relatively low at dc and increases inductively at high frequencies. For this development, we will only consider that Z_i is a small-valued resistive impedance, R_i. The intent of the buffer is to simultaneously force the inverting node voltage to follow the noninverting input voltage while also providing a low impedance path for an error current to flow. Any small signal error current flowing in the inverting node, i_{err}, is passed through the buffer to a high transimpedance gain stage and on to the output pin as a voltage. This transimpedance gain, $Z(s)$, senses i_{err} and generates an output voltage proportional to it. Thus $Z(s)$ has a very high dc value, a dominant low frequency pole, and higher order poles. When the loop is closed, the feedback loop drives i_{err} to zero much like a voltage feedback amplifier will drive the delta voltage across its inputs to zero. Hence $Z(s)$ ideally transforms the error current into a 0 Ω output impedance voltage source.

The following analysis neglects the impact of a finite output impedance from $Z(s)$ to the output, output loading interactions with that output impedance, and the effect of stray capacitance shunting R_g. Starting by summing currents at the V^- node of Figure 8.20,

$$i_{err} + \frac{V_o - V^-}{R_f} = \frac{V^-}{R_g} \tag{8.67}$$

Also,

$$V^- = \alpha(s)V^+ - i_{err}R_i \tag{8.68}$$

$$i_{err}Z(s) = V_o, \qquad \text{then } i_{err} = \frac{V_o}{Z(s)} \tag{8.69}$$

Multiply (8.69) through by R_f and isolating V^-,

$$i_{err}R_f + V_o = V^+\left(1 + \frac{R_f}{R_g}\right) \tag{8.70}$$

Now substitute in for i_{err} and V^- from previously

$$\frac{R_fV_o}{Z(s)} + V_o = \alpha(s)V^+ - \frac{R_fV_o}{Z(s)}\left(1 + \frac{R_f}{R_g}\right) \tag{8.71}$$

Gather V_o terms and solve for V_o/V^+

$$\frac{V_o}{V^+} = \frac{\alpha(s)\left(1 + \frac{R_f}{R_g}\right)}{1 + \frac{R_f + R_i(1 + R_f/R_g)}{Z(s)}} \tag{8.72}$$

Under ideal conditions, $\alpha(s) = 1$ and $Z(s) = \infty$, so the voltage gain becomes

$$\frac{V_o}{V^+} = \left(1 + \frac{R_f}{R_g}\right) \tag{8.73}$$

which is identical to equation (8.6) for the voltage op-amp case. In general, all the equations derived previously for the voltage op-amp topologies also apply to current op-amps with one important difference. With the current op-amp, there is ample opportunity for independent control of both the signal gains and the frequency response by using only resistive elements. Selection of the exact resistor values must be deferred to the specifications accompanying any particular current amplifier, however, the resistor values will usually range in the tens or hundreds of ohms.

8.13 SUMMARY

This chapter gives a variety of topologies to select for designing active filter sections. The poles and zeros of the filter are obtained first, and then the voltage transfer function is factored into single, quadratic, or cubic terms. Each section is realized independently, using the op-amp's near zero output impedance to provide isolation between one stage and the next. Dynamic range optimization techniques can guarantee that none of the op-amps saturate and that the signal-to-noise ratio is acceptable. Techniques for realizing high-Q sections of an active filter include the state-variable circuit and the high-Q insensitivity topology. As a final example, Figure 8.21 shows a fourth-order Tchebyscheff Type I filter with 1 dB passband ripple between 1000 Hz to 1200 Hz.

The methods in this chapter have been used to produce active realizations with as many as 44 op-amps between the input and the output. This particular state-variable realization with 44 op-amps was gain optimized at the output of each individual op-amp and produced a full undistorted output of ±10 V peak to peak with less than 3 mV (peak to peak) output noise for an approximate signal-to-noise ratio approaching 70 dB.

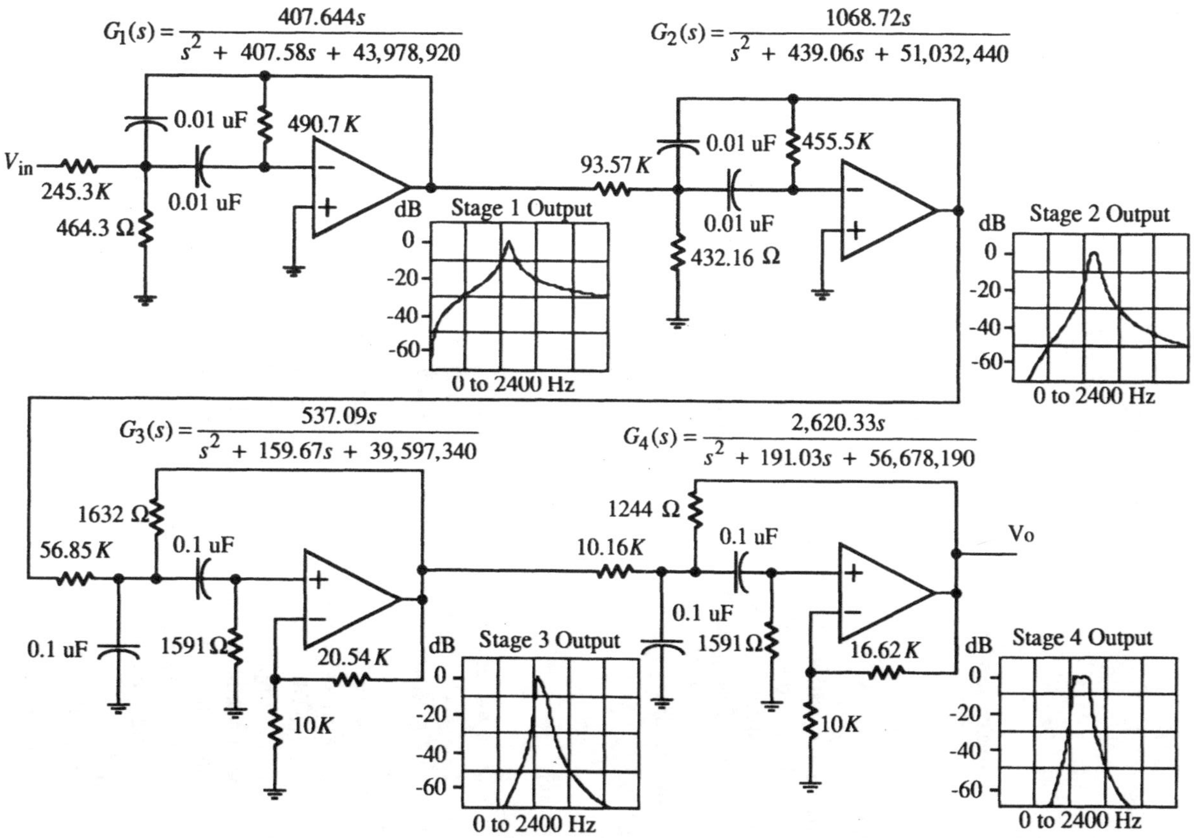

Figure 8.21 Tchebyscheff Type I bandpass filter, fourth-order, with 1 dB passband ripple between 1000 Hz and 1200 Hz, derived from a normalized low-pass prototype with normalized low-pass pole locations at $-0.3368698 \pm j0.4073291$ and $-0.139536 \pm j0.9833795$.

REFERENCES

[1] Graeme, Jerald G., *Operational Amplifiers, Design and Applications,* New York: McGraw-Hill Book Company, 1971.

[2] Kerwin, W. J., and Huelsman, L. P., "The Design of High Performance Active RC Band-Pass Filters," *IEEE Int. Conv. Rec.,* Vol. 14, part 10, 1966, pp. 74–80.

[3] Mitra, Sanjit K., *Active Inductorless Filters,* New York: IEEE Press, 1971.

[4] Williams, Arthur B., *Electronic Filter Design Handbook,* New York: McGraw-Hill, 1981.

[5] Tarmy, R., and Ghausi, Mohammed, "Very High-*Q* Insensitive Active RC Networks," *IEEE Trans. Circuit Theory,* Vol. CT-17, August 1970, pp. 358–366.

[6] Steffes, Michael, "Current Feedback Amplifier Loop Gain Analysis and Performance Enhancements," Application Note OA-13, Comlinear Corporation.

PROBLEMS

8.1 If $Z_f = 100K$ and $Z_{in} = 10K$, what is the voltage gain of the circuit in (a) Figure 8.1 and (b) Figure 8.2? Assume ideal op-amps.

8.2 Decompose the transfer function

$$G(s) = \frac{s^2}{s^4 + 4s^3 + 6s^2 + 4s + 1}$$

into the product of

(a) a first-order high-pass and a third-order transfer function,
(b) a second-order low-pass quadratic and a second-order high-pass quadratic,
(c) two first-order low-pass transfer functions and two first-order high-pass ones,
(d) two second-order bandpass quadratics.

8.3 Implement a second-order bandpass with a $Q = 10$, $H = 1$, and $\omega_o = 1000$, using

(a) a MFB topology,
(b) a VCVS topology,
(c) a state variable topology,
(d) a high-*Q* insensitive topology.

8.4 What are the coefficients *A, B,* and *C* of the transfer function

$$G(s) = \frac{As^2}{s^2 + Bs + C}$$

with $Q = 2$, a high-frequency gain of 5, and $f_o = 1000$ Hz?

8.5 Derive equation (8.21) from (8.13) and (8.20).

8.6 Derive the transfer function for the MFB high pass in Figure 8.5 using the transfer function for the MFB band pass as a starting point and making the substitution $1/(sC_1)$ for R_1.

8.7 Derive equations (8.43) and (8.44) for the state variable topology.

8.8 The poles of a Tchebyscheff Type I low pass with 1 dB passband ripple and a passband cutoff frequency of 5000 Hz are

$-7290.466 \pm j8362.408$, $-5336.992 \pm j22{,}846.53$ and $-1953.474 \pm j31{,}208.93$.

Implement this filter as a cascade of three VCVS stages.

8.9 Implement the function

$$\frac{V_o}{V_{in}} = \frac{0.0163s^2 + 1.754 \times 10^8}{s^2 + 16{,}782.3s + 1.789 \times 10^8}$$

with the elliptic topologies in Figure 8.13. Which one would you recommend for an actual implementation and why?

8.10 Derive the sensitivity equations (8.48) using the definition for sensitivity given in (8.47).

8.11 Derive equation (8.57) assuming an ideal op-amp.

8.12 For Figure 8.15, if $K_3 = 1$, show that the magnitude of V_2 is exactly equal to the magnitude of V_1 at all frequencies.

8.13 For the MFB low pass in Figure 8.5, if $C_1 = C_2$ and $H = 1$ find the value of Q that causes R_1 to be unrealizable.

8.14 Modify the state variable equations in Figure 8.10 so that a circuit with a Q of 0.1 is realizable.

8.15 Modify the circuit in Figure 8.13(a) to realize the transfer function given by

$$G(s) = \frac{s^2 + 78{,}956{,}830}{s^2 + 62{,}516.91s + 78{,}956{,}830}$$

without any negative resistors. The final circuit should use five resistors, four capacitors, and one op-amp.

8.16 Find a value of R_2 and C_1 in the all pass in Figure 8.12 so that the output voltage is 90° out of phase with the input voltage at 1000 Hz.

8.17 If the MFB bandpass in Figure 8.5 has $R_1 = R_2 = R_3 = 10K$ and $C_1 = C_2 = 0.001$ uF, what is α and ω_o?

8.18 Implement

$$\frac{s^2}{s^2 + 151{,}512s + 1.413 \times 10^{10}} \, \frac{s^2}{s^2 + 17{,}774s + 4.000 \times 10^9}$$

using a two-stage state variable topology but with only six op-amps.

8.19 In example 8.4, modify the values of R_1 and R_2 so that the center frequency is 5000 Hz with a Q of 725.

8.20 Assume that a nonideal op-amp has an open loop gain of 10,000. Derive the voltage gains for the circuits in Figures 8.1 and 8.2 using $Z_f = 100K$ and $Z_{in} = Z_i = 10K$ and compare the voltage gains to the ideal case.

8.21 Use equation (8.57) to determine the gain of the circuit in Figure 8.16 given R_a = $1K$, $R_b = 1K$, $R_c = 2K$, $R_d = 4K$, $V_{in1} = 1$ and $V_{in2} = 0$. Verify the results using Spice or an equivalent circuit simulator.

8.22 Write a computer program to perform gain optimization over 0 to 5000 Hz and repeat problem 8.8 assuming that V_{in} is $10\cos(\omega t)$ and that the op-amps are powered from a bipolar ±15 V supply.

Chapter 9

Infinite Impulse Response Digital Filters

9.1 DIGITAL FILTER CONCEPTS

Digital filters can be categorized into two broad classes known as finite impulse response and infinite impulse response filters. The IIR filter is similar to an analog filter in that the response to a unit impulse decays exponentially over time but never disappears completely. The properties of both FIR and IIR filters are summarized in Figure 9.1, where z^{-1} represents a unit sample delay.

The usual approach to the design of IIR filters involves the transformation of an analog filter into a digital filter meeting the prescribed specifications. This is the most often used method because the art of analog filter design is highly advanced, and it is advantageous to utilize the design procedures already developed for analog filters. The closed circuit formulas given in previous chapters can be used for IIR realizations of Butterworth, Tchebyscheff, and elliptic filters. In many applications, it is necessary to use a digital filter to simulate the performance of an analog linear filter. Except for warping of the frequency axis, the IIR filter performs exactly like its analog counterpart.

An analog signal is to be converted into a sampled data signal in preparation for digital filtering. Nyquist sampling theory states that the sample rate must be at least twice the highest baseband frequency component of the input signal to be processed without error. A low-pass analog filter is normally used as an antialiasing preprocessor on the input to ensure that no frequency components are greater than 1/2 the sample rate by the time the analog signal reaches the input of the analog-to-digital (A/D) converter (see Figure 9.2).

The function of sample-and-hold circuit is to ensure that the voltage at the input to the A/D converter does not change during the finite time that it takes to perform the conversion from analog to digital. The output of the A/D converter is often an 8-bit, or a 16-bit, digital word that represents the amplitude of the sampled analog voltage.

Infinite impulse response

$$y(n) = \sum_{i=0}^{M} b(i)\, x(n-i) - \sum_{i=1}^{M} a(i)\, y(n-i)$$

Finite impulse response

$$y(n) = \sum_{i=0}^{M} b'(i)\, x(n-i)$$

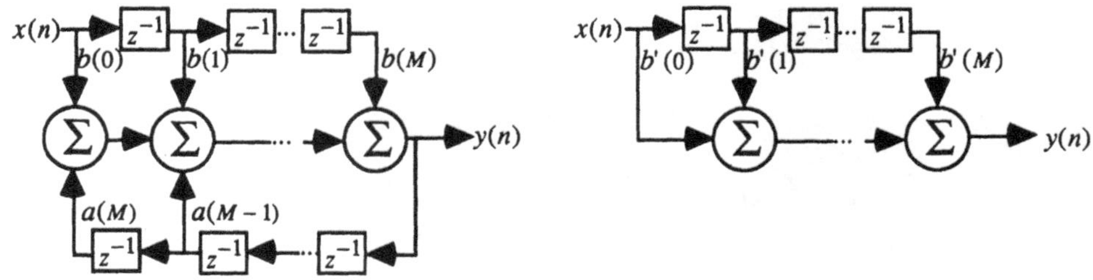

Figure 9.1 A comparison of IIR and FIR digital design approaches.

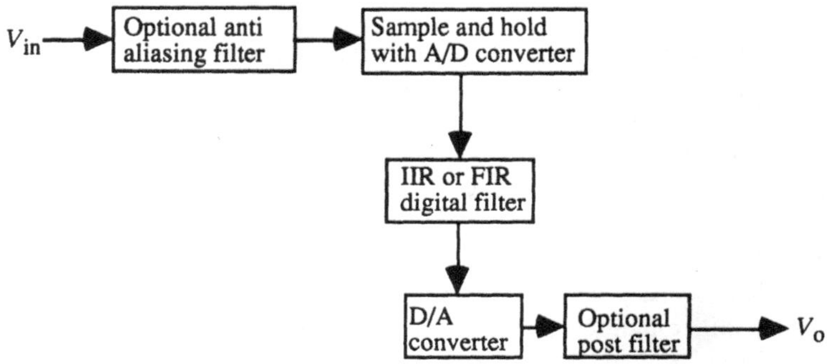

Figure 9.2 Typical digital filter block diagram.

The digital filter can be implemented with a series of cascaded difference equations, each similiar to equation (1.12). At the end of every sample period, the digital filter produces an 8-bit, or 16-bit, word to the digital-to-analog (D/A) converter, which reproduces an analog amplitude corresponding to the value of the digital word. A low-pass filter is sometimes used as a postprocessing smoothing filter to eliminate the step transitions in the reconstructed analog output.

9.2 THE ANTIALIASING FILTER

A 0–4000 Hz analog signal, sampled at 8000 samples per second, would require an antialiasing filter with a flat passband from dc to 4000 Hz and high attenuation starting immediately for frequencies above 4000 Hz. In general, the requirements of an antialiasing filter for a system with a sample rate of exactly twice the highest baseband frequency of interest is prohibitive because the transition band between the passband and the stopband is infinitesimal. It is always preferable to use a sampling frequency greater than the minimum specified by Nyquist's theorem. For the previous example, if the sampling rate were increased to 12,000 samples per second, then the antialiasing filter could have an approximately flat bandpass from dc to 4000 Hz, with high attenuation for frequencies above 6000 Hz. (The exact attenuation required would depend on the anticipated spectral content of the input signal above 6000 Hz.) A very high sampling frequency is often used in switched capacitor filters so that a simple one-pole low pass more than satisfies the requirements for antialiasing.

Example 9.1

A digital preemphasis circuit for an FM radio station uses a sampling rate of 50,000 samples per second with quantization noise of –96 dB. The analog signal to be processed has a bandwidth of 50–15,000 Hz, however any frequency content above 25,000 Hz is already attenuated by 20 dB due to roll-off in the finite frequency response of the audio mixers in the studio. Design an antialiasing filter suitable for this application.

Solution

The antialiasing filter should be a low pass with approximately constant amplitude between 50–15,000 Hz and 76 dB (96–20) attenuation for frequencies above 25,000 Hz. A 7th-order elliptic, or 11th-order Tchebyscheff, filter with 0.1 dB passband ripple would satisfy these requirements.

9.3 THE SAMPLE-AND-HOLD CIRCUIT

The analog signal is sampled and held constant for the duration of the analog-to-digital conversion. The analog input signal, V_{in}, is sampled just prior to the start of the conversion cycle. At the end of the sampling period, the capacitor (see Figure 9.3) is discharged in preparation for the next input sample.

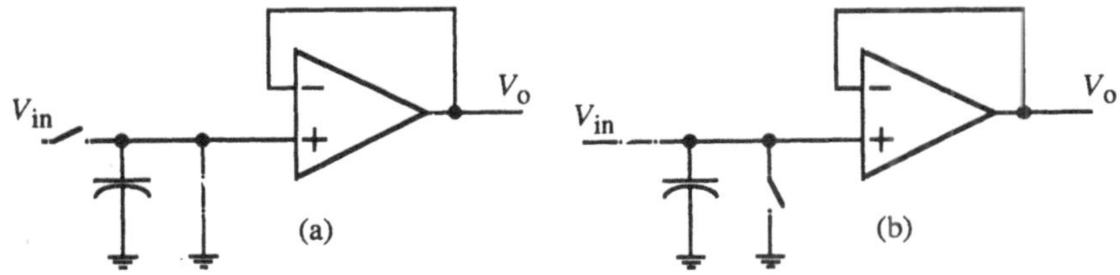

Figure 9.3 Sample-and-hold circuit with sampling frequency of $1/T$ Hz.

The sample-and-hold circuit has a frequency response characteristic that may be significant in some cases. Because the Laplace transform of a unit step is $1/s$ and the Laplace transform of a delay is e^{-sT}, where T represents the delay time, the transfer function for a zero-order analog-to-digital converter can be written as

$$\frac{V_o}{V_{in}} = \frac{1}{T}\left(\frac{1}{s} - \frac{e^{-sT}}{s}\right) = \frac{1 - e^{-sT}}{Ts} = \frac{1 - z^{-1}}{Ts} \tag{9.1}$$

The notation z^{-1} is simply a shorthand expression for e^{-sT}. The constant, $1/T$, is required since the conversion occurs $1/T$ times per second. Rewriting equation (9.1) as

$$e^{-j\omega T/2}\frac{e^{j\omega T/2} - e^{-j\omega T/2}}{j\omega T} \tag{9.2}$$

allows the magnitude and phase of the sample-and-hold to be expressed conveniently as

$$|F(s)| = \left|\frac{V_o}{V_{in}}\right| = \left|\frac{2\sin(\omega T/2)}{\omega T}\right| = \left|\frac{2}{\omega T}\sin\left(\frac{\omega T}{2}\right)\right| \tag{9.3}$$

$$\angle F(s) = \angle\frac{V_o}{V_{in}} = -\frac{\omega T}{2} \tag{9.4}$$

To compensate for this spectrum shaping by the A/D converter, the digital sequence is often preprocessed by a digital filter with a magnitude response approximating $1/|F(s)|$. The digital compensation network can be combined with any other digital processing in the system and need not be realized as a separate filter.

Figure 9.4 plots the response of a sample-and-hold circuit showing a gain of 0.636 at the frequency corresponding to 1/2 the sampling rate. If frequencies higher than $1/(2T)$ were to be present at the input of the sample-and-hold, these frequencies would be aliased, or heterodyned, into the lower frequency portion (usable region) of the spectrum, interfering with the desired signal.

9.4 ANALOG-TO-DIGITAL CONVERSION

Figure 9.5 shows an 8-bit successive approximation A/D converter [1]. This type of converter requires eight clock bits to complete a conversion, regardless of the magnitude of the input signal. For general purpose situations, the successive approximation converter usually has adequate conversion speed. The conversion process begins with a *start conversion* pulse and ends eight clock periods later (relative to the A/D master clock in the digital control logic). The input signal, V_{in}, remains constant during the conversion process since it is preceded with a sample-and-hold circuit.

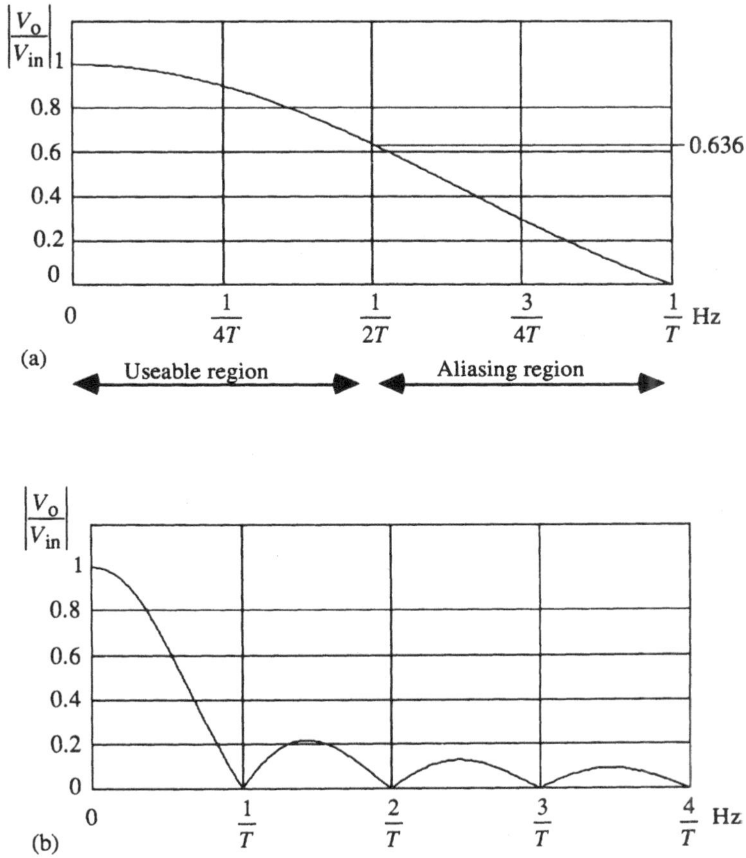

Figure 9.4 A/D frequency response from equation (9.3).

For a given input voltage, V_{in}, the A/D converter finds the maximum digital word that sets the comparator output voltage, V_c, at 0 V. For an n-bit converter, this can always be done in n steps. Initially, all 8 bits are set to zero. During step 1, the most significant bit (MSB) is set high (+5 V). If V_c is still zero, then step 1 is completed and bit 6 is set high beginning with step 2. If V_c had been high in step 1, then the MSB would have been set back to zero (low), and bit 6 set high beginning with step 2. The process is repeated until all eight bits have been set.

From Figure 9.5 the voltage V^- at the negative input of the comparator can be seen to be a linear sum of the eight binary voltages given by

$$V^- = \frac{1}{2}\left[V_7 + \frac{1}{2}V_6 + \frac{1}{4}V_5 + \frac{1}{8}V_4 + \frac{1}{16}V_3 + \frac{1}{32}V_2 + \frac{1}{64}V_1 + \frac{1}{128}V_0\right] \tag{9.5}$$

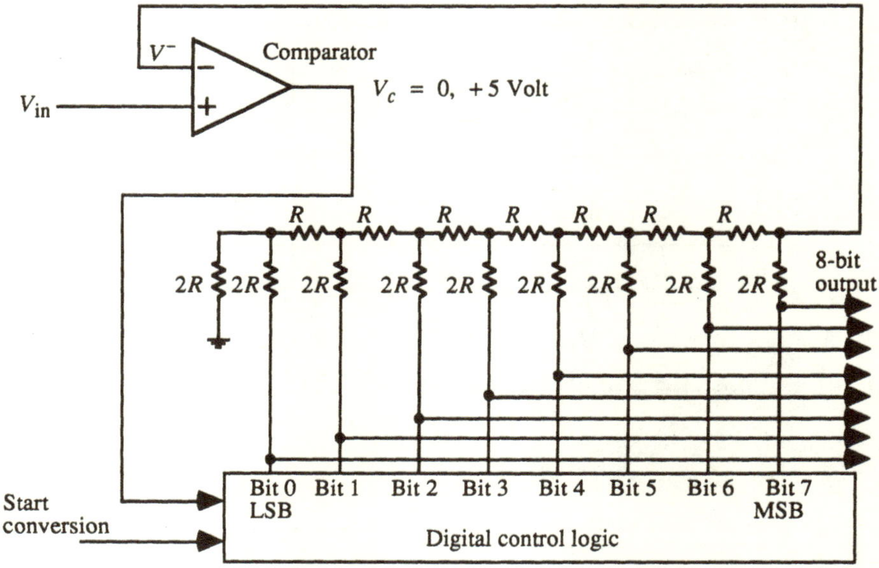

Figure 9.5 Successive approximation A/D converter.

where V_7 is the MSB and V_0 is the LSB. The resistive ladder network is actually a digital-to-analog converter that reconstructs the analog voltage so that a comparison can be made between the input voltage and the reconstructed voltage to determine the closest bit settings. The maximum speed of the conversion depends upon the time required for the comparator to change levels, plus the time required for switching transients to settle. Voltage and current offsets in the comparator will affect the comparison point and thus the accuracy of the A/D converter. After the conversion is completed, the digital output will be accurate to ±1/2 LSB.

9.5 THE DIGITAL FILTER

The transposed direct form II structure [2] of Figure 9.6 implements an infinite impulse response function:

$$\frac{V_o}{V_{\text{in}}} = \frac{b_2 z^{-2} + b_1 z^{-1} + b_o}{a_2 z^{-2} + a_1 z^{-1} + 1} \tag{9.6}$$

with a canonical form using a minimum number of delay elements. These second-order quadratic stages may be cascaded to construct higher order composite filters in much the same way that an active filter is realized as a cascade of second-order stages. With digital

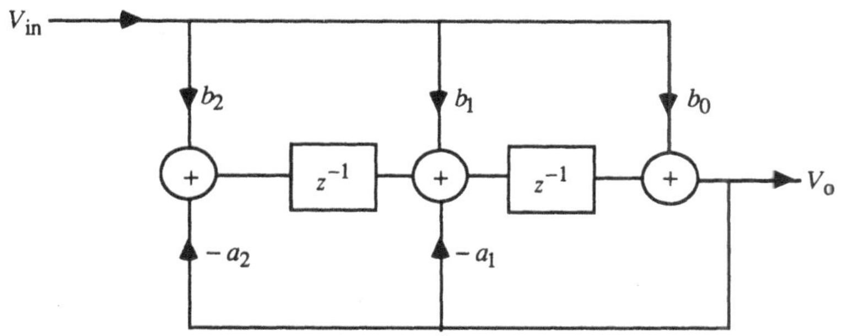

Figure 9.6 Transposed direct form II.

filters, both numeric overflow and underflow need to be avoided or minimized. As with active filters, gain normalization at the output of each second-order stage will optimize the dynamic range while preventing overflow.

9.6 THE BILINEAR TRANSFORM

Because multiplication by the Laplace variable s in the frequency domain is equivalent to differentiation in the time domain, a difference expression approximation to the variable s can be derived as

$$s \rightarrow \frac{1}{T} \frac{1 - z^{-1}}{\left(\dfrac{1 + z^{-1}}{2}\right)} \tag{9.7}$$

where $1 - z^{-1}$ represents the difference between the present voltage sample and the previous sample, $(1 + z^{-1})/2$ represents the average of the present and the previous sample, and $1/T$ is used to normalize with respect to the sampling time, T. Rearranging terms, the desired algebraic form of the bilinear transformation becomes

$$s \rightarrow \frac{2}{T} \frac{1 - z^{-1}}{1 + z^{-1}} \tag{9.8}$$

If ω_a is used to represent the frequency domain of the analog filter, and ω_d represents the frequency domain of the digital filter, then equation (9.8) becomes

$$j\omega_a \rightarrow \frac{2}{T} \frac{1 - e^{-j\omega_d T}}{1 + e^{j\omega_d T}} \tag{9.9}$$

which can be written as

$$j\omega_a \rightarrow \frac{2}{T}\frac{e^{j\omega_d T/2} - e^{-j\omega_d T/2}}{e^{j\omega_d T/2} + e^{j\omega_d T/2}} \tag{9.10}$$

or

$$\omega_a \rightarrow \frac{2}{T}\tan\left(\frac{\omega_d T}{2}\right) \tag{9.11}$$

For small values of ω_a, the mapping is almost linear. The frequency warping expressed in (9.11) does not mean that an analog input frequency of ω_a is converted to a digital frequency of ω_d, but that the response of the digital filter at ω_d is the same as the response of the analog filter at ω_a.

Equation (9.8) maps the s-domain into the z-domain. (See Figure 9.7.) The entire left-hand plane is mapped into the unit circle with the $j\omega_a$ axis mapping into the boundary of the unit circle in the z-domain. Because poles in the right-half s-plane are a sign of instability, poles outside the unit circle in the z-plane would also indicate instability in the digital filter. The location of the z-domain poles can be obtained by writing the digital transfer function as a function of z (not z^{-1}) and finding the roots of the denominator. If the magnitude of all roots is ≤ 1, then the IIR digital filter is stable.

Example 9.2

Derive the transfer function for a gain optimized IIR digital bandpass filter with a bandpass region of 1000 Hz to 2000 Hz, stopband frequencies of 500 Hz and 2500 Hz, a 1 dB

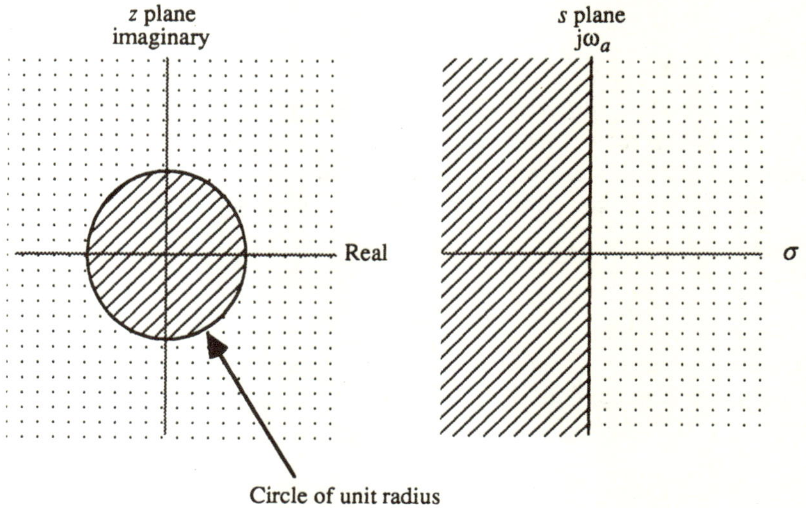

Figure 9.7 Mapping the s-domain into the z-domain.

passband ripple, 50 dB stopband attenuation, and an A/D sampling frequency of 10 KHz. Base the filter on a Tchebyscheff Type II design.

Solution

In Step 1, using equation (9.11), the analog prototype must be designed for bandpass frequencies of 1034.25 Hz and 2312.66 Hz. Stopband frequencies become 504.15 Hz and 3183.10 Hz.

In Step 2, calculating a transition ratio from the lower band edge yields

$$\omega_{x1} = \left|\frac{1034.25 \cdot 2312.66 - 504.15 \cdot 504.15}{504.15 \cdot 2312.66 - 504.15 \cdot 1034.25}\right| = 3.32$$

whereas

$$\omega_{x2} = \left|\frac{1034.25 \cdot 2312.66 - 3183.10 \cdot 3183.10}{3183.10 \cdot 2312.66 - 3183.10 \cdot 1034.25}\right| = 1.9.$$

In Step 3, using $\omega_x = 1.9$ for design with a Tchebyscheff Type II filter, the analog low-pass prototype parameters become $\epsilon = 0.5088$, $[\text{Mag}]^2 = 0.00001$, $N = 5.66$, so a sixth-order low-pass filter will be used (reference Table 6.3).

In Step 4, from equations (4.16), (4.17), and (4.19) the normalized low-pass analog pole and zero locations are

Low-Pass Poles	*Low-Pass Zeros*
$-1.285976 \pm j0.4354488$	$0 \pm j1.855197$
$-0.7539852 \pm j0.9528279$	$0 \pm j2.534246$
$-0.2301577 \pm j1.085487$	$0 \pm j6.923689$

resulting in a low-pass transfer function (with $\omega_c = 1$ rad/sec) of

$$G(s) = \frac{s^2 + 3.4417559}{s^2 + 2.5720s + 1.84335} \, \frac{s^2 + 6.4224028}{s^2 + 1.5089s + 1.47637} \, \frac{s^2 + 47.937469}{s^2 + 0.46032s + 1.23125}$$

In Step 5, from Table 6.1, replace s with $[s^2 + (\omega_u \cdot \omega_L)]/[s \cdot (\omega_u - \omega_L)]$, where $\omega_L = 2\pi(1034.25)$ and $\omega_u = 2\pi(2316.66)$.

The analog bandpass poles and zeroes become

Poles	*Zeros*
−4099.886 ± j6733.155	0 ± 19,696.05
−6229.691 ± j10,230.89	0 ± j24,250.16
−1876.415 ± j6234.435	0 ± j4794.22
−4179.956 ± j13,888	0 ± j3893.879
−544.8056 ± j6257.518	0 ± j1648.998
−1303.931 ± j14,976.67	0 ± j57,263.38

The resulting denormalized analog bandpass transfer function is

$$G(s) = \prod_{i=1}^{6} \frac{A_i s}{s^2 + B_i s + C_i} = \frac{s^2 + 3.8793 \times 10^8}{s^2 + 8199.77s + 6.2144 \times 10^7} \frac{s^2 + 5.8807 \times 10^8}{s^2 + 12459.38s + 1.4348 \times 10^8}$$
$$\times \frac{s^2 + 1.5162 \times 10^7}{s^2 + 8.359.91s + 2.1035 \times 10^8} \frac{s^2 + 2.29846 \times 10^7}{s^2 + 3752.83s + 4.2389 \times 10^7}$$
$$\frac{s^2 + 3.2791 \times 10^9}{s^2 + 2607.86s + 2.2600 \times 10^8} \frac{s^2 + 2.7192 \times 10^7}{s^2 + 1089.61s + 3.9453 \times 10^7}$$

In Step 6, making the substitution

$$s \rightarrow \frac{2}{T} \frac{1 - z^{-1}}{1 + z^{-1}}$$

where $T = 0.0001$, and performing the gain optimization technique described in Chapter 8 results in an IIR digital filter transfer function given by

$$\prod_{i=1}^{6} \frac{A_i z^{-2} + B_i z^{-1} + C_i}{D_i z^{-2} + E_i z^{-1} + F_i}$$

where

$A_1 = 0.1770810 = 16AAH$ $B_1 = -0.00542 = FF4EH$ $C_1 = 0.1770810 = 16AAH$
$D_1 = 0.4412384 = 387AH$ $E_1 = -1.0 = 8000H$ $F_1 = 0.9266384 = 769BH$
$A_2 = 0.2969890 = 2603H$ $B_2 = 0.11305670 = E79H$ $C_2 = 0.296989 = 2603H$
$D_2 = 0.3712701 = 2F85H$ $E_2 = -0.6472337 = AD28H$ $F_2 = 1.0 = 7FFFH$
$A_3 = 0.5609648 = 47CDH$ $B_3 = -1.0 = 8000H$ $C_3 = 0.56096480 = 47CDH$
$D_3 = 0.4179093 = 357EH$ $E_3 = -0.8136973 = 97DAH$ $F_3 = 0.58869060 = 4B5AH$
$A_4 = 0.5393995 = 450BH$ $B_4 = -1.0 = 8000H$ $C_4 = 0.5393995 = 450BH$
$D_4 = 0.1970954 = 193AH$ $E_4 = -0.168699 = EA68H$ $F_4 = 0.34582070 = 2C44H$

$A_5 = 0.2617405 = 2180H \quad B_5 = -0.5164117 = BDE7H \quad C_5 = 0.2617405 = 2180H$

$D_5 = 0.5792058 = 4A23H \quad E_5 = -1.0 = 8000H \quad F_5 = 0.6396479 = 51DFH$

$A_6 = 0.2971008 = 2607H \quad B_6 = 0.4649947 = 3B84H \quad C_6 = 0.2971008 = 2607H$

$D_6 = 0.8461802 = 6C4FH \quad E_6 = -0.5131546 = BE51H \quad F_6 = 1.0 = 7FFFH$

The 16-bit hexadecimal values in example 9.2 are produced by multiplying the decimal values by 2^{15} and then converting to a hexadecimal integer. For implementation in a digital signal processor (DSP), it is important that no coefficient be greater than unity and that gain optimization be performed on each successive quadratic stage to prevent overflow in the DSP.

Figure 9.8 shows the response of the IIR digital filter of example 9.2 relative to its corresponding s-domain analog filter. The analog frequency range of $0 < f_a < \infty$ Hz is mapped into the digital frequency range of $0 < f_d < 5000$ Hz. Both filters exceed the original design specification. In practice, however, the sampling frequency is usually taken to be at least an order of magnitude higher than the bandpass region of the filter being transformed.

The bilinear transform can be applied to analog transfer functions using the identities

$$\frac{As + B}{Cs + D} \rightarrow \frac{(BT - 2A)z^{-1} + (BT + 2A)}{(DT - 2C)z^{-1} + (DT + 2C)} \tag{9.12}$$

$$\frac{As^2 + Bs + C}{Ds^2 + Es + F} \rightarrow \frac{(4A - 2BT + CT^2)z^{-2} + (2CT^2 - 8A)z^{-1} + (4A + 2BT + CT^2)}{(4D - 2ET + FT^2)z^{-2} + (2FT^2 - 8D)z^{-1} + (4D + 2ET + FT^2)} \tag{9.13}$$

The left-hand poles in the s plane are mapped inside the unit circle on the z plane. Pole-zero locations for both the analog bandpass filter and the IIR bandpass filter of example 9.2 are shown in Figure 9.9. The z-plane poles and zeros are derived by taking the reciprocals of the z^{-1} plane poles and zeros for the quadratic functions given in example 9.2.

9.7 DIGITAL-TO-ANALOG CONVERSION

The D/A conversion process has already been described as part of the successive approximation circuit presented earlier. Figure 9.10 shows a D/A converter based on a resistive ladder. The conversion process is very fast because it depends primarily on the settling time of a single op-amp. The binary voltages at the D/A input must be referenced very accurately to specific high and low levels for precise reproduction of the analog voltage.

Analog bandpass filter for Example 9.2

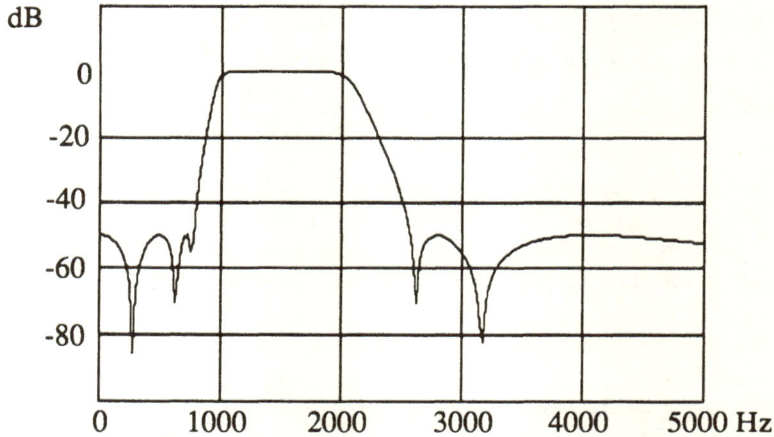

(a)

Digital bandpass filter of Example 9.2

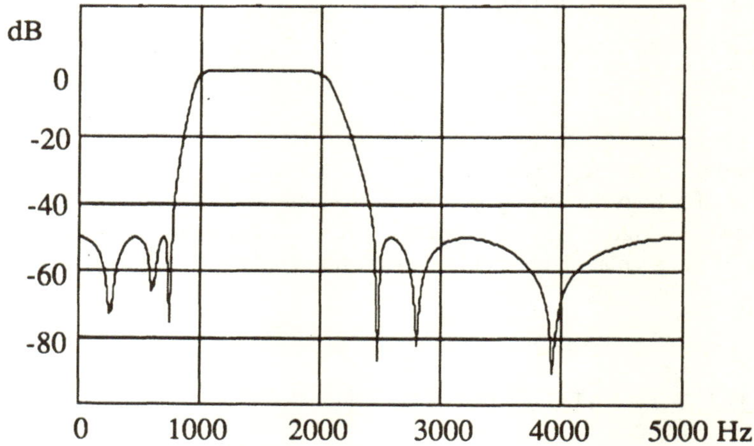

(b)

Figure 9.8 (a) Analog bandpass filter and (b) its corresponding z-transform implementation.

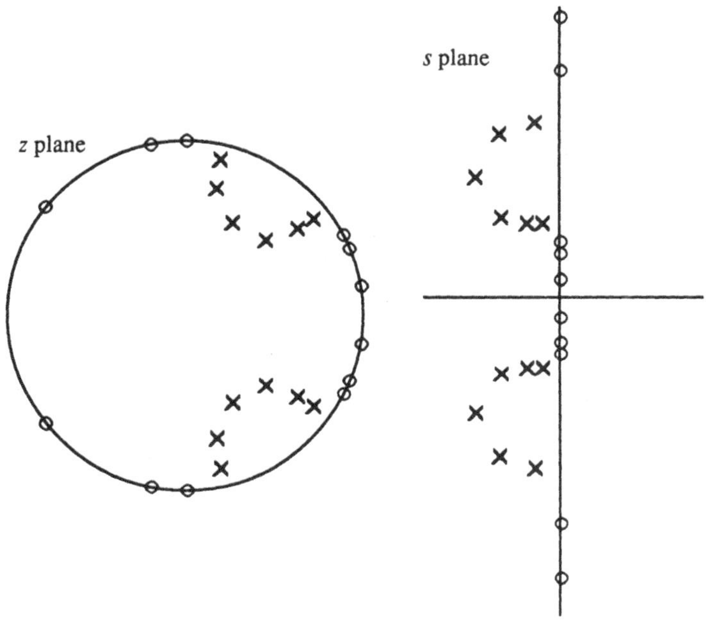

Figure 9.9 Pole-zero locations for the bandpass filters in example 9.2.

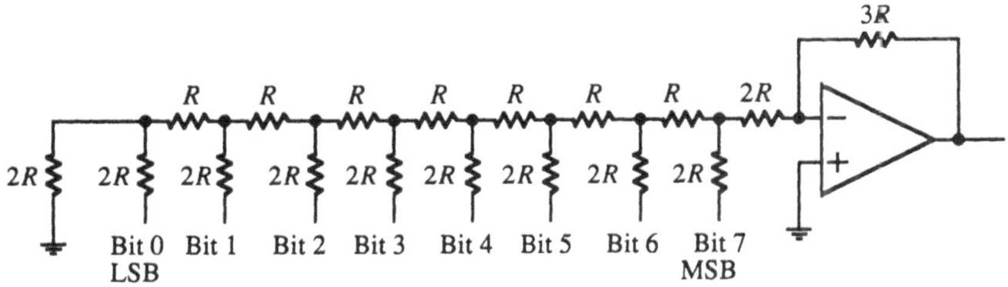

Figure 9.10 Digital-to-analog converter.

REFERENCES

[1] Graeme, Jerald G., Tobey, Gene E., and Huelsman, Lawrence P., *Operational Amplifiers*, New York: McGraw-Hill, 1971, pp. 335–352.

[2] Williams, Arthur B., and Taylor, Fred J., *Electronic Filter Design Handbook*, New York: McGraw-Hill, 1988, pp. 15-1 to 15-16.

PROBLEMS

9.1 An IIR bandpass filter has bandpass edge frequencies at 1000 Hz and 2000 Hz and stopband edge frequencies at 900 Hz and 2200 Hz. What are the corresponding passband and stopband frequencies for the analog prototype filter if the IIR filter is to use a sampling frequency of (a) 20,000 Hz and (b) 25,000 Hz?

9.2 Compute the analog output voltage for Figure 9.8 given

Bit 0 LSB	Bit 1	Bit 2	Bit 3	Bit 4	Bit 5	Bit 6	Bit 7 MSB
5 V	0 V	0 V	5 V	0 V	5 V	0 V	5 V

9.3 If the poles of a prewarped analog filter are given as

Real	*Imaginary*
−1881.242	0
−1521.957	±j 3976.496
− 581.3357	±j 6434.106

what is the corresponding transfer function in the z-domain if a sampling frequency of 10,000 Hz is to be used?

9.4 Show that the bilinear transform maps the entire $j\omega$ axis of the s-domain into the circumference of the unit circle in the z-domain.

9.5 Find the coefficients corresponding to equation (9.6) for the following IIR filters:
   (a) a fifth-order low-pass Tchebyscheff Type I IIR filter with 0.1 dB passband ripple, cutoff frequency of 5,000 Hz, and a sampling frequency of 100,000 Hz,
   (b) A sixth-order low-pass Tchebyscheff Type II IIR filter with 0.1 dB passband ripple, 60 dB minimum stopband attenuation, a passband cutoff frequency of 5,000 Hz, and a sampling frequency of 50,000 Hz,
   (c) a fifth-order bandpass elliptic IIR filter with 0.1 dB passband ripple, 60 dB minimum stopband attenuation, passband edge frequencies of 2,000 Hz and 3000 Hz, and a sampling frequency of 50,000 Hz,
   (d) a ninth-order high-pass Butterworth IIR filter with 3 dB attenuation at a cutoff frequency of 5,000 Hz and a sampling frequency of 50,000 Hz.

9.6 Implement a gain-normalized seventh-order Tchebyscheff Type I filter with 0.1 dB passband ripple, cutoff frequency of 5,000 Hz, and a sampling frequency of 50,000 Hz, using the transposed direct form II of Figure 9.6 such that all intermediate calculations are less than, or equal to, unity, for a sine wave input that has a maximum peak-to-peak variation of ±1 V.

Chapter 10

Finite Impulse Response Window Filter Design

10.1 FIR WINDOWED DIGITAL FILTER CONCEPTS

From Figure 9.1, a FIR filter is represented by a sampled data expression of the form

$$y(n) = \sum_{i=0}^{M} b'(i)x(n - i) \tag{10.1}$$

Window designs are based on taking the z transform of equation (10.1) as described in Chapter 1, yielding

$$y(n) = \sum_{i=0}^{M} b'(i)x(n - i) = \sum_{i=0}^{M} h(i)x(n)z^{-i}, \qquad \text{with } h(i) = b'(i) \tag{10.2}$$

or

$$\frac{y(e^{j\omega})}{x(e^{j\omega})} = H(e^{j\omega}) = \sum_{i=0}^{M} h(i)z^{-i} \tag{10.3}$$

Using the notation of Chapter 1 and extending the impulse response for both positive and negative time yields

$$H(e^{j\omega}) = \sum_{n=-\infty}^{n=\infty} h(n)z^{-n} \tag{10.4}$$

which provides a means of determining the impulse response, $h(n)$, from the desired frequency response, $H(e^{j\omega})$, by using the inverse Fourier transform.

There are many reasons for designing FIR filters. It is easy to design a FIR filter that approximates an arbitrary magnitude but retains a linear phase. The importance of the linear phase arises in applications that must preserve the shape of the original waveform, such as speech processing and data transmissions. Also FIR filters realized nonrecursively are always stable. The round-off noise associated with FIR realizations can easily be made small. The major disadvantage of FIR filters is that a large value of N is required to approximate sharp cutoff filters. The required value of N can be in the hundreds or thousands of taps for an equivalent IIR filter consisting of only ten quadratic sections.

Figure 9.2 can also be used for FIR realizations by replacing the recursive structure of Figure 9.6 with a nonrecursive structure similar to Figure 10.1. The rules for FIR window design follow:

1. Find the inverse Fourier transform of the desired response.
2. Apply a window to smooth out ripples in the frequency response.

The particular four-tap arrangement in Figure 10.1 requires eight memory locations: four of which store the four most recent input samples, and four locations for the impulse response coefficients. For each sample period, the output is generated by the sum of the multiplication of the impulse response coefficients by the digitized sample voltages with the most recent sample being multiplied by $h(0)$, and so forth. In practice, a circular buffer would hold the input samples and the oldest sample would be replaced by the most recent A/D conversion.

10.2 FOURIER SERIES EXPANSION

The impulse response, $h(n)$, is related to the frequency response, $H(e^{j\omega})$ by the following Fourier expansions.

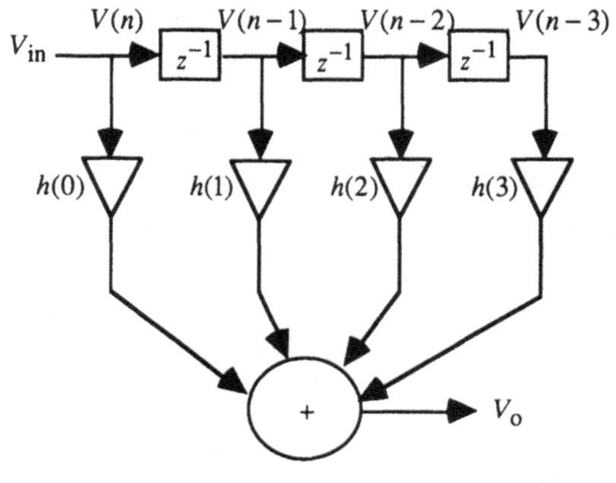

Figure 10.1 Direct form FIR filter shown with four taps.

Fourier transform:

$$\boldsymbol{H(e^{j\omega}) = \sum_{n=-\infty}^{\infty} h(n)\ e^{-j\omega n}} \tag{10.5}$$

Inverse Fourier transform:

$$\boldsymbol{h(n) = \frac{1}{2\pi}\int_{-\pi}^{\pi} H(e^{j\omega})\ e^{j\omega n} d\omega} \tag{10.6}$$

Equation (10.6) will be used to find the z-domain transfer function of all FIR filters in this chapter. Integration of (10.6), assuming flat passband regions, yields the formulas for various filters given in Table 10.1. For example, if a three-tap low-pass FIR filter is desired with a sampling frequency, f_s, of 10,000 Hz, and a cutoff frequency of 1000 Hz ($0.1f_s$), as shown in Figure 10.2, then the expression for the impulse response becomes

$$h(n) = \frac{1}{2\pi}\int_{-0.2\pi}^{0.2\pi} e^{j\omega n} d\omega = \frac{1}{2\pi}\frac{1}{jn} j2\ \sin(0.2\pi n) = \frac{\sin(0.2\pi n)}{\pi n} \tag{10.7}$$

For $n = -1$, 0, and 1, $h(n)$ is 0.1870979, 0.2, and 0.1870979, respectively, resulting in a z-domain transfer function of

$$H(e^{j\omega}) = 0.1870979z + 0.2 + 0.1870979z^{-1} \tag{10.8}$$

There are two minor difficulties with the use of (10.10) for designing FIR filters. First, the filter impulse response, $h(n)$, is infinite in duration because the summation in (10.10) extends to $\pm\infty$. Second, the filter is unrealizable (noncausal) because the impulse response begins at $-\infty$ and no finite amount of delay can make the impulse response realizable. Therefore (10.8) is an approximation of the desired frequency response obtained by truncating the infinite Fourier series. Direct truncation of the series leads to the well-known Gibbs phenomenon, which manifests itself as a fixed percentage overshoot and ripple before and after an approximated discontinuity in the frequency response. Thus in the approximation of standard filters, such as the ideal low-pass or bandpass filter, the largest ripple in the frequency response is about 9% of the size of the discontinuity, and its amplitude does not decrease with increasing impulse response duration; that is, including more and more terms in the Fourier series does not decrease the amplitude of the largest ripple. Instead, the overshoot is confined to a smaller and smaller frequency range as n is increased.

Because the Fourier series was truncated to three terms in our example, it is possible to add a delay to make the filter realizable as

$$H(e^{j\omega}) = 0.1870979 + 0.2z^{-1} + 0.1870979z^{-2} \tag{10.9}$$

Table 10.1
Normalized FIR Design Equations

$f_s = 1, \quad f_c =$ low-pass, or high-pass, cutoff frequency with $0 \le f_c \le 0.5$
$f_L =$ lower frequency bandpass edge for bandpass or stopband realization
$f_u =$ upper frequency bandpass edge for bandpass or stopband realization
$H(e^{j\omega}) = 1$ (for passband) and 0 (for stopband)

Low pass:
$$h(n) = \frac{1}{2\pi}\int_{-2\pi f_c}^{2\pi f_c} e^{j\omega n} d\omega = \frac{1}{2\pi}\frac{1}{jn} 2\,j\,\sin(2n\pi f_c) = \frac{\sin(2n\pi f_c)}{\pi n} \tag{10.10a}$$

for an odd number of taps, $n = \ldots -2, -1, 0, -1, -2, \ldots$
for an even number of taps, $n = \ldots -1.5, -0.5, 0.5, 1.5, \ldots$

Bandpass:
$$h(n) = \frac{1}{2\pi}\left[\int_{-2\pi f_u}^{-2\pi f_L} e^{j\omega n} d\omega + \int_{2\pi f_L}^{2\pi f_u} e^{j\omega n} d\omega\right] = \frac{\sin(2n\pi f_u) - \sin(2n\pi f_L)}{\pi n} \tag{10.10b}$$

for an odd number of taps, $n = \ldots -2, -1, 0, -1, -2, \ldots$
for an even number of taps, $n = \ldots -1.5, -0.5, 0.5, 1.5, \ldots$

Bandstop:
$$h(n) = \frac{\sin(2n\pi f_L) - \sin(2n\pi f_u) + \sin(n\pi)}{\pi n} \tag{10.10c}$$

for an odd number of taps, $n = \ldots -2, -1, 0, -1, -2, \ldots$

High pass:
$$h(n) = \frac{1}{2\pi}\left[\int_{-\pi}^{-2\pi f_c} e^{j\omega n} d\omega + \int_{2\pi f_c}^{\pi} e^{j\omega n} d\omega\right] = \frac{\sin(n\pi) - \sin(2n\pi f_c)}{\pi n} \tag{10.10d}$$

for an odd number of taps, $n = \ldots -2, -1, 0, -1, -2, \ldots$

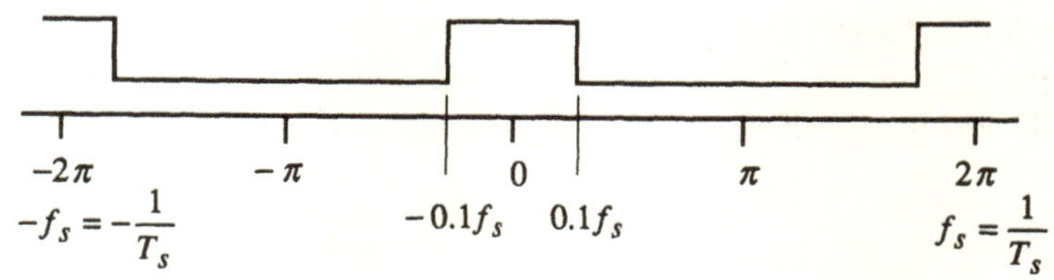

Figure 10.2 Example of a low-pass frequency spectrum.

10.3 CONDITIONS FOR LINEAR PHASE

By definition, a linear phase filter has both a constant phase delay, $\phi/\omega = K$, and a constant group delay, $d\phi/d\omega = K$. The output waveform of a linear phase filter is simply a time-delayed version of the input waveform.

The requirement for linear phase can be obtained by expanding equation (10.5) into the series,

$$H(e^{j\omega}) = \ldots + h(-1)e^{-j\omega(-1)} + h(0) + h(1)e^{-j\omega(1)} + \ldots \tag{10.11}$$

or

$$H(e^{j\omega}) = \ldots + h(-1)[\cos(\omega) + j\sin(\omega)] + h(0) + h(1)[\cos(\omega) - j\sin(\omega)] + \ldots \tag{10.12}$$

If N is the total number of terms in the truncated Fourier series, then setting

$$h(n) = h(N - 1 - n), \qquad 0 \le n \le N - 1 \tag{10.13}$$

results in the cancellation of all imaginary terms in equation (10.12) and produces a linear phase filter, as shown in Figure 10.3. The phase shift and group delay of a causal implementation are given by

$$\phi = -\frac{\omega}{f_s}\frac{N-1}{2} \text{ rad}, \qquad f_s = \text{sampling frequency} \tag{10.14}$$

$$\text{Group delay} = \frac{1}{f_s}\frac{N-1}{2} \text{ sec}, \qquad f_s = \text{sampling frequency} \tag{10.15}$$

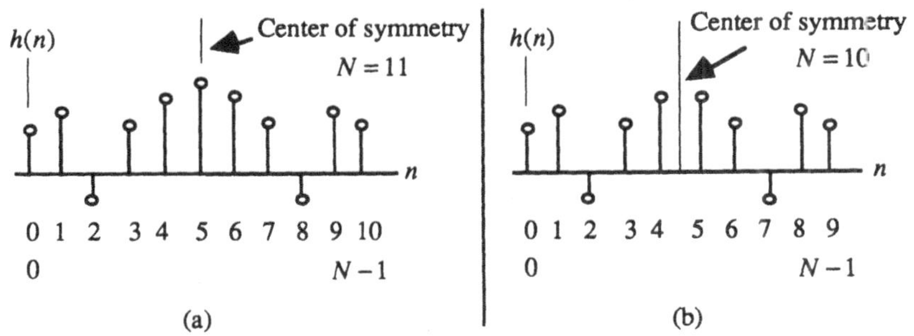

Figure 10.3 (a) Typical impulse response for N odd, even symmetry; (b) typical impulse response for N even, even symmetry.

A second type of "linear phase" filter can be defined in which the phase of $H(e^{j\omega})$ is a piecewise linear function. If

$$h(n) = -h(N-1-n) \qquad 0 \le n \le N-1 \tag{10.16}$$

then the real components of equation (10.12) cancel leaving an expression that has a constant phase of ±90°.
The phase shift and group delay of a causal implementation are given by

$$\phi = -\frac{\omega}{f_s}\frac{N-1}{2} \pm \frac{\pi}{2} \text{ rad}, \qquad f_s = \text{sampling frequency} \tag{10.17}$$

$$\text{Group delay} = \frac{1}{f_s}\frac{N-1}{2} \text{ sec}, \qquad f_s = \text{sampling frequency} \tag{10.18}$$

Filters that satisfy equation (10.16) have an antisymmetric impulse response as shown in Figure 10.4. It should be noted that for odd values of N, equation (10.16) requires $h[(N-1)/2] = 0$, because the filter can have only imaginary components. Bandstop and high-pass filters can be designed directly from (10.6) or from bandpass and low-pass prototypes, respectively. To convert (10.9) into a high-pass filter, the delay corresponding to the center of symmetry is simply subtracted from (10.9), resulting in a high-pass transfer function of

$$H(z^{-1}) = 0.1870979 - 0.8z^{-1} + 0.1870979z^{-2} \tag{10.19}$$

From Figures 10.3(b) and 10.4(b), it can be seen that even-order bandstop and high-pass filters are not possible using the design techniques in this chapter because the center of symmetry does not coincide with an integer multiple of the sampling time.

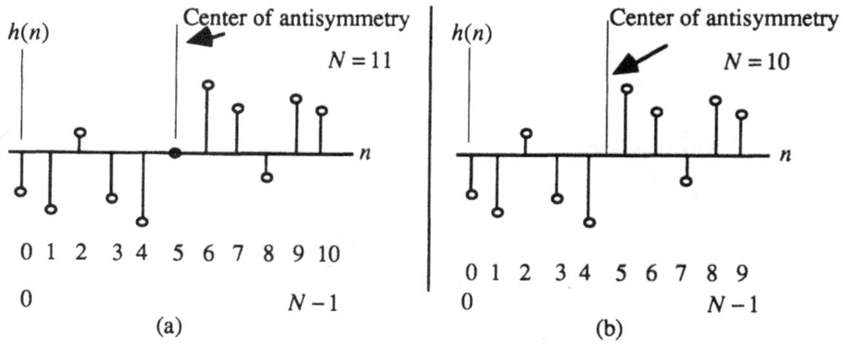

Figure 10.4 (a) Typical impulse response for N odd, odd symmetry; (b) typical impulse response for N even, odd symmetry.

10.4 WINDOWS

Direct truncation of the Fourier series is not a reasonable way of designing FIR filters. A window function can be applied to the truncated Fourier coefficients to reduce the effects of the Gibbs phenomenon.

Example 10.1

Find the impulse response for eighth- and ninth-order FIR low-pass, bandpass, bandstop, and high-pass filters with a sampling frequency of 10,000 Hz and a cutoff frequency of 1000 Hz for the low-pass and high-pass filters, and cutoff frequencies of 1000 Hz and 2000 Hz for the bandpass and bandstop filters.

Solution

Using equation (10.10) the coefficients of these filters follow.

Low Pass	*Bandpass*	*Bandstop*	*High Pass*
For eight-tap filters:			
$h(0) = 0.0735765$	$h(0) = -0.1600711$	Not applicable	Not applicable
$h(1) = 0.127324$	$h(1) = -0.127324$		
$h(2) = 0.1716787$	$h(2) = 0.03014171$		
$h(3) = 0.1967264$	$h(3) = 0.1774694$		
$h(4) = 0.1967264$	$h(4) = 0.1774694$		
$h(5) = 0.1716787$	$h(5) = 0.03014171$		
$h(6) = 0.127324$	$h(6) = -0.127324$		
$h(7) = 0.0735765$	$h(7) = -0.1600711$		
For nine-tap filters:			
$h(0) = 0.0467744$	$h(0) = -0.1224571$	$h(0) = 0.1224571$	$h(0) = -0.04677$
$h(1) = 0.1009102$	$h(1) = -0.1632762$	$h(1) = 0.1632762$	$h(1) = -0.10091$
$h(2) = 0.1513653$	$h(2) = -0.05781644$	$h(2) = 0.0578264$	$h(2) = -0.15136$
$h(3) = 0.1870979$	$h(3) = 0.1156328$	$h(3) = -0.1156328$	$h(3) = -0.18709$
$h(4) = 0.2000000$	$h(4) = 0.2000000$	$h(4) = 0.8000000$	$h(4) = 0.80000$
$h(5) = 0.1870979$	$h(5) = 0.1156328$	$h(5) = -0.1156328$	$h(5) = -0.18709$
$h(6) = 0.1513653$	$h(6) = -0.05781644$	$h(6) = 0.0578164$	$h(6) = -0.15136$
$h(7) = 0.1009102$	$h(7) = -0.1632762$	$h(7) = 0.1632752$	$h(7) = -0.10091$
$h(8) = 0.0467744$	$h(8) = -0.1224571$	$h(8) = 0.1224571$	$h(8) = -0.04677$

The effects of a Kaiser window are shown in Figure 10.5(a) using a 32-tap low-pass FIR filter with the coefficients obtained by direct evaluation of equation (10.10a). In

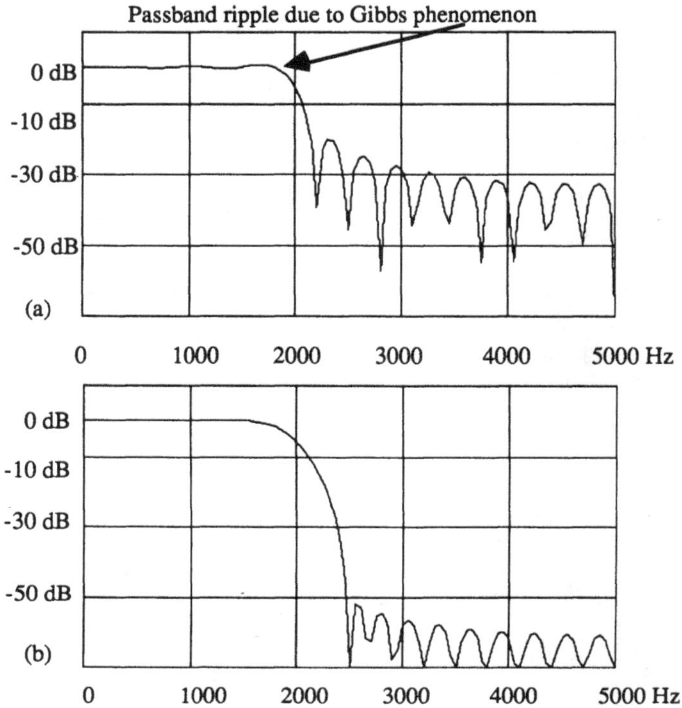

Figure 10.5 (a) 32-tap FIR filter with rectangular window, (b) with Kaiser window.

Figure 10.5(b), these coefficients have been modified by a Kaiser window, which has eliminated the passband ripple and reduced the maximum stopband gain to less than –50 dB.

Several issues often impede the use of windows. One problem is the need for a closed form expression for the Fourier series coefficients, $h(n)$, where $h(n)$ is defined by (10.6). When $H(e^{j\omega})$ is complicated or cannot easily be put into a closed form mathematical expression, (10.6) is often cumbersome or difficult to evaluate. Another issue with windows is that the passband edge frequency generally cannot be specified exactly because the window "smears" the discontinuity in frequency.

10.4.1 Rectangular Window

The rectangular window corresponds to direct truncation of the Fourier series and is equivalent to no window at all. From equation (10.5), the frequency response of the rectangular window (Figure 10.6) is

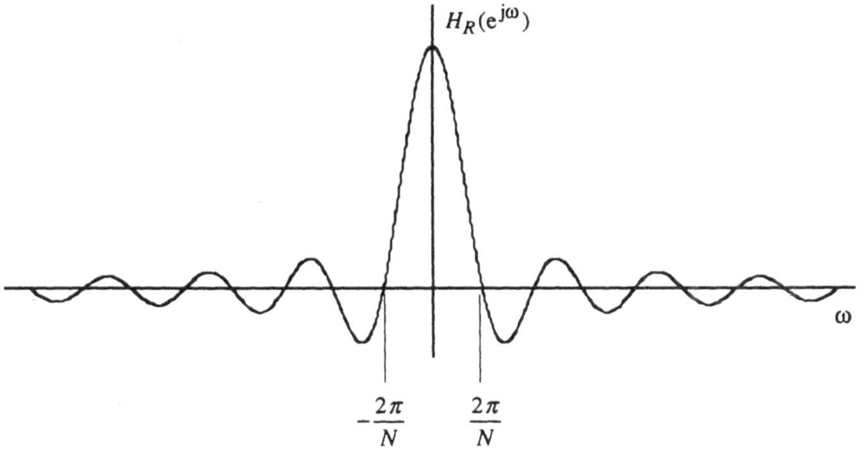

Figure 10.6 Frequency response of a rectangular window.

$$H_R(e^{j\omega}) = \sum_{n=-(N-1)/2}^{(N-1)/2} e^{-j\omega n} = \frac{e^{j\omega[(N-1)/2]}(1 - e^{-j\omega N})}{(1 - e^{j\omega})} = \frac{\sin(\omega N/2)}{\sin(\omega/2)} \tag{10.20}$$

The final frequency response of the filter is the circular convolution of the ideal frequency response with the window's frequency response. The ripple from the side lobes of the rectangular window produces approximation errors (ripples in the passband) for all frequencies and is referred to as the *Gibbs phenomenon* seen in Figure 10.5(a).

10.4.2 Generalized Hamming Window

The generalized Hamming window is often given with the form

$$\omega_H(n) = \begin{cases} \alpha + (1 - \alpha)\cos\left(\frac{2\pi n}{N}\right) & -\left(\frac{N-1}{2}\right) \le n \le \left(\frac{N-1}{2}\right) \\ 0.0 & \text{elsewhere} \end{cases} \tag{10.21}$$

where α is in the range $0 \le \alpha \le 1$. If $\alpha = 0.54$, the window is called a *Hamming window*; if $\alpha = 0.5$, it is called a *Hanning window*.

The frequency response of this window can be obtained from equation (10.5) and is shown in Figure 10.7 with $\alpha = 0.54$ and $N = 25$. Although the overall frequency response of the Hamming window appears to have no ripples beyond $\omega = \pm 4\pi/N$, the ripples are still present but very small in amplitude. A comparison of Figure 10.6 and Figure 10.7 shows the main lobe of the frequency response of the Hamming window is twice the

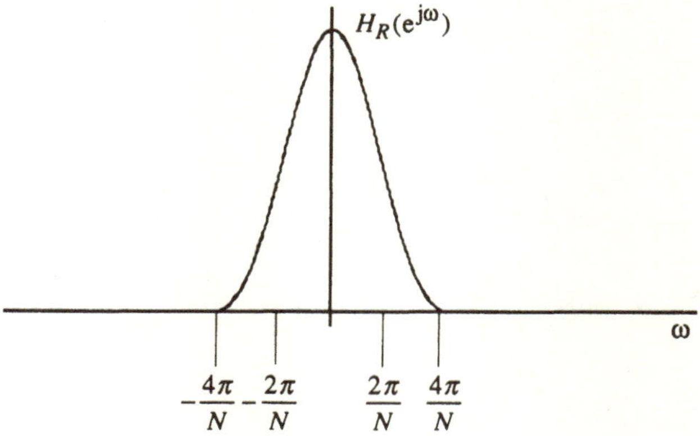

Figure 10.7 Frequency response of a Hamming window for $\alpha = 0.54$.

width of the main lobe of the frequency response of the rectangular window. The side lobe amplitudes of the Hamming window frequency response are considerably smaller, however, than the side lobe amplitudes of the rectangular window frequency response. For $\alpha = 0.54$ (the conventional Hamming window), 99.96% of the spectral energy is in the main lobe and the peak side lobe ripple is down about 40 dB from the main lobe peak. In contrast, for the rectangular window, the spectral side lobes are down only about 14 dB from the main lobe peak.

For a low-pass filter, the increased main lobe width corresponds to the increased transition bandwidth between the passband and the stopband, whereas decreased side lobe levels correspond to smaller ripples (more rejection) in the filter stopband and passband.

10.4.3 Kaiser Window

The problem of designing windows can be reduced to the mathematical problem of finding a time-limited function whose Fourier transform best approximations a band-limited function; that is, a time-limited function that has the minimum energy outside some selected frequency interval. For continuous-time functions, this problem has been solved in closed form to give a class of functions called *prolate speroidal wavefunctions*. The form of these functions is very complicated. Kaiser has proposed a relatively simple approximation to these functions as an ideal candidate for a window. The approximation, called the *Kaiser window*, is given in equation (10.28).

The constant β in equation (10.28) specifies a frequency response trade-off between the peak height of the side lobe ripples and the width or energy of the main lobe, and $I_o(x)$ is the modified zero-order Bessel function.

The Kaiser window is an optimum window in that it is a finite duration sequence that has the minimum spectral energy beyond some specified frequency. Another optimum window is the Dolph-Tchebyscheff window, which has the minimum width of the main lobe of its frequency response for a fixed peak value of side lobe ripple. This window has the property that all its spectral side lobes are of equal amplitude. Neither of these windows can be used to design optimum (mini-max) approximations to any ideal frequency response function since they are convolved with the desired ideal response to give the actual filter response. Thus although there are optimum windows, there are no optimum window designed filters.

The function $I_o(x)$ has the power series expansion [1] given by

$$I_o(x) = 1 + \sum_{k=1}^{\infty} \left[\frac{(x/2)^k}{k!} \right]^2 \tag{10.22}$$

The Kaiser window coefficients and the function $I_o(x)$ can be computed by the routine, B6, in Appendix B.

10.4.4 Window Summary

The exact equations of each of the following windows will vary slightly depending on their source of reference. The following forms are frequently used in computer programs that synthesize digital filters [2].

Rectangular Window

$$\begin{aligned} &\text{For } N \text{ odd:} \quad \omega(n) = 1 \quad -(N-1)/2 \le n \le (N-1)/2 \\ &\text{For } N \text{ even:} \quad \omega(n) = 1 \quad -(N/2) \le n \le (N/2 - 1) \end{aligned} \tag{10.23}$$

Triangular (Bartlett) Window

$$\begin{aligned} &\text{For } N \text{ odd:} \quad \omega(n) = \quad 1 - |2n|/(N+1) \quad -(N-1)/2 \le n \le (N-1)/2 \\ &\text{For } N \text{ even:} \quad \omega(n) = \text{some } 1 - |2n+1|/N \quad -(N/2) \le n \le (N/2 - 1) \end{aligned} \tag{10.24}$$

Hamming Window

$$\begin{aligned} &\text{For } N \text{ odd:} \quad \omega(n) = 0.54 + 0.46 \cos[2\pi n/(N-1)] \quad -(N-1)/2 \le n \le (N-1)/2 \\ &\text{For } N \text{ even:} \quad \omega(n) = 0.54 + 0.46 \cos\left[\frac{2\pi(2n+1)}{2(N-1)}\right] \quad -(N/2) \le n \le (N/2 - 1) \end{aligned} \tag{10.25}$$

Hanning Window

$$\begin{aligned} &\text{For } N \text{ odd:} \quad \omega(n) = 0.5 + 0.5 \cos[2\pi n/(N+1)] \quad -(N-1)/2 \le n \le (N-1)/2 \\ &\text{For } N \text{ even:} \quad \omega(n) = 0.5 + 0.5 \cos\left[\frac{2\pi(2n+1)}{2(N+1)}\right] \quad -(N/2) \le n \le (N/2 - 1) \end{aligned} \tag{10.26}$$

Generalized Hamming Window

For N odd: $\omega(n) = \alpha + (1 - \alpha)\cos[2\pi n/(N - 1)] \qquad -(N - 1)/2 \le n \le (N - 1)/2$

For N even: $\omega(n) = \alpha + (1 - \alpha)\cos\left[\dfrac{2\pi(2n + 1)}{2(N - 1)}\right] \qquad -(N/2) \le n \le (N/2 - 1)$

(10.27)

where α is a variable parameter.

Kaiser (I_o-Sinh) Window

$$\text{For } N \text{ odd:} \quad \omega(n) = \frac{I_o\left[\beta\sqrt{1 - \dfrac{4n^2}{(N-1)^2}}\right]}{I_o(\beta)} \qquad -(N - 1)/2 \le n \le (N - 1)/2$$

(10.28)

$$\text{For } N \text{ even:} \quad \omega(n) = \frac{I_o\left[\beta\sqrt{1 - \dfrac{4(n + 1/2)^2}{(N-1)^2}}\right]}{I_o(\beta)} \qquad -(N/2) \le n \le (N/2 - 1)$$

Blackman Window

$$\text{For } N \text{ odd:} \quad \omega(n) = 0.42 - 0.5\cos\left[\frac{2\pi n + \pi(N - 1)}{(N - 1)}\right] + 0.08\cos\left[\frac{4\pi n + 2\pi(N - 1)}{(N - 1)}\right]$$

$$-(N - 1)/2 \le n \le (N - 1)/2 \qquad (10.29)$$

$$\text{For } N \text{ even:} \quad \omega(n) = 0.42 - 0.5\cos\left[\frac{2\pi n + \pi N}{(N - 1)}\right] + 0.08\cos\left[\frac{4\pi n + 2\pi N}{(N - 1)}\right]$$

$$-(N/2) \le n \le (N/2 - 1)$$

Example 10.2

For the eight-tap low pass in example 10.1, modify the coefficients with a Kaiser window using an attenuation of 40 dB.

Solution

Using program B6, $\beta = 3.399974$. The final filter coefficients are as follows.

Unwindowed Coefficients		*Kaiser Window*		*Final Coefficients*
0.0735765	×	0.1473911	=	0.0108445
0.127324	×	0.4425494	=	0.0563471
0.1716787	×	0.7626705	=	0.1309342
0.1967264	×	0.9713108	=	0.1910425
0.1967264	×	0.9713108	=	0.1910425
0.1716787	×	0.7626705	=	0.1309342
0.127324	×	0.4425494	=	0.0563471
0.0735765	×	0.1473911	=	0.0108445

10.5 THE HILBERT TRANSFORM

The existence of antisymmetric FIR filters with a constant phase of 90° is useful for generating a class of filters that approximate a Hilbert transform. The frequency response of a Hilbert transform is given by

$$H(e^{j\omega}) = \begin{cases} -j & 0 < \omega \leq \pi \\ j & -\pi \leq \omega < 0 \end{cases} \tag{10.30}$$

Using equation (10.6) the impulse response can be obtained from

$$h(n) = \frac{1}{2\pi}\left(\int_0^{\pi} -j\ e^{j\omega n} d\omega + \int_0^{\pi} j\ e^{j\omega n} d\omega\right) \tag{10.31}$$

or equivalently

$$h(n) = \begin{cases} \dfrac{1 - \cos(\pi n)}{\pi n} & n \neq 0 \\ 0 & n = 0 \end{cases} \tag{10.32}$$

The impulse response is plotted in Figure 10.8.

Two significant uses for Hilbert transformer approximations are in single sideband modulators and in image rejecting mixers [3]. The image rejection in dB can be derived from Figure 10.9(a). The single sideband modulator will accept an audio input frequency

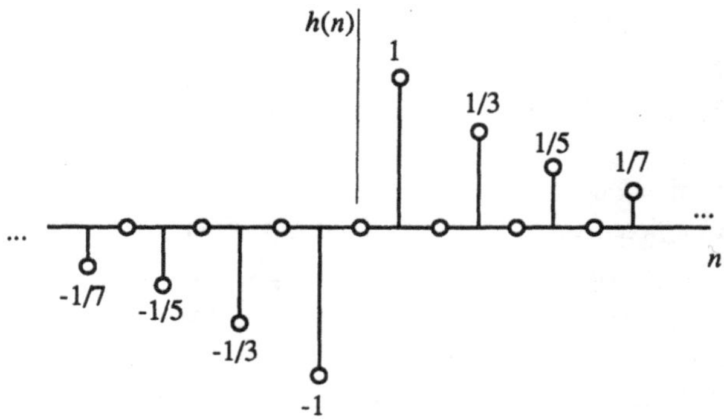

Figure 10.8 Time response of the ideal Hilbert transformer.

range, and up convert it into a single sideband RF signal. The carrier frequency is denoted as ω_o. A Hilbert transformer approximation is used to phase shift the components of the audio spectrum (say 50 Hz to 3000 Hz) by a constant 90°. (The other 90° phase shift is done on the local oscillator at a constant frequency of ω_o.) Without loss of generality, if

$$V_{\text{in}} = 2\cos(\omega_m t) \tag{10.33}$$

then

$$V_1 = 2\cos(\omega_m t + 90°) \tag{10.34}$$

$$V_2 = \cos(\omega_o t) \tag{10.35}$$

$$V_3 = \cos(\omega_o t + 90°) \tag{10.36}$$

Using the identity $\cos(a) \cdot \cos(b) = 0.5 \cdot \cos(a - b) + 0.5 \cdot \cos(a + b)$ yields

$$V_4 = \cos[(\omega_o - \omega_m)t] + \cos[(\omega_o + \omega_m)t] \tag{10.37}$$

$$\begin{aligned} V_5 &= \cos[(\omega_o - \omega_m)t] + \cos[(\omega_o + \omega_m)t + 180°] \\ &= \cos[(\omega_o - \omega_m)t] - \cos[(\omega_o + \omega_m)t] \end{aligned} \tag{10.38}$$

V_o is produced from either the sum or difference of V_4 and V_5, resulting in the lower or upper sideband, respectively.

If the Hilbert transformer has exactly unity gain, but only approximately 90° phase shift with a phase error of $\phi_e(\omega)$, then the addition, or subtraction, of V_4 and V_5 will not completely cancel the undesired sideband. In this case V_5 can be rewritten as

$$V_5 = \cos[(\omega_o - \omega_m)t - \phi_e(\omega)] - \cos[(\omega_o + \omega_m)t + \phi_e(\omega)] \tag{10.39}$$

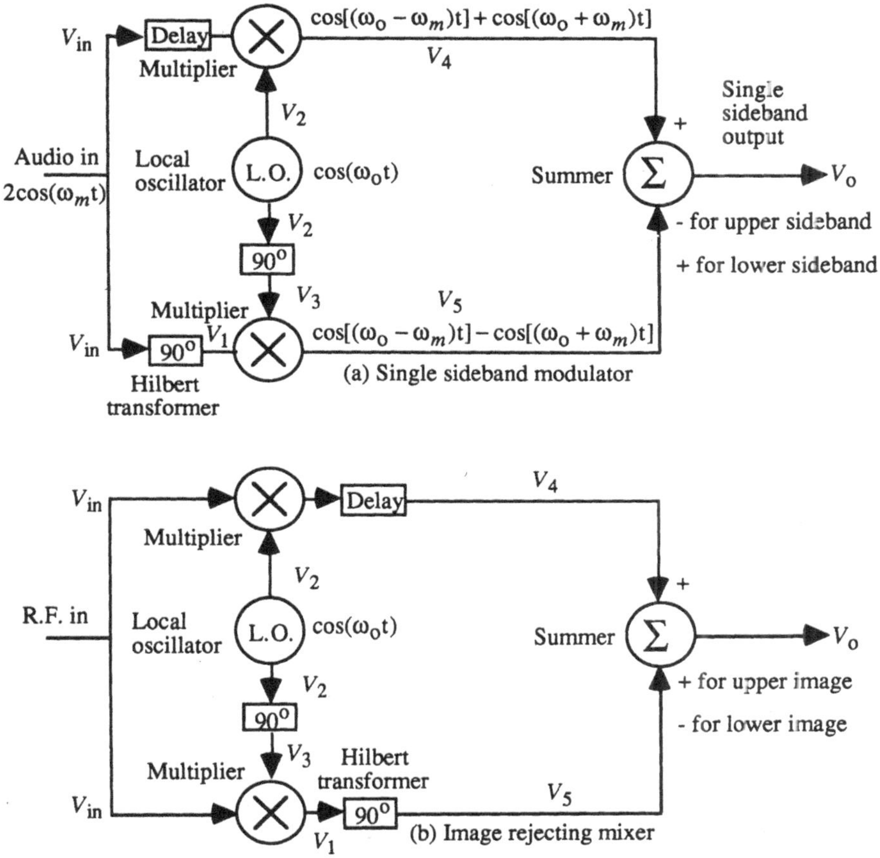

Figure 10.9 Applications for Hilbert transform: (a) single-sideband generation, and (b) image rejecting mixer.

The dB rejection can be calculated from the ratio of the energy in the undesired sideband to the energy in the desired sideband and integrating over a full cycle with $x = (\omega_o + \omega_m)t$ and $y = (\omega_o - \omega_m)t$, or

$$\text{dB rejection} = -10 \log \left[\frac{\int_0^{2\pi} [\cos(x) - \cos(x + \phi_e(\omega))]^2 dx}{\int_0^{2\pi} [\cos(y) + \cos(y + \phi_e(\omega))]^2 dy} \right] \tag{10.40}$$

$$= -20 \log \left[\frac{\sin[\phi_e(\omega)]}{1 + \cos[\phi_e(\omega)]} \right]$$

If the phase is exactly 90° for all frequencies, but a gain error, $G_e(\omega)$, causes the gain of the Hilbert transformer to deviate from unity, then the dB rejection can be calculated as

$$\text{dB rejection} = -20\ \log\left[\frac{G_e(\omega)}{2 + G_e(\omega)}\right] \tag{10.41}$$

Similarly for the image rejecting mixer in Figure 10.9(b), if

$$V_{\text{in}} = 2\cos\{[(\omega_o + \omega_{m1})t] + \cos[(\omega_o - \omega_{m2})t]\}, \qquad \omega_{m1} = \omega_{m2} \tag{10.42}$$

then

$$V_4 = \cos[(2\omega_o + \omega_{m1})t] + \cos(\omega_{m1}t) + \cos[(2\omega_o - \omega_{m2})t] + \cos(\omega_{m2}t) \tag{10.43}$$

Assuming that the double frequency terms will be eliminated by low-pass filtering of the output, V_o, then

$$V_4 = \cos(\omega_{m1}t) + \cos(\omega_{m2}t) \tag{10.44}$$

Similarly,

$$V_5 = \cos(\omega_{m1}t) - \cos(\omega_{m2}t) \tag{10.45}$$

The modulating frequency, ω_{m1} or ω_{m2}, can now be recovered by adding or subtracting V_4 and V_5, respectively. Both circuits in Figure 10.9 require delays in the main branch to match the midpoint delay of the Hilbert transformer for proper cancellation of the undesired signals.

10.6 Σ-Δ MODULATOR

Analog-to-digital converters with 8-bit accuracy are usually implemented with a successive approximation design similar to Figure 9.5. If higher conversion speed is required, then a flash converter constructed from 256 voltage comparators and decoding logic can provide conversion rates as high as 100 million samples per second. In both cases precision resistors are required with the precision of each resistor on the order of 0.4% (1/256), for 8-bit accuracy. If a continuous analog voltage is sampled with an A/D converter, the conversion accuracy is limited to ±1/2 the least significant bit (LSB). This inaccuracy in conversion is referred to as quantization noise. The quantization noise associated with an A/D converter is 6 dB per bit because each bit of increased accuracy in the A/D converter decreases the quantization noise by 1/2, or by –20 log(1/2) = 6 dB. Based on quantization noise alone, an 8-bit converter will have a noise floor at –48 dB, and a 16-bit converter will have a noise floor at –96 dB.

An A/D converter with 16-bit accuracy would require precision resistors on the order of 0.0015% accuracy, which would neither be feasible nor cost effective. The oversampled Σ-Δ modulator has been widely used in recent years to provide 16-bit analog-to-digital conversion without the need for any precision components [4]. It operates by converting an analog input voltage to a binary bit stream (consisting of 1-bit bytes) followed by a FIR low-pass filter using 16-bit arithmetic. The output of the FIR filter is the 16-bit digital word. The circuit in Figure 10.10 is a commonly used second-order Σ-Δ modulator.

The voltage, V_5, at the output of the comparator has the exact spectral makeup as the input signal, V_{in}, except that high-frequency noise components are added. The FIR low-pass can be implemented with a Kaiser window filter and is usually of very high order (> 256). Delays shown by the z^{-1} blocks are implemented with analog sample-and-hold circuits similar to Figure 9.3. Although the parallel outputs of the shift register are 1-bit binary levels, the FIR filter can be implemented with 16-bit arithmetic providing a 16-bit digital output word, V_o, at the end of every sample period, representing the digital conversion of the input signal, V_{in}. For example, the design of a sigma-delta A/D converter for frequencies between 0 and 5000 Hz could be implemented using an (over)sampling rate of 500 KHz and a FIR Kaiser window filter with 500 taps (Figure 10.11). The following equations, used iteratively, would provide a simulated output response:

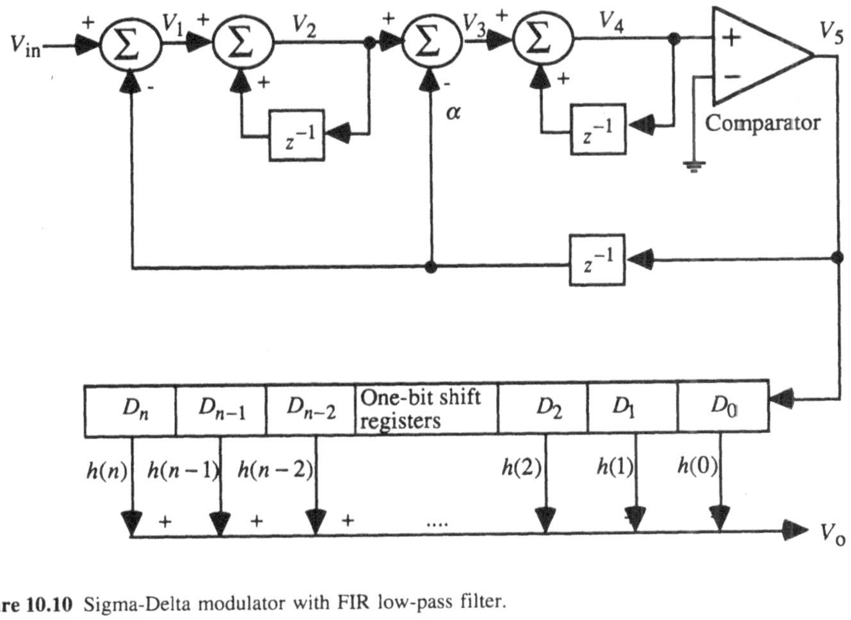

Figure 10.10 Sigma-Delta modulator with FIR low-pass filter.

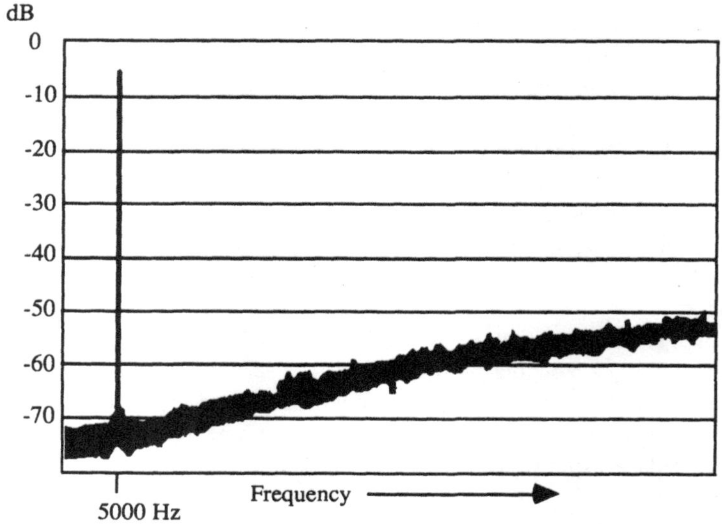

Figure 10.11 Typical Σ-Δ spectral output at location V_5 before low-pass filtering, for a single sine wave input of 5,000 Hz (adapted from [5] by permission of the publisher, © 1992 IEEE).

$$V_{\text{in}} = \sin(2 \cdot 3.14159 \cdot 5000 \cdot i/f_s), \qquad i = 0, 1, \ldots, 4999; \qquad f_s = 5000 \tag{10.46}$$

$$V_1 = V_{\text{in}} - V_5 \tag{10.47}$$

$$V_2 = V_2 + V_1 \tag{10.48}$$

$$V_3 = V_2 - V_5 \qquad \text{(implies } \alpha = 1) \tag{10.49}$$

$$V_4 = V_3 + V_4 \tag{10.50}$$

$$V_5 = \begin{cases} 2 \text{ if } V_4 \text{ is greater than or equal to } 0 \\ -2 \text{ otherwise} \end{cases} \tag{10.51}$$

The response of the Σ-Δ modulator can be analyzed with the Fourier transform by taking 500 continuous samples of the intermediate output signal, V_5. Using equations (1.35) and (1.36), the component frequencies are given by

$$m\text{th harmonic energy in dB} = 10 \log_{10}(a_m^2 + b_m^2) \tag{10.52}$$

where

$$a_m = \frac{1}{\pi m} \sum_{i=1}^{N} V_5(i) \left[\sin\left(\frac{2\pi m i}{N}\right) - \sin\left(\frac{2\pi m(i-1)}{N}\right) \right] \tag{10.53}$$

$$b_m = -\frac{1}{\pi m} \sum_{i=1}^{N} V_5(i) \left[\cos\left(\frac{2\pi m i}{N}\right) - \cos\left(\frac{2\pi m(i-1)}{N}\right) \right] \tag{10.54}$$

with $N = 500$ for the previous example. The spectral analysis is important in determining if the length of the FIR filter length is sufficient to give the desired signal-to-noise ratio. The FIR low-pass filter is designed to pass the desired low frequency components while rejecting the higher frequency noise due to quantization in the Σ-Δ comparator.

REFERENCES

[1] Rabiner, Lawrence R., and Gold, Bernard, *Theory and Application of Digital Signal Processing*, Englewood Cliffs, NJ: Prentice-Hall, 1975, pp. 102–103.

[2] Digital Signal Processing Committee, *Programs for Digital Signal Processing*, New York: IEEE Press, 1979, pp. 5.2-1 to 5.2-3.

[3] Schwartz, Mischa, *Information Transmission, Modulation, and Noise*, New York: McGraw-Hill, 1970, p. 221.

[4] Johns, David A., and Lewis, David M., "Design and Analysis of Delta-Sigma Based IIR Filters," *IEEE Trans. Circuits and Systems*, April 1993, pp. 233–240.

[5] Huang, Qiuting, and Moschytz, George S., "Analog FIR Filters with an Oversampled Σ-Δ Modulator," *IEEE Trans. Circuits and Systems*, September 1992, pp. 658–663.

PROBLEMS

10.1 Derive equation (10.10 c) from (10.6).

10.2 Implement a 33-tap FIR filter with sampling frequency of 10,000 Hz for
- (a) a low pass with cutoff frequency of 1000 Hz,
- (b) a bandpass with cutoff frequencies of 2000 and 3000 Hz,
- (c) a bandstop with cutoff frequencies of 2000 and 3000 Hz,
- (d) a high pass with cutoff frequency of 1000 Hz.

10.3 Implement each of the filters in problem 10.2 with the following windows:
- (a) rectangular,
- (b) Blackman,
- (c) Kaiser with attenuation of 40 dB,
- (d) Hanning.

10.4 Integrate equation (10.6) numerically to design a 64-tap FIR filter using a rectangular window and 10,000 Hz sampling rate such that the passband increases linearly in amplitude from $|H(e^{j\omega_1})| = 1$ to $|H(e^{j\omega_2})| = 2$ for $f_1 = 1000$ Hz, $f_2 = 2000$ Hz, and $|H(e^{j\omega})| = 0$ outside the passband range of 1000 Hz to 2000 Hz.

10.5 Derive an expression for the dB rejection of the unwanted sideband for a single sideband modulator if both the phase and the amplitude response of the Hilbert transformer are allowed to be in error.

10.6 Use (10.46) through (10.51) to find the signal-to-noise ratio of V_5 in Figure 10.10 for (a) a 256-tap FIR filter, (b) a 512-tap FIR filter, and (c) a 1024-tap FIR filter, using a 5,000 Hz sine wave as the input with a sampling frequency of 500 KHz. For each case, generate a plot similar to Figure 10.11.

10.7 Design a 128-tap FIR low-pass filter with cutoff frequency of 1,000 Hz, and a sampling frequency of 10 KHz, using a Kaiser window with an attenuation factor of 50 dB. Plot the amplitude response if the impulse response is (a) truncated to 16-bit accuracy and (b) truncated to 8-bit accuracy.

Chapter 11
Equiripple Approximation of FIR Filters

11.1 THE EXCHANGE METHOD OF MINI-MAX APPROXIMATION

In the previous chapter, the use of the Fourier series led to the design of FIR filters from the relatively simple formulas in Table 10.1. It is a fundamental result of the theory of Fourier series that the use of the rectangular window in these designs provides the best least-squares approximation to a desired frequency response for a given value of N. A disadvantage of the least-squares approximation is that it may not be satisfactory at all points.

Mini-max approximation uses a different error criteria that minimizes the absolute error in both the passband and the stopband, resulting in equiripple error in both bands. The exchange algorithm operates like the simplex optimization algorithm of linear algebra but is significantly faster since it performs multiple exchanges per iteration [1]. The Remez exchange algorithm is an adaptation of the exchange method that allows the edges of the passbands and stopbands to be precisely defined. It is based on a reformulation, by Parks and McClellan [2], of a powerful approximation theorem. This theorem, called the *alternation theorem,* is stated as follows.

Alternation Theorem

If $P(e^{j\omega})$ is a linear combination of L cosine functions such that

$$P(e^{j\omega}) = \sum_{n=0}^{L-1} \alpha(n) \cos(\omega n)$$

then a necessary and sufficient condition that $P(e^{j\omega})$ be the unique, best weighted Tchebyscheff approximation to a continuous function $\overline{D}(e^{j\omega})$ on A, a compact subset of $(0, \pi)$, is that the weighted error function $E(e^{j\omega})$ exhibit at least $(L + 1)$ extremal frequencies in A; that is, there must exist $(L + 1)$ points ω_i such that $\omega_1 < \omega_2 < \ldots < \omega_L < \omega_{L+1}$, and such that $E(e^{j\omega_i}) = -E(e^{j\omega_{i+1}})$, $i = 1, 2, \ldots L$ and $|E(e^{j\omega})| = \max_{\omega \in A}[E(e^{j\omega})]$.

The Remez exchange algorithm is the starting point for many advanced approximation techniques in filter design. McGee [3] proposed an adaptation of the Remez exchange for s-domain filters with both poles and zeros to approximate arbitrary passband and stopband characteristics. Temes and Calahan [4] described the use of the Remez exchange both for amplitude equalization and group delay equalization. Temes and Gyi [5] used the exchange method to adjust the location of zeros in an arbitrary s-domain filter to provide equiripple stopband attenuation. The algorithm will be used in this chapter to design linear phase equiripple FIR filters as an alternative to windows.

11.2 THE BASIC EXCHANGE ALGORITHM

Mini-max approximation is often called *equiripple*, or Tchebyscheff, *approximation* because it produces a function, $p(x)$, that misses the desired data points, $f(x_i)$, by an error that alternates between $\pm h$. The exchange method finds the unique polynomial, $p(x)$, that minimizes the value of h. Lagrange interpolation is used to compute a polynomial that takes on the prescribed extremal values $[f(x) \pm h]$ at the initial estimates of the extremum points. The extrema of the resulting polynomial are computed by evaluating the polynomial on a finely spaced set of points and searching for the local maxima and minima. If the maxima and minima have the prescribed values, the procedure is terminated, otherwise a new polynomial is computed with the new estimate of extremum locations being the extrema of the previous polynomial. This process is summarized in the following and example 11.1.

1. For a set of M data points defined at (x_i, y_i) for $i = 1, 2, \ldots M$, choose an interpolating polynomial of degree N that satisfies the $N + 2$ simultaneous equations (with $N + 2 < M$) given by

$$\begin{bmatrix} a_0 + a_1x_1 + a_2x_1^2 + \ldots + a_nx_1^n - y_1 = -h \\ a_0 + a_1x_2 + a_2x_2^2 + \ldots + a_nx_2^n - y_2 = +h \\ a_0 + a_1x_3 + a_2x_3^2 + \ldots + a_nx_3^n - y_3 = -h \\ \cdot \\ \cdot \\ \cdot \\ a_0 + a_1x_{n+2} + \ldots + a_nx_{n+2}^n - y_{n+2} = (-1)^{(n+2)}h \end{bmatrix} \tag{11.1}$$

2. Identify the $N + 2$ values of x_i from the set of M data points such that the error alternates in sign and is at its maximum amplitude.

3. Construct a new interpolating polynomial of degree N using the new values of x_i corresponding to the locations of maximum alternating error.
4. Repeat steps 2 and 3 until the absolute value of h no longer changes. Then the process has converged.

Example 11.1

Let the function, $f(x)$, be defined as

$$f(x) = 1 \text{ for } 0 \le x \le 1$$
$$f(x) = x \text{ for } 1 \le x \le 2$$

Find a fifth-order equiripple approximating polynomial using the exchange method for the range $0 \le x \le 2$.

Solution

In Step 1, for a fifth-degree polynomial of the form $p(x) = a_0 + a_1x + a_2x^2 + a_3x^3 + a_4x^4 + a_5x^5$, the initial values of $x_i = \{0, 0.333333, 0.6666667, 1.00000, 1.333333, 1.666667, 2\}$ are arbitrarily chosen corresponding to equally spaced intervals over the range $0 \le x \le 2$.

The set of seven linear simultaneous equations becomes

$$
\begin{aligned}
&a_0 + && -h = 1\\
&a_0 + 0.3333a_1 + 0.1111a_2 + 0.037037a_3 + 0.012345a_4 + 0.004115a_5 + h = 1\\
&a_0 + 0.6667a_1 + 0.4444a_2 + 0.296296a_3 + 0.197531a_4 + 0.131687a_5 - h = 1\\
&a_0 + 1.0000a_1 + 1.0000a_2 + 1.000000a_3 + 1.000000a_4 + 1.000000a_5 + h = 1\\
&a_0 + 1.3333a_1 + 1.7777a_2 + 2.370370a_3 + 3.160493a_4 + 4.213990a_5 - h = 4/3\\
&a_0 + 1.6667a_1 + 2.7778a_2 + 4.629628a_3 + 7.716047a_4 + 12.86008a_5 + h = 5/3\\
&a_0 + 2.0000a_1 + 4.0000a_2 + 8.000000a_3 + 16.00000a_4 + 32.00000a_5 - h = 2
\end{aligned}
$$

Solving for the coefficients yields

$$p_1(x) = 0.9687497 + 0.750012x - 2.375051x^2 + 2.250066x^3 - 0.562531x^4 + 0x^5$$

with $h = -0.03125032$.

In Step 2, the polynomial $p_1(x)$ has extrema at the following:

x	$f(x) - p_1(x)$
0	−0.0312507
0.22	0.04144084
0.73	−0.0338413
1.00	0.03124547
1.27	−0.03385174
1.78	0.0413624
2.00	−0.3139532

In Step 3, the coefficients for a new interpolating polynomial $p_2(x)$ are obtained by solving the set of linear simultaneous equations that follows:

$$a_0 + \qquad -h = 1$$

$$a_0 + 0.2200a_1 + 0.0484a_2 + 0.01065a_3 + 0.0023426a_4 + 0.000515a_5 + h = 1$$

$$a_0 + 0.7300a_1 + 0.5329a_2 + 0.389017a_3 + 0.283983a_4 + 0.207307a_5 - h = 1$$

$$a_0 + 1.0000a_1 + 1.0000a_2 + 1.000000a_3 + 1.000000a_4 + 1.000000a_5 + h = 1$$

$$a_0 + 1.2700a_1 + 1.6129a_2 + 2.048383a_3 + 2.601446a_4 + 3.303837a_5 - h = 1.27$$

$$a_0 + 1.7800a_1 + 3.1684a_2 + 5.639752a_3 + 10.03876a_4 + 17.86899a_5 + h = 1.78$$

$$a_0 + 2.0000a_1 + 4.0000a_2 + 8.000000a_3 + 16.00000a_4 + 32.00000a_5 - h = 2$$

Solving for the coefficients yields

$$p_2(x) = 0.9662585 + 0.6998696x - 2.22973x^2 + 2.129779x^3 - 0.532431x^4 + 0x^5$$

with $h = -0.0337415$.

In Step 4, finding the extrema of $p_2(x)$ and repeating the process yields

$$p_3(x) = 0.9661971 + 0.700743x - 2.231421x^2 + 2.131034x^3 - 0.5327449x^4 + 0x^5$$

with $h = -0.03380291$. The solution, $p_3(x)$, is sufficiently close to the previous iteration that the process may be stopped. The exact curve, $f(x)$, and the approximating polynomial, $p(x)$, are plotted in Figure 11.1.

11.3 FOUR TYPES OF FINITE IMPULSE RESPONSE FILTERS

The application of the exchange algorithm to the design of FIR filters is only slightly more difficult than the polynomial approximation in example 11.1. It is first necessary

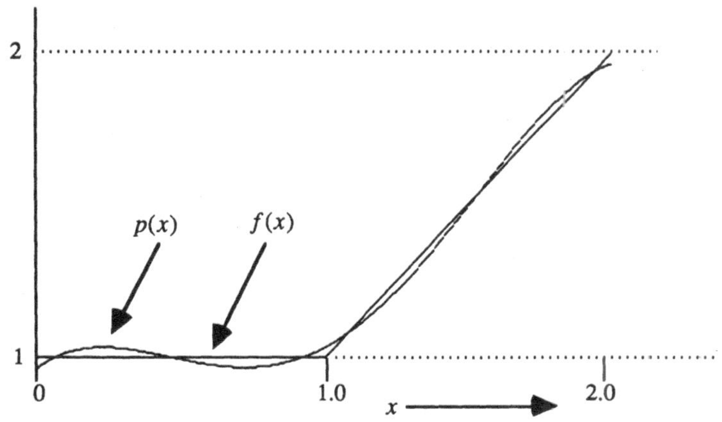

Figure 11.1 Polynomial approximation of a piecewise linear function using the exchange method. The maximum error alternates between ±0.03380291 at seven locations.

to express the FIR filter as a sum of cosine terms to use the alternation theorem as stated by Parks and McClellan. The four types of FIR filters were shown previously in Figures 10.3 and 10.4. As an example, an odd-order (N odd), symmetric FIR filter is usually written as

$$H(e^{j\omega}) = h(0) + h(1)z^{-1} + h(2)z^{-2} + \ldots + h(N-1)z^{-(N-1)} \tag{11.2}$$

The objective of the approximation is to choose the impulse response, $h(i)$, such that the resulting filter has an equiripple response. For the odd-order symmetric filter, $H(e^{j\omega})$ can be rewritten as the product of $z^{-(N-1)/2}$ and

$$[h(0)z^{(N-1)/2} + h(1)z^{(N-3)/2} + \ldots + h[(N-1)/2] + \ldots + h(N-2)z^{-(N-3)/2} + h(N-1)z^{-(N-1)/2}]$$

Note that the term $z^{-(N-1)/2}$ has been factored out of the previous expression and that it has unity magnitude at all frequencies. Because our objective is the approximation of the magnitude of $H(e^{j\omega})$, the phase term can be dropped, yielding

$$H'(e^{j\omega}) = h(0)z^{(N-1)/2} + h(1)z^{(N-3)/2} + \ldots + h[(N-1)/2] + \ldots + h(N-2)z^{-(N-3)/2} + h(N-1)z^{-(N-1)/2}$$

or, using equations (10.11) through (10.13),

$$H'(e^{j\omega}) = [h(0) + h(N-1)] \cos[\omega(N-1)/2] + [h(1) + h(N-2)] \cos[\omega(N-3)/2] + \ldots + h[(N-1)/2]$$

or

$$H'(e^{j\omega}) = \sum_{n=0}^{(N-1)/2} a(n)\cos(n\omega) \tag{11.3}$$

with

$$a(0) = h[(N-1)/2] \tag{11.4}$$

$$a(n) = h\left[\frac{N-1}{2} - n\right] + h\left[\frac{N-1}{2} + n\right], \qquad n = 1, \ldots, (N-1)/2 \tag{11.5}$$

Each of the four cases is summarized in Table 11.1. Each expression for $H'(e^{j\omega})$ in Table 11.1 must be rewritten as a product of a fixed function of ω, $Q(e^{j\omega})$, and an expansion

Table 11.1

$H'(e^{j\omega}) = h(0) + h(1)z^{-1} + \ldots + h(N-1)z^{-(N-1)}$ (from [1], pp. 125–126, adapted by permission of Prentice-Hall, Englewood Cliffs, New Jersey)

	$H'(e^{j\omega})$
Case 1: N odd, symmetric impulse response	$\sum_{n=0}^{(N-1)/2} a(n)\cos(n\omega)$
$a(0) = h\left(\frac{N-1}{2}\right)$	
$a(n) = h\left[\frac{N-1}{2} - n\right] + h\left[\frac{N-1}{2} + n\right], \quad n = 1, \ldots, (N-1)/2$	
Case 2: N even, symmetric impulse response	$\sum_{n=1}^{N/2} b(n)\cos\left[\omega\left(n - \frac{1}{2}\right)\right]$
$b(n) = h\left[\frac{N}{2} - n\right] + h\left[\frac{N}{2} + n - 1\right], \quad n = 1, 2, \ldots, N/2$	
Case 3: N odd, anti-symmetric impulse response	$\sum_{n=1}^{(N-1)/2} c(n)\sin(n\omega)$
$c(n) = h\left[\frac{N-1}{2} - n\right] - h\left[\frac{N-1}{2} + n\right], \quad n = 1, 2, \ldots, (N-1)/2$	
Case 4: N even, anti-symmetric impulse response	$\sum_{n=1}^{N/2} d(n)\sin\left[\omega\left(n - \frac{1}{2}\right)\right]$
$d(n) = h\left[\frac{N}{2} - n\right] - h\left[\frac{N}{2} + n - 1\right], \quad n = 1, 2, \ldots, N/2$	

of cosine terms, $P(e^{j\omega})$ before the alternation theorem can be applied. This is done in Table 11.2.

To complete the formulation of the equiripple approximation problem, it is necessary to define the desired frequency response, $D(e^{j\omega})$, and a weighting function on the error that can be used to specify the relative size of the error in different frequency bands. The approximation error, $E(e^{j\omega})$, is defined as

Table 11.2

$H'(e^{j\omega})$ as a Product of a Fixed Function of ω and an Expansion of Cosine Terms (from [1], pp. 125–126, adapted by permission of Prentice-Hall, Englewood Cliffs, New Jersey)

	$Q(e^{j\omega})$	$P(e^{j\omega})$
Case 1	1	$\sum_{n=0}^{(N-1)/2} \bar{a}(n)\cos(n\omega)$
Case 2	$\cos\left(\frac{\omega}{2}\right)$	$\sum_{n=0}^{(N/2)-1} \bar{b}(n)\cos(n\omega)$
Case 3	$\sin(\omega)$	$\sum_{n=0}^{(N-3)/2} \bar{c}(n)\cos(n\omega)$
Case 4	$\sin\left(\frac{\omega}{2}\right)$	$\sum_{n=0}^{(N/2)-1} \bar{d}(n)\cos(n\omega)$

Equating the coefficients to those in Table 11.1 yields the following:

Case 1: $\bar{a}(n) = a(n)$	Case 3: $\bar{c}\left(\frac{N-3}{2}\right) = 2c\left(\frac{N-1}{2}\right)$ $\bar{c}\left(\frac{N-5}{2}\right) = 2c\left(\frac{N-3}{2}\right)$ $\begin{cases}\bar{c}\left(\frac{N-3}{2} - n\right) = 2c\left(\frac{N-1}{2} - n\right) + \bar{c}\left(\frac{N-3}{2} - (n-2)\right)\\ n = 2, 3, \ldots, (N-5)/2\end{cases}$ $\bar{c}(0) = c(1) + 0.5\bar{c}(2)$
Case 2: $\bar{b}\left(\frac{N}{2} - 1\right) = 2b\left(\frac{N}{2}\right)$ $\begin{cases}\bar{b}\left(\frac{N}{2} - (n+1)\right) = 2b\left(\frac{N}{2} - n\right) - \bar{b}\left(\frac{N}{2} - n\right)\\ n = 1, 2, \ldots, (N/2) - 2\end{cases}$ $\bar{b}(0) = b(1) - 0.5\bar{b}(1)$	Case 4: $\bar{d}\left(\frac{N}{2} - 1\right) = 2d\left(\frac{N}{2}\right)$ $\begin{cases}\bar{d}\left(\frac{N}{2} - (n+1)\right) = 2d\left(\frac{N}{2} - n\right) + \bar{d}\left(\frac{N}{2} - n\right)\\ n = 1, 2, \ldots, (N/2) - 2\end{cases}$ $\bar{d}(0) = d(1) + 0.5\bar{d}(1)$

$$E(e^{j\omega}) = W(e^{j\omega})[D(e^{j\omega}) - H'(e^{j\omega})] \qquad (11.6)$$

Substituting $H'(e^{j\omega}) = P(e^{j\omega})Q(e^{j\omega})$ yields

$$E(e^{j\omega}) = W(e^{j\omega})[D(e^{j\omega}) - P(e^{j\omega})Q(e^{j\omega})] \qquad (11.7)$$

The term $Q(e^{j\omega})$ is a fixed function of frequency and normally factored out of the previous equation, giving

$$E(e^{j\omega}) = W(e^{j\omega})Q(e^{j\omega})\left[\frac{D(e^{j\omega})}{Q(e^{j\omega})} - P(e^{j\omega})\right] \qquad (11.8)$$

or

$$E(e^{j\omega}) = \overline{W}(e^{j\omega})[\overline{D}(e^{j\omega}) - P(e^{j\omega})] \qquad (11.9)$$

with $\overline{W}(e^{j\omega}) = W(e^{j\omega})Q(e^{j\omega})$ and $\overline{D}(e^{j\omega}) = D(e^{j\omega})/Q(e^{j\omega})$.

The well-known alternation theorem can now be applied to find the coefficients, $\overline{a}(n)$, $\overline{b}(n)$, $\overline{c}(n)$, or $\overline{d}(n)$, to minimize the absolute value of $E(e^{j\omega})$.

11.4 THE NUMBER OF EXTREMA IN A LINEAR PHASE FIR FILTER

Consider a symmetric FIR filter of length N, where N is odd. The z-domain expression can be written as a series of a nonzero constant term plus $(N - 1)/2$ cosine terms, where N is the total number of taps. It can be proven that the maximum number of extrema for $H'(e^{j\omega})$ in a case 1 filter is

$$\text{Case 1:} \quad N_e \leq \frac{(N + 1)}{2} \qquad (11.10a)$$

because extrema can exist at both 0 and π on the closed interval $0 \leq \omega \leq \pi$. For cases 2 to 4, the number of extrema, N_e, of $H'(e^{j\omega})$ is given as follows:

$$\text{Case 2:} \quad N_e \leq \frac{N}{2} \qquad (11.10b)$$

$$\text{Case 3:} \quad N_e \leq \frac{(N - 1)}{2} \qquad (11.10c)$$

$$\text{Case 4:} \quad N_e \leq \frac{N}{2} \qquad (11.10d)$$

Equations (11.10) show only the constraints for the extrema of $H'(e^{j\omega})$. The approximation problem is usually solved over a union of disjoint frequency bands, and the error function, $E(e^{j\omega})$, can add an extremum at each band edge where ω is not equal to 0 or π. A case 1 low-pass filter is a two-band approximation problem and can have a maximum of $(N + 1)/2 + 2$ extrema. A case 1 stopband filter is a three-band approximation problem and can have a maximum of $(N + 1)/2 + 4$ extrema.

Two well-known techniques design optimal mini-max FIR filters with only the maximum number of extrema. These techniques will not be discussed here because they allow the design of only a subset of optimal filters allowed by the alternation theorem. Instead the Remez exchange algorithm will be presented, which allows the design of the full set of optimal filters.

11.5 THE REMEZ EXCHANGE ALGORITHM

As shown earlier, the optimal linear phase FIR design problem can be formulated as an expansion of L independent cosine functions. An initial guess of the $(L + 1)$ extrema is made for which the error function is forced to have magnitude h with alternating signs. The following set of equations given by

$$\overline{W}(e^{j\omega k})[\overline{D}(e^{j\omega k}) - P(e^{j\omega k})] = (-1)^k h \qquad k = 0, 1, \ldots, L \tag{11.11}$$

with

$$P(e^{j\omega k}) = \sum_{n=0}^{L-1} \alpha(n) \cos(\omega n) \tag{11.12}$$

In matrix form, the set of simultaneous equations is

$$\begin{bmatrix} 1 & \cos(\omega_o) & \cos(2\omega_o) \ldots \cos[(L-1)\omega_o] & \dfrac{(-1)^0}{\overline{W}(e^{j\omega_o})} \\ 1 & \cos(\omega_1) & \ldots & \\ \cdot & & & \\ \cdot & & & \\ \cdot & & & \\ 1 & \cos(\omega_L) & \cos(2\omega_L) \ldots \cos[(L-1)\omega_L] & \dfrac{(-1)^L}{\overline{W}(e^{j\omega_L})} \end{bmatrix} \begin{bmatrix} \alpha(0) \\ \alpha(1) \\ \cdot \\ \cdot \\ \cdot \\ \alpha(L-1) \\ h \end{bmatrix} = \begin{bmatrix} \overline{D}(e^{j\omega_o}) \\ \overline{D}(e^{j\omega}) \\ \cdot \\ \cdot \\ \cdot \\ \overline{D}(e^{j\omega_L}) \end{bmatrix} \tag{11.13}$$

The solution can proceed in exactly the same manner as example 11.1; however, it is not necessary to solve equation (11.13) directly because a solution for h can be found more quickly using the analytical formula

$$h = \frac{a_0\overline{D}(e^{j\omega_o}) + a_1\overline{D}(e^{j\omega_1}) + \ldots + a_L\overline{D}(e^{j\omega_L})}{a_0/\overline{W}(e^{j\omega_o}) - a_1/\overline{W}(e^{j\omega_1}) + \ldots + (-1)^L a_L/\overline{W}(e^{j\omega_L})} \tag{11.14}$$

where

$$a_k = \prod_{\substack{i=0 \\ i\neq k}}^{L} \frac{1}{(x_k - x_i)} \tag{11.15}$$

$$x_i = \cos(\omega_i) \tag{11.16}$$

The Lagrange formula in the barycentric form is used to interpolate $P(e^{j\omega})$ on the L points $\omega_o, \omega_1, \ldots, \omega_L$ to the values

$$C_k = \overline{D}(e^{j\omega_k}) - (-1)^k \frac{h}{\overline{W}(e^{j\omega_k})} \qquad k = 0, 1, \ldots, L-1 \tag{11.17}$$

$$P(e^{j\omega}) = \frac{\sum_{k=0}^{L-1}\left(\frac{\beta_k}{x - x_k}\right)C_k}{\sum_{k=0}^{L-1}\left(\frac{\beta_k}{x - x_k}\right)} \tag{11.18}$$

where

$$\beta_k = \prod_{\substack{i=0 \\ i\neq k}}^{L-1} \frac{1}{(x_k - x_i)} \tag{11.19}$$

and $x = \cos(\omega)$.

$E(e^{j\omega})$ must be evaluated on the dense set of frequencies. If $|E(e^{j\omega})| > h$ for some frequencies in the dense set, then a new set of $(L + 1)$ frequencies must be chosen as candidates for the extremal frequencies. The new points are chosen as the peaks of the last error curve. The value of h increases with each iteration and ultimately converges to an upper bound, which corresponds to the solution to the problem. If more than $(L + 1)$ extrema occur in $E(e^{j\omega})$ in any iteration, then the $(L + 1)$ frequencies at which $|E(e^{j\omega})|$ is largest are retained as the new estimates for the extremal frequencies.

The last step is to obtain the filter impulse response by evaluating $P(e^{j\omega})$ at 2^M equally spaced frequencies (with $2^M \geq N$) and using the inverse discrete Fourier transform to obtain the coefficients $\alpha(n)$ [$\overline{a}(n)$, $\overline{b}(n)$, $\overline{c}(n)$, $\overline{d}(n)$]. Depending on which case linear phase filter is used, a recursive formula can be written for obtaining $h(n)$[$a(n)$, $b(n)$, $c(n)$, or $d(n)$].

Figure 11.2 shows the results of a typical Remez exchange design.

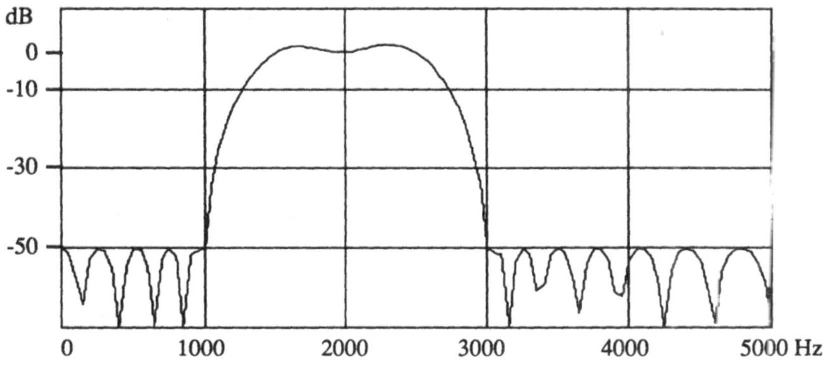

Figure 11.2 A 32-tap FIR equiripple filter. Bandpass regions are ±0.89326 dB from 1500 Hz to 2500 Hz, and 50.049 dB minimum attenuation for $f < 1$ KHz and $f > 3$ KHz.

Example 11.2

Formulate the Remez exchange problem for an eight-tap FIR equiripple low-pass filter with a ±1 dB passband ripple from 0 to 1000 Hz, and 20 dB attenuation from 2 KHz to 5 KHz, with a sampling frequency of 10 KHz.

Solution

The roll-off requirements cannot be met with a FIR filter of length 8; however, the exchange algorithm will nevertheless produce the closest equiripple response to a weighted error function. The normalized bandpass frequency range is $0 \leq \omega \leq \pi/5$, and the stopband range is $2\pi/5 \leq \omega \leq \pi$.

Step 1 is to determine the initial parameters. For an even-order symmetric FIR filter, the expression

$$E(e^{j\omega}) = W(e^{j\omega})\left[D(e^{j\omega}) - \sum_{n=1}^{N/2} b(n)\cos[\omega(n - 1/2)]\right]$$

becomes

$$E(e^{j\omega}) = W(e^{j\omega})\cos\left(\frac{\omega}{2}\right)\left[\frac{D(e^{j\omega})}{\cos\left(\frac{\omega}{2}\right)} - \sum_{n=0}^{(N/2)-1} \bar{b}(n)\cos(\omega n)\right]$$

with

$$\overline{W}(e^{j\omega}) = W(e^{j\omega}) \cos\left(\frac{\omega}{2}\right) = W(e^{j\omega})Q(e^{j\omega}) \quad \text{and} \quad \overline{D}(e^{j\omega}) = \frac{D(e^{j\omega})}{\cos\left(\frac{\omega}{2}\right)}$$

Step 2 is to identify $\overline{W}(e^{j\omega})$ and $\overline{D}(e^{j\omega})$. For simplicity, assume $W(e^{j\omega}) = 1$ for all ω. The desired response is $D(e^{j\omega}) = \{1, 1, 0, 0, 0\}$ for unity gain at two passband frequencies and zero gain at three stopband frequencies. Because a 5×5 matrix of equations must be solved, five equally spaced frequency samples will be assumed according to Figure 11.3. A ± 1 dB passband response corresponds to $\pm[1 - 10^{-1/20}]$, or extrema of 1 ± 0.108749 whereas a 20 dB stopband attenuation corresponds to extrema of ± 0.1. Five equal samples at $\{0, \pi/5, 2\pi/5, 3\pi/5, \text{and } 4\pi/5\}$ will be used for the initial iteration. Note that the gain at $\omega = \pi$ is necessarily zero so that this point cannot be used as a sample point in the exchange. Then $Q(e^{j\omega})$ becomes $\{1, 0.951056, 0.8090173, 0.5877859, 0.309018\}$ whereas $\overline{D}(e^{j\omega}) = \{1, 1.0514621, 0, 0, 0\}$ and $\overline{W}(e^{j\omega}) = W(e^{j\omega})Q(e^{j\omega}) = Q(e^{j\omega})$.

Step 3 is to write the system of equations to be solved. A system of five linear simultaneous equations can be written using

$$\left[\sum_{n=0}^{N/2-1} \overline{b}(n) \cos[\omega n]\right] + \frac{(-1)^n h}{\cos\left(\frac{\omega}{2}\right)} = \frac{D(e^{j\omega})}{\cos\left(\frac{\omega}{2}\right)}$$

at the specified frequency points. The system of equations is

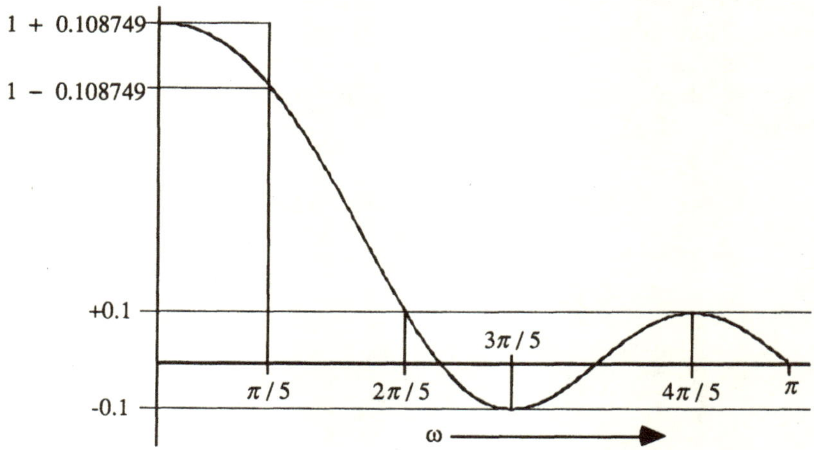

Figure 11.3 Initial placement of extrema for Remez exchange.

$$\begin{bmatrix} 1 & 1 & 1 & 1 & 1 \\ 1 & \cos(\pi/5) & \cos(2\pi/5) & \cos(3\pi/5) & -1/0.9510566 \\ 1 & \cos(2\pi/5) & \cos(4\pi/5) & \cos(6\pi/5) & 1/0.8090173 \\ 1 & \cos(3\pi/5) & \cos(6\pi/5) & \cos(9\pi/5) & -1/0.5877859 \\ 1 & \cos(4\pi/5) & \cos(8\pi/5) & \cos(12\pi/5) & 1/.309018 \end{bmatrix} \begin{bmatrix} \bar{b}_o \\ \bar{b}_1 \\ \bar{b}_2 \\ \bar{b}_3 \\ h \end{bmatrix} = \begin{bmatrix} 1.0 \\ 1.0514621 \\ 0 \\ 0 \\ 0 \end{bmatrix}$$

Step 4 is to perform the exchange. The first iteration yields $h = -0.1428805$ with extrema at $\omega = \{0, \pi/5, 2\pi/5, 1.771868, 2.670377\}$. The sample points at $\pi/5$ and $2\pi/5$ are never exchanged because these represent the desired edges of the passband and the stopband, respectively. The iterations are continued exactly as shown in example 11.1 until the value of the equiripple error, h, converges.

An empirical formula, taken from a program by Ulrich Heute [6], is given in program B7 (Appendix B) for both the length of an equiripple FIR filter to meet desired passband and stopband characteristics and for the required word length in bits.

REFERENCES

[1] Rabiner, Lawrence R., and Gold, Bernard, *Theory and Application of Digital Signal Processing*, Englewood Cliffs, NJ: Prentice-Hall, 1975, p. 140.

[2] Parks, T. W., and McClellan, J. H., "Chebyshev Approximation for Non-Recursive Digital Filters with Linear Phase," *IEEE Trans. Circuit Theory*, Vol. CT-19, March 1972, pp. 189–194.

[3] McGee, W. F., "Numerical Approximation Technique for Filter Characteristic Functions," *IEEE Trans. Circuit Theory*, Vol. CT-14, March 1967, pp. 92–94.

[4] Temes, G. C., and Calahan, D. A., "Computer-Aided Network Optimization—The State of the Art," *Proc. IEEE*, Vol. 53, November 1967, pp. 1832–1863.

[5] Temes, G. C., and Gyi, M., "Design of Filters with Arbitrary Passband and Chebyshev Stopband Attenuation," *IEEE Int. Conv. Rec.*, Vol. 15, 1967, pp. 184–189.

[6] Digital Signal Processing Committee, *Programs for Digital Signal Processing*, New York: John Wiley & Sons, 1979, pp. 5.4-1 to 5.4-20.

PROBLEMS

11.1 Complete the first iteration of example 11.2 by finding the coefficients $\{\bar{b}_0, \bar{b}_1, \bar{b}_2, \bar{b}_3\}$, $\{b_1, b_2, b_3, b_4\}$ and $\{h(0), h(1), h(2), h(3), h(4), h(5), h(6), h(7)\}$.

11.2 Show that the value of $h = -0.1428805$, obtained in the first iteration of example 11.2, can be found from equation (11.14).

11.3 Complete example 11.2 by iterating until h converges. What are the final values of $\{h(0), h(1), h(2), h(3), h(4), h(5), h(6), h(7)\}$?

11.4 Using program B7, find the number of taps necessary to realize an equiripple FIR lowpass filter with ±1 dB passband ripple from 0 to 1000 Hz, and 20 dB attenuation from 2 KHz to 5 KHz, with a sampling frequency of 10 KHz.

11.5 Use the Remez exchange to find the impulse response of

(a) a 32-tap bandpass FIR filter with less than ±1 dB passband ripple from 1500 Hz to 2500 Hz, and greater than 50 dB attenuation for $f < 1$ KHz and $f > 3$ KHz (use a sampling frequency of 10 KHz),

(b) a 16-tap low-pass FIR filter with ±1 dB passband ripple from 0 to 1000 Hz and greater than 40 dB attenuation for $f > 2$ KHz, with a sampling frequency of 10 KHz.

11.6 Explain how to use the Remez exchange to realize FIR equiripple filters with arbitrary passband shapes.

Chapter 12
Switched Capacitor Filters

12.1 SWITCHED CAPACITOR EQUIVALENT RESISTORS

The objective for this chapter is to develop z-domain filters by replacing the resistors of an RC active network with switched capacitor simulated resistors. The switched capacitor resistors are exactly equivalent to resistors by themselves; however, such an equivalence does not always hold true when these components are used to replace the resistors of a RC active filter. The theory of switched capacitors dates back to the 1960s, when several methods were proposed that used switches and capacitors to simulate resistors in active filters. It was shown that the filter transmittances had the important property of depending only on capacitor ratios. By 1972, MOS technology was determined to be applicable for the construction of switched capacitor filters in integrated circuits. The tight tolerance with which capacitor ratios can be implemented allows the performance of the IC filter to approach the ideal. It is not required that the MOS technology be able to accurately implement exact capacitor values or exact resistor values, but only that the ratio of the capacitor values be precisely controlled.

The charge stored by a capacitor is given by

$$Q = CV \tag{12.1}$$

where Q is charge in coulombs, C is capacitance in farads, and V is the voltage across the capacitor in volts. For the circuit in Figure 12.1(a), the two switches close on different phases of a master clock and are never closed at the same time. A two-phase switch configuration, as shown in Figure 12.2, is assumed throughout this chapter with the phases denoted as ϕ_1 and ϕ_2. The capacitor, C, in Figure 12.1(a) charges to V_{in} volts during clock phase 1 (ϕ_1). Without loss of generality it can be assumed that the capacitor discharges into a short circuit when the ϕ_2 switch is closed. The charge transferred during phase 2

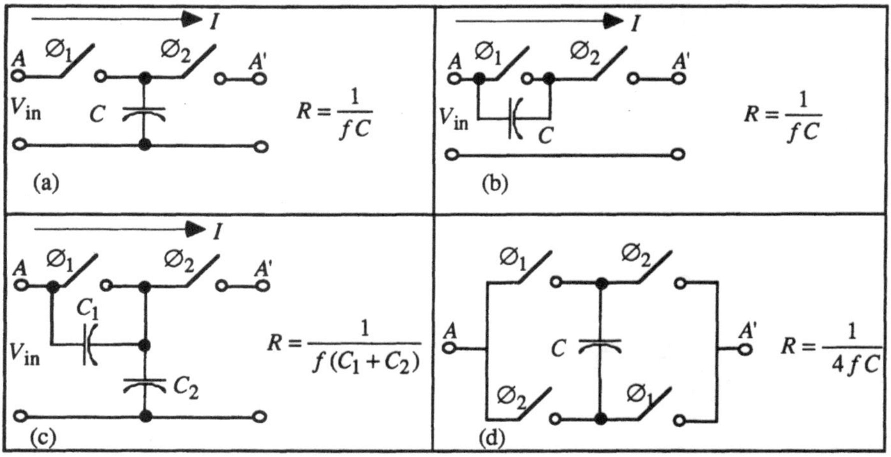

Figure 12.1 Resistor equivalents using switched capacitors.

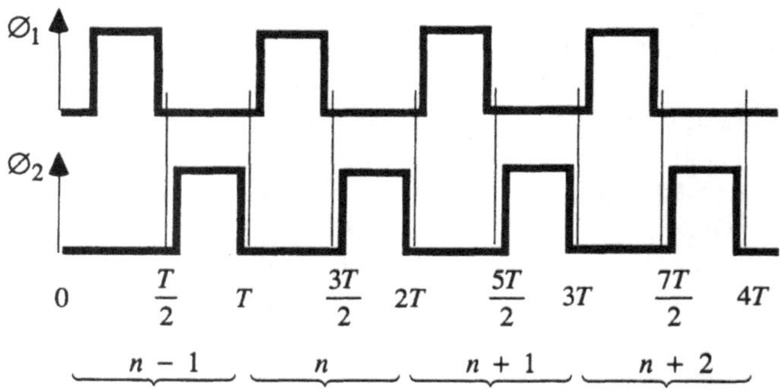

Figure 12.2 Clock waveforms for switched capacitor realizations.

is CV_{in} coulombs. If this cycle repeats with a frequency of $f = 1/T$ Hz, then the total charge transferred in one second is fCV_{in} coulombs/sec or fCV_{in} amps. The equivalent resistance is the total current flow in one second divided by the input voltage, or

$$R_{eq} = \frac{1}{fC} \tag{12.2}$$

The equivalent resistances for the other switch capacitor configurations can be derived in a similar manner and are given in Figure 12.1.

Each of the configurations in Figure 12.1 is sensitive to any parasitic capacitance from the switching node to ground. The effect of parasitic capacitance is described by

Figure 12.3. Parasitic capacitance always exists between every circuit node and ground. In Figure 12.3(b), the parasitic capacitors C_{p1} and C_{p3} have no effect on the circuit because C_{p1} is driven by a low-impedance voltage source, V_1, and C_{p3} connects virtual ground at the inverting node of the op-amp to actual ground; therefore, no current flows through C_{p3}. However the actual value of C_1 is changed by C_{p2}. In the configuration in Figure 12.3(c), C_{p2} and C_{p3} are either driven by a low-impedance voltage source or shorted to ground depending on the clock phase. Many parasitic insensitive topologies can be derived in which the equivalent resistance is not affected by any parasitic capacitance from any of the switching nodes to ground. Two such parasitic insensitive configurations are shown in Figure 12.4. The switched capacitor circuits in Figures 12.1 and 12.4 will measure exactly the equivalent resistance if an ohm meter is applied from A to A' in each of the circuits. The behavior of each switched capacitor substitution may behave differently in an actual circuit, as shown in Figure 12.5.

The switched capacitor circuit in Figure 12.5(b) tries to simulate the continuous summer of Figure 12.5(a) but has no feedback for both the ϕ_1 and ϕ_2 phase periods. Under this condition, no null port can exist. Therefore, the voltage at the inverting input

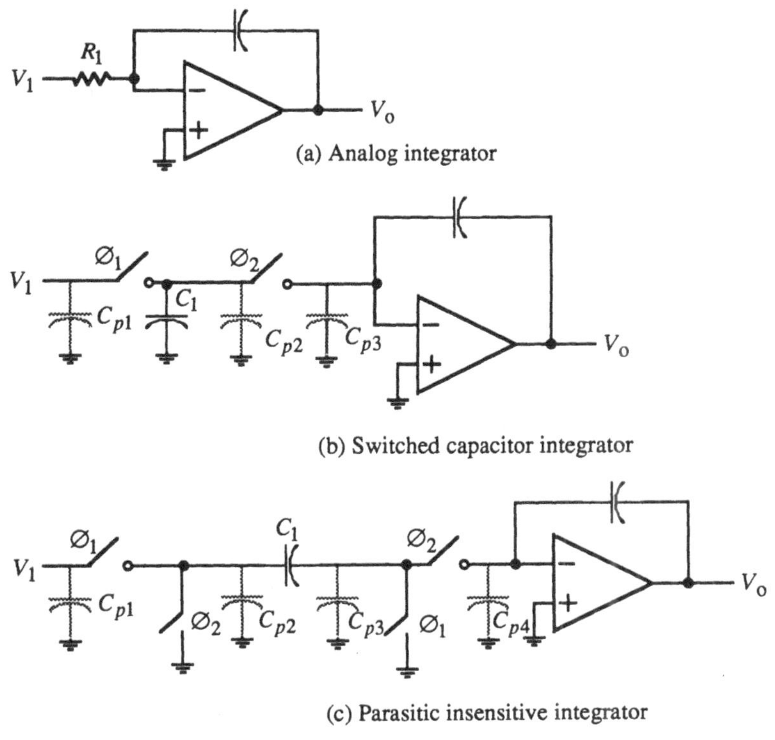

Figure 12.3 Illustration of parasitic capacitance in a switched capacitor circuit.

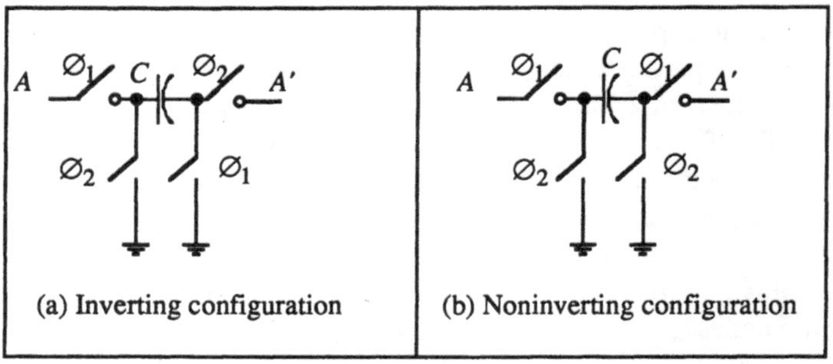

Figure 12.4 Parasitic insensitive topologies for implementing (a) a negative resistor and (b) a positive resistor.

is unlikely to have a nonzero value of voltage. Because this voltage is amplified by the open loop gain of the op-amp, the output voltage, V_o, will be at one of its limits.

Figure 12.5(c) solves the feedback problem by providing a stable feedback path for both of the clock phases. Neglecting any parasitic capacitance relative to C_1 and C_2 and the time delay due to sampling, the transfer function can be written as

$$V_o = -\frac{C_1}{C_o}V_1 - \frac{C_2}{C_o}V_2 \tag{12.3}$$

It should be noticed that the output of the op-amp is valid only for ϕ_2 in this example.

Figure 12.5(d) uses all parasitic insensitive switches with inverting voltage configurations. Again, neglecting the time delay due to sampling, the output voltage is

$$V_o = +\frac{C_1}{C_o}V_1 + \frac{C_2}{C_o}V_2 \tag{12.4}$$

The analysis of any switched capacitor circuit can be done using a charge conservation approach. We begin by using Figure 12.6 and assuming that the entire system is inactive. That is, both switches are open and the capacitors are completely discharged. The first clock pulse, ϕ_1, which will occur during the first phase period (or odd phase), will close the switches labeled ϕ_1. At this time, the capacitor in Figure 12.6(a) will be charged to V_{in} volts. In practice, a finite resistance, R, is associated with the switch. This resistance is connected in series with the capacitor and prevents instantaneous charging of the capacitor. Obviously, the RC time constant must be much less than the width of ϕ_1 for the charge to be transferred.

For Figure 12.6(a), all capacitors are assumed to be discharged. The charge conservation analysis will begin with Figure 12.6(b) and continue until the circuit has reached a steady state. Because the inverting input of the op-amp has a very high input impedance, the entire charging current I_1 will flow through both C_1 and C_2. The inverting input of

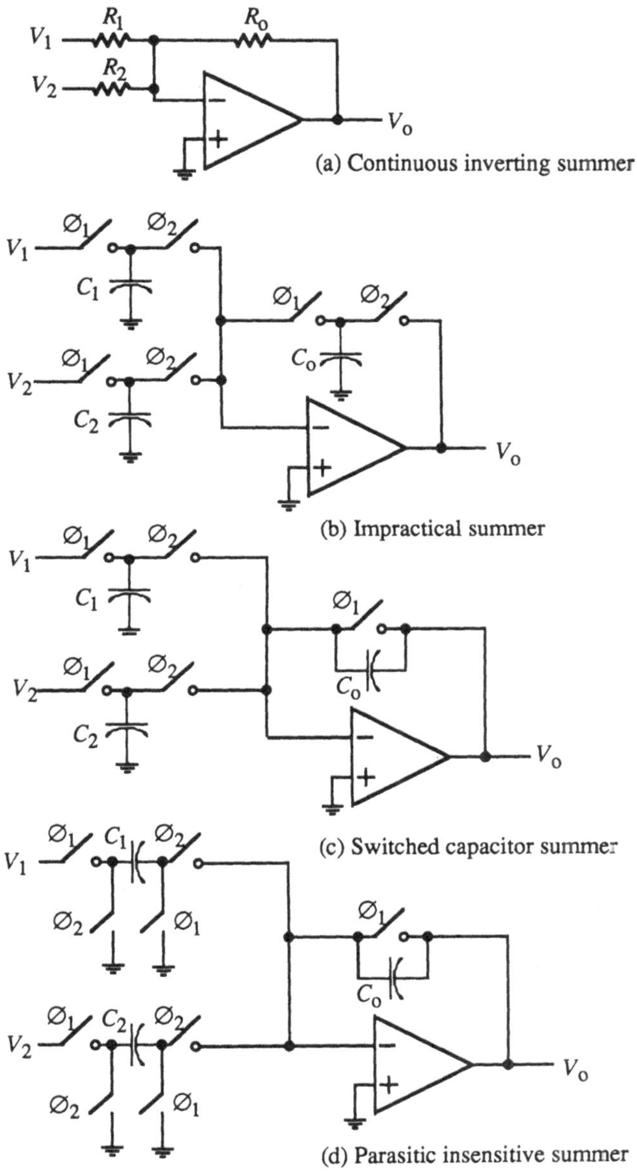

(a) Continuous inverting summer

(b) Impractical summer

(c) Switched capacitor summer

(d) Parasitic insensitive summer

Figure 12.5 Switched capacitor summers.

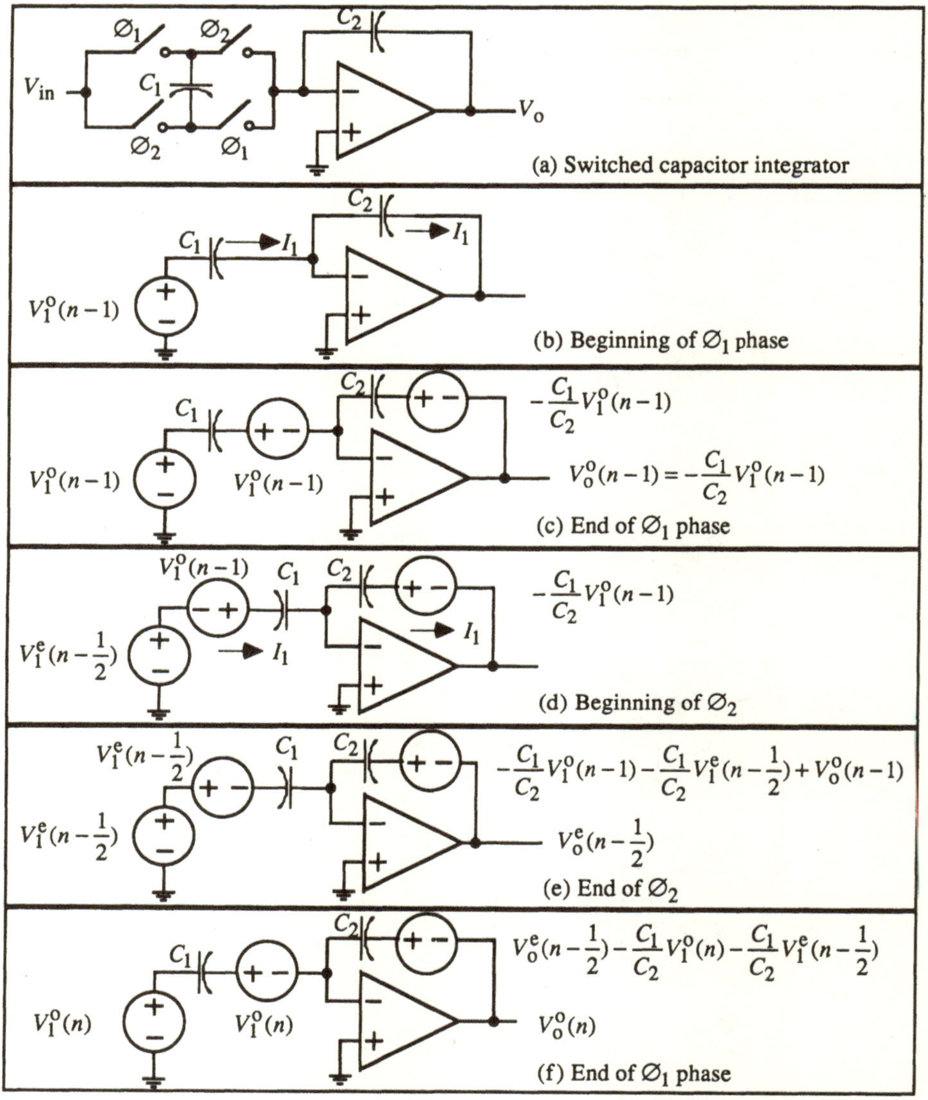

Figure 12.6 Analysis of a bilinear switched capacitor integrator.

the op-amp is virtual ground, so capacitor C_1 eventually charges to $V_1^o(n-1)$ volts. The notation $V_1^o(n-1)$ denotes the input voltage during the odd clock phase (ϕ_1) during the $(n-1)$ clock interval relative to Figure 12.2. The charge accumulated on C_1 is $Q_1 = C_1 V_1^o(n-1)$. Because I_1 flow charges both C_1 and C_2, the capacitor C_2 must also accumulate the same charge. The charge on C_2 becomes $Q_2 = Q_1 = C_2 V_{C2}^{o}(n-1)$. The voltage across C_2 is the output voltage of the op-amp in Figure 12.6(b), written as

$$V_o^o(n-1) = \frac{-C_1}{C_2} V_1^o(n-1) \tag{12.5}$$

In Figure 12.6(c), the final accumulated charges for capacitors C_1 and C_2 are shown by representing each charged capacitor as an uncharged capacitor in series with a voltage source that represents the voltage across each capacitor. No more current flows through either C_1 or C_2. The input voltage, $V_1^o(n-1)$, is assumed to be constant during the entire ϕ_1 interval.

In Figure 12.6(d), the ϕ_1 switches have opened and the ϕ_2 switches have closed. Capacitor C_1 charges to $V_1^e(n-1/2)$ volts and the accumulated voltage across C_2 becomes

$$V_o^e\left(n-\frac{1}{2}\right) = -\frac{C_1}{C_2} V_1^o(n-1) - \frac{C_1}{C_2} V_1^e\left(n-\frac{1}{2}\right) + V_o^o(n-1) \tag{12.6}$$

The analysis for the bilinear switched capacitor realization using the substitution shown in Figure 12.1(d) requires that the even-phase input voltage be a different sample than the odd-phase input voltage. Thus the input voltage is sampled twice during each clock cycle. The final capacitor voltages at the end of the ϕ_2 period are shown in Figure 12.6(e).

Figure 12.6(f) shows the output during the end of the next ϕ_1 period. The output voltage is

$$V_o^o(n) = V_o^e\left(n-\frac{1}{2}\right) - \frac{C_1}{C_2} V_1^o(n) - \frac{C_1}{C_2} V_1^e\left(n-\frac{1}{2}\right) \tag{12.7}$$

The circuit reached steady state in Figure 12.6(d) so either equation (12.6) or (12.7) can be used to derive the voltage transfer function. Because the even and odd phases differ in time by a delay of half a clock cycle, this delay may be represented using z transforms by $z^{-1/2}$. From equation (12.6), the voltage transfer function is

$$\frac{V_o}{V_1} = \frac{-C_1}{C_2} \frac{1 + z^{-1/2}}{1 - z^{-1/2}} \tag{12.8}$$

It should be emphasized that all switches change position twice every T seconds, thereby doubling the effective sample rate. Therefore (12.8) can be rewritten as

$$\frac{V_o}{V_1} = \frac{-C_1}{C_2} \frac{1 + z'^{-1}}{1 - z'^{-1}} \tag{12.9}$$

if a new period is defined with $T/2 = T'$. In general, the bilinear switched capacitor substitution implements the bilinear transform regardless of circuit topology.

The circuit in Figure 12.6(a) is a switched capacitor representation of the RC integrator in Figure 12.7. If the equivalent resistance given in Figure 12.1(d) is substituted for R_1 into the voltage transfer function given in Figure 12.7, then the voltage transfer function of the integrator becomes

$$\frac{V_o}{V_1} = \frac{-2C_1}{sT'C_2} \tag{12.10}$$

Comparing (12.10) with (12.9), it can be seen that

$$s = \frac{2}{T'}\frac{1 - z'^{-1}}{1 + z'^{-1}} \tag{12.11}$$

which is the bilinear transform of Chapter 9 with $z'^{-1} = e^{-j\omega T'}$.

The design of the active switched capacitor filter based on an element-by-element resistor replacement in analog RC filters via the bilinear transform can be described [1] generically as follows:

1. The desired passband and stopband frequencies, ω_d, of the switched capacitor filter are prewarped to obtain the corresponding ω_a values of the active RC prototype filter, using the relation

$$\omega_a = \frac{2}{T'}\tan\left(\frac{\omega_d T'}{2}\right) \tag{12.12}$$

2. The active RC prototype is designed from the prewarped frequencies of ω_a.
3. Each resistor, R_m, in the active RC circuit is replaced by a switched capacitor simulation from Figure 12.1(d) of value

$$C_{rm} = \frac{T'}{2R_m} \tag{12.13}$$

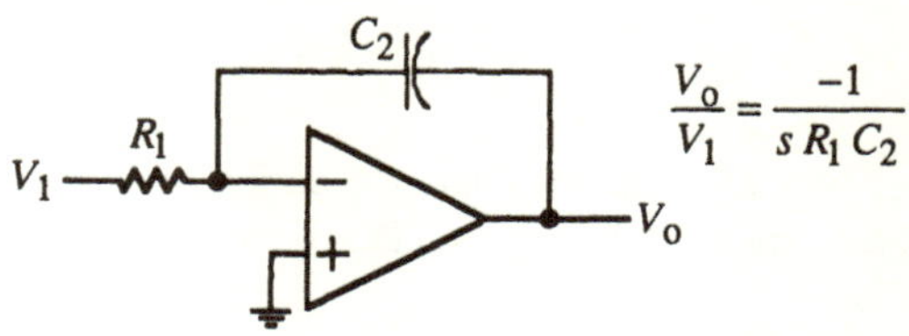

Figure 12.7 RC integrator.

Because all switches change position twice every T seconds, $T' = T/2$, equation (12.13) yields

$$C_{rm} = \frac{T}{4R_m} \tag{12.14}$$

4. A sample-and-hold circuit, operating at twice the sampling rate ($1/T$), is added at the input. A simple smoothing low-pass filter is usually connected at the output of the low-pass filter.

Example 12.1

Given the low-pass prototype shown in Figure 12.8, design the equivalent bilinear switched capacitor version. The desired center frequency, ω_o, and the quality factor, Q, are 2π krad/sec and 0.707106, respectively. Consider two cases of sampling frequency: (a) $T = 0.00002$ sec and (b) $T = 0.000125$ sec.

Solution

The transfer function of the prototype is given by

$$G(s) = \frac{\dfrac{-1}{C_1C_2R_1R_3}}{s^2 + s\left(\dfrac{1}{C_1}\right)\left(\dfrac{1}{R_1} + \dfrac{1}{R_2} + \dfrac{1}{R_3}\right) + \dfrac{1}{C_1C_2R_2R_3}} \tag{12.15}$$

The design equations are given in Figure 8.5. For part (a),

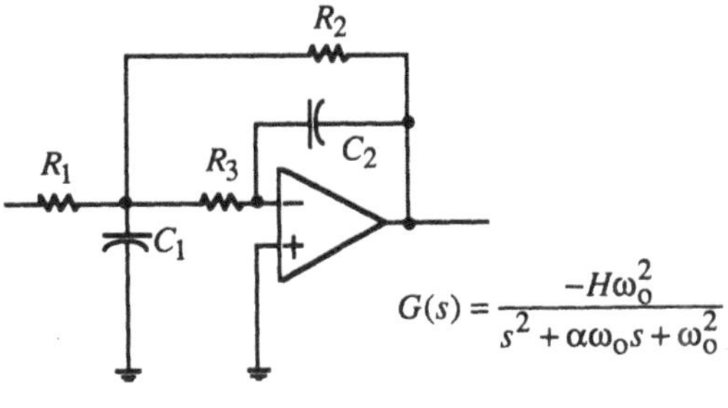

Figure 12.8 Second-order multiple-feedback active filter.

$$T' = T/2 = 0.00001 \text{ sec}$$

$$H = 1, \qquad \alpha = 1/Q = 1.414, \qquad \omega_d = 2\pi(1000) \text{ rad/sec}$$

$$\omega_a = \frac{2}{T'}\tan\left(\frac{\omega_d T'}{2}\right) = 2\pi(1000.3291) \text{ rad/sec.}$$

Thus the value of ω_o used in the design of the analog prototype is $2\pi(1000.3291)$. Choosing $C_1 = 100$ pF, the value of C_2 can be determined from

$$C_1 \geq \frac{4(H+1)C_2}{\alpha^2} = 4C_2 \tag{12.16}$$

If C_2 is arbitrarily set to 25 pF, then (12.16) is satisfied. The value of R_1 will be determined using (8.21) with a minus sign before the radical because this gives a lower value resistance than the equation in Figure 8.5. Therefore

$$R_1 = \frac{\alpha\omega_o C_1 H \pm \sqrt{(\alpha\omega_o C_1 H)^2 - 4(H+1)H^2\omega_o^2 C_1 C_2}}{2H^2\omega_o^2 C_1 C_2} = 4.498 \text{ M}\Omega \tag{12.17}$$

The other resistors become $R_2 = HR_1$ and $R_3 = 1/(\omega_o^2 C_1 C_2 R_2) = 2.251$ MΩ.

Using equation (12.14), the remaining capacitor values are $C_{r1} = C_{r2} = 1.11$ pF, and $C_{r3} = 2.22$ pF. The capacitor values can be summarized as

$$C_{r1} = 1.11 \text{ pF} \rightarrow 1.0\ Cu$$
$$C_{r2} = 1.11 \text{ pF} \rightarrow 1.0\ Cu$$
$$C_{r3} = 2.22 \text{ pF} \rightarrow 2.0\ Cu$$
$$C_1 = 100 \text{ pF} \rightarrow 90.09\ Cu$$
$$C_2 = 25 \text{ pF} \rightarrow 22.52\ Cu$$

where *Cu* represents the smallest unit capacitance that can be implemented in an integrated circuit. The total capacitance is about 116.6 *Cu*, and if *Cu* = 1 pF, then the total capacitance is 116.6 pF. The value of *Cu* can be set to the minimum that can be manufactured with precision on a silicon die because the scaling of the capacitors by an arbitrary constant does not change the transfer function of a switched capacitor filter. The most important consideration for the implementation of switched capacitor filters in integrated circuits is the minimization of the total capacitance required so that the die size is kept as small as possible. Typically capacitor ratios of less than 10 are very desirable.

If equations (12.11) and (12.13) are substituted into (12.15) then the transfer function, $G(z)$, of the switched capacitor filter can be written as

$$\frac{\frac{-C_{r1}C_{r3}}{C_1C_2}(1 + z'^{-1})^2}{\left(1 + \frac{C_{r1} + C_{r2} + C_{r3}}{C_1} + \frac{C_{r2}C_{r3}}{C_1C_2}\right)z'^{-2} + \left(\frac{2C_{r2}C_{r3}}{C_1C_2} - 2\right)z'^{-1} + \left(1 - \frac{C_{r1} + C_{r2} + C_{r3}}{C_1} + \frac{C_{r2}C_{r3}}{C_1C_2}\right)}$$

with $z'^{-1} = e^{-j\omega T'}$.

Now for (b) where $f = 1/T = 8000$ Hz, $\omega_o = 2\pi(1013.052)$ rad/sec. Therefore $R_1 = R_2 = 4.44$ MΩ, $C_{r1} = C_{r2} = 7.035$ pF, $R_3 = 2.22$ MΩ, and the corresponding $C_{r3} = 14.06$ pF. Finally, C_1 and C_2 are as before. Thus,

$$\begin{aligned} C_{r1} &= 7.035 \text{ pF} \rightarrow 1.0\ Cu \\ C_{r2} &= 7.035 \text{ pF} \rightarrow 1.0\ Cu \\ C_{r3} &= 14.06 \text{ pF} \rightarrow 2.0\ Cu \\ C_1 &= 100 \text{ pF} \rightarrow 14.21\ Cu \\ C_2 &= 25 \text{ pF} \rightarrow 3.55\ Cu \end{aligned}$$

The total capacitance is about 22 *Cu*.

Comparing both cases, the largest capacitor ratios are nearly 90 and 14 for cases (a) and (b), respectively. Case (b) needs less area (about 1/6) to be implemented, although the antialising filter associated with it involves more stringent requirements than the one associated with case (a). The switched realization is shown in Figure 12.9.

12.2 FIRST-ORDER PARASITIC INSENSITIVE SC FILTERS

The bilinear substitution shown in Figure 12.9 has the advantage that it can be applied to any RC active filter regardless of topology, but it is sensitive to parasitic capacitance from the switching nodes to ground. In Figure 12.10, two parasitic insensitive configurations are shown for first-order topologies. The bilinear substitution of Chapter 9 is applied to a prewarped s-domain filter to generate a z-domain transfer function. The capacitor values are found simply by matching coefficients between the z-domain transfer function and the design equations in Figure 12.10.

The transfer functions for Figure 12.10 can be found by either using a charge conservation analysis or by inspection; however, it will be necessary to assume that the input voltage changes only during the odd phase (ϕ_1). The even-phase output voltage is given by

$$V_2^e\left(n + \frac{1}{2}\right) = \left[V_2^e\left(n - \frac{1}{2}\right) - \frac{C_3}{C_4}V_1^e\left(n + \frac{1}{2}\right) + \frac{C_3}{C_4}V_1^e\left(n - \frac{1}{2}\right) + \frac{C_1}{C_4}V_1^e\left(n + \frac{1}{2}\right)\right]\frac{C_4}{C_2 + C_4} \tag{12.18}$$

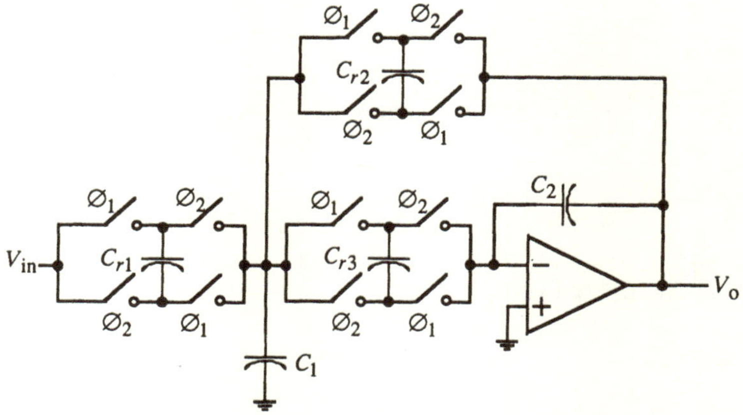

Figure 12.9 An active switched capacitor low-pass filter.

Because the input voltage changes only on the odd phase, $V_2^e(n - 1/2)$ is equal to $V_2^o(n - 1)$, allowing equation (12.18) to be written solely in terms of the even-phase input and output voltages. The first term on the right-hand side of the equation, $V_2^e(n - 1/2)$, is the contribution of the previous even phase output voltage to the present output. The middle terms, $-(C_3/C_4)V_1^e(n + 1/2) + (C_3/C_4)V_1^e(n - 1/2)$, are due to the change in charge on C_3 resulting from the difference between the present and the previous even-phase input voltages; and the final term, $(C_1/C_4)V_1^e(n + 1/2)$, is the contribution of charge from C_1 due to the present even-phase input voltage. The scale factor, $C_4/(C_2 + C_4)$, is necessary because C_2 was discharged during the odd phase and now shunts C_4 during the even phase, causing the output voltage to be reduced by the ratio $C_4/(C_2 + C_4)$. Rewriting equation (12.18) yields

$$V_2^e\left(n + \frac{1}{2}\right) = \frac{C_4}{C_2 + C_4} V_2^e\left(n - \frac{1}{2}\right) - \frac{C_3 - C_1}{C_2 + C_4} V_1^e\left(n + \frac{1}{2}\right) + \frac{C_3}{C_2 + C_4} V_1^e\left(n - \frac{1}{2}\right)$$

or

$$V_2 = V_2 \frac{C_4}{C_2 + C_4} z^{-1} - V_1 \frac{C_3 - C_1}{C_2 + C_4} + V_1 \frac{C_3}{C_2 + C_4} z^{-1} \tag{12.19}$$

which rearranges to

$$\frac{V_2}{V_1} = -\frac{(C_3 - C_1) - C_3 z^{-1}}{(C_2 + C_4) - C_4 z^{-1}} \tag{12.20}$$

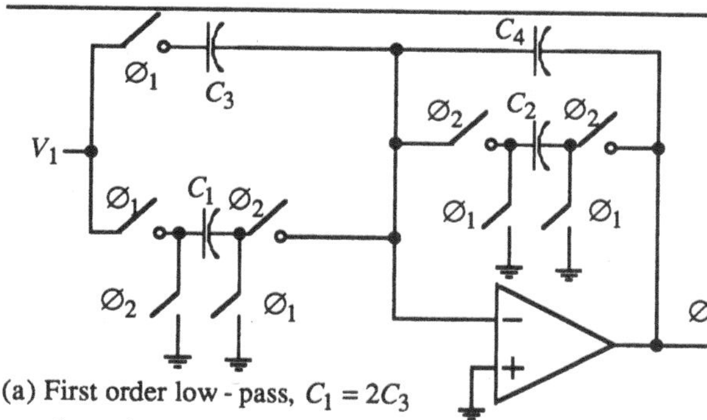

$$\frac{V_2}{V_1} = -\frac{(C_3 - C_1) - C_3 z^{-1}}{(C_2 + C_4) - C_4 z^{-1}}$$

(a) First order low - pass, $C_1 = 2C_3$

$$V_2^e\left(n+\frac{1}{2}\right) = \frac{C_4}{C_2 + C_4} V_2^e\left(n-\frac{1}{2}\right) - \frac{C_3 - C_1}{C_2 + C_4} V_1^e\left(n+\frac{1}{2}\right) + \frac{C_3}{C_2 + C_4} V_1^e\left(n-\frac{1}{2}\right)$$

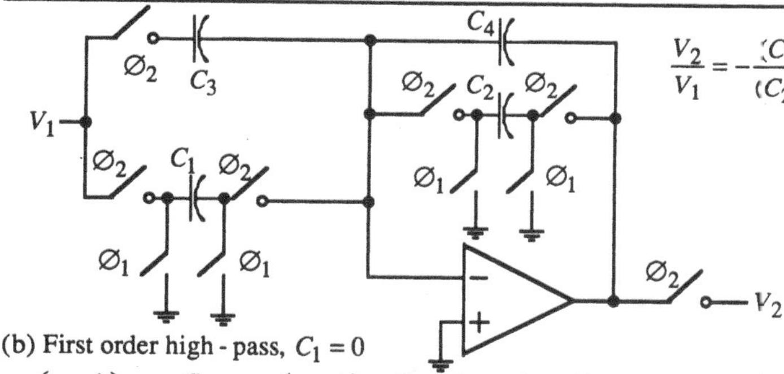

$$\frac{V_2}{V_1} = -\frac{(C_1 + C_3) - C_3 z^{-1}}{(C_2 + C_4) - C_4 z^{-1}}$$

(b) First order high - pass, $C_1 = 0$

$$V_2^e\left(n+\frac{1}{2}\right) = \frac{C_4}{C_2 + C_4} V_2^e\left(n-\frac{1}{2}\right) - \frac{C_1 + C_3}{C_2 + C_4} V_1^e\left(n+\frac{1}{2}\right) + \frac{C_3}{C_2 + C_4} V_1^e\left(n-\frac{1}{2}\right)$$

Figure 12.10 SC parasitic insensitive first-order filters and design equations.

12.3 EXACT DESIGN OF SECOND-ORDER SWITCHED CAPACITOR STAGES

An exact design method for second-order stages will be based upon the use of the bilinear transformation. The bilinear transformation has the advantage of mapping the entire $j\omega$ axis in the continuous-time domain onto the entire unit circle in the discrete-time domain. After the s-domain filter has been prewarped by the bilinear z-transform, then one can either replace the integrators of the prewarped s-domain by bilinear integrators or transform the prewarped filter specifications to the z-domain and solve for the parameters of the z-domain through a set of nonlinear equations.

The approach now to be described uses the bilinear z-transform and only the stray-insensitive configurations of the integrators. Figure 12.11 shows a second-order biquad circuit capable of realizing the bilinear z-transform of any second-order transfer function [2]. It assumes that the output is sampled at the end of clock phase ϕ_2 and that the input

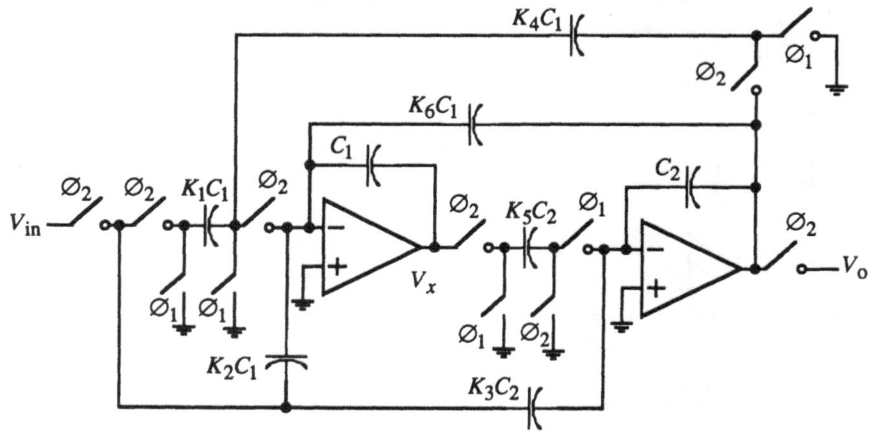

Figure 12.11 Martin and Sedra biquad for realizing any second-order transfer function.

voltage is allowed to change only during the ϕ_2 interval. The z-transform transfer function can be found by writing the difference equations for the charge transfer.

For Figure 12.11, the charge transfer equations can be written down by inspection as, for $\phi_1(n)$,

$$V_x^o(n) = V_x^e(n - 1/2) - K_5K_6V_x^e(n - 1/2) \tag{12.21}$$

$$V_o^o(n) = V_o^e(n - 1/2) + K_5V_x^e(n - 1/2) \tag{12.22}$$

for $\phi_2(n + 1/2)$,

$$V_x^e(n + 1/2) = V_x^o(n) - K_1V_{\text{in}}^e(n + 1/2) - K_2V_{\text{in}}^e(n + 1/2) + K_2V_{\text{in}}^e(n - 1/2) - K_6V_o^e(n + 1/2) + K_6V_o^o(n) - K_4V_o^e(n + 1/2) \tag{12.23}$$

$$V_o^e(n + 1/2) = V_o^o(n) - K_3V_{\text{in}}^e(n + 1/2) + K_3V_{\text{in}}^e(n - 1/2) \tag{12.24}$$

Using (12.22) and (12.24) to solve for V_o^e,

$$V_o^e = z^{-1}V_o^e - K_3V_{\text{in}}^e + z^{-1}K_3V_{\text{in}}^e + z^{-1}K_5V_x^e \tag{12.25}$$

Equations (12.21) and (12.23) can be used to solve for V_x in terms of V_o and V_{in}, yielding

$$V_x^e = \frac{V_o(-K_6 + z^{-1}K_6 - K_4) + V_{\text{in}}(-K_1 - K_2 + K_2z^{-1})}{1 - z^{-1}} \tag{12.26}$$

Substituting equation (12.26) into (12.25) provides the final voltage transfer function for the Martin and Sedra biquad in Figure 12.11.

$$\frac{V_o}{V_{in}} = \frac{-K_3 + z^{-1}(2K_3 - K_1K_5 - K_2K_5) + z^{-2}(-K_3 + K_2K_5)}{1 + z^{-1}(-2 + K_5K_6 + K_4K_5) + z^{-2}(1 - K_5K_6)} \quad (12.27)$$

The one available degree of freedom in (12.27) may be used to maximize the dynamic range of the biquad. A near-optimum solution is obtained by choosing equal time constants for the two integrators such that

$$K_4 = K_5 \quad (12.28)$$

Example 12.2

Design a SC low-pass filter for minimum total capacitance with a cutoff frequency of 5000 Hz, sampling frequency of 100,000 Hz, 3 dB maximum passband attenuation, and 50 dB minimum stopband attenuation for frequencies greater than 10,000 Hz. Maximize the dynamic range at the output of each op-amp. Use $Cu = 1$ pF.

Solution

From the design requirements, the order and total capacitance needed to meet the specifications are as follows:

Filter Type	*Order Required*	*Total Capacitance*
Butterworth	9	300.33 pF
Tchebyscheff I	5	256.96 pF
Tchebyscheff II	5	45.62 pF
Elliptic	4	108.96 pF

The cascaded quadratic sections for the Tchebyscheff II filter are developed from bilinear substitution of the prewarped s-domain functions, given as

Stage 1	Stage 2	Stage 3
$\dfrac{-0.2261z^{-1} - 0.2261}{z^{-1} - 1.4523}$	$\dfrac{0.4137z^{-2} - 0.6688z^{-1} + 0.4137}{z^{-2} - 2.5369z^{-1} + 1.6954}$	$\dfrac{0.1262z^{-2} - 0.1428z^{-1} + 0.1262}{z^{-2} - 2.0741z^{-1} + 1.1837}$

The parasitic insensitive implementation is shown in Figure 12.12.

12.4 EFFECTS OF FINITE OP-AMP GAIN AND BANDWIDTH IN SC FILTERS

If the SC integrator in Figure 12.13 is modeled with a finite gain and finite bandwidth with a single dominate pole at frequency ω_1, then the op-amp open loop gain, $A(s)$, can be written as

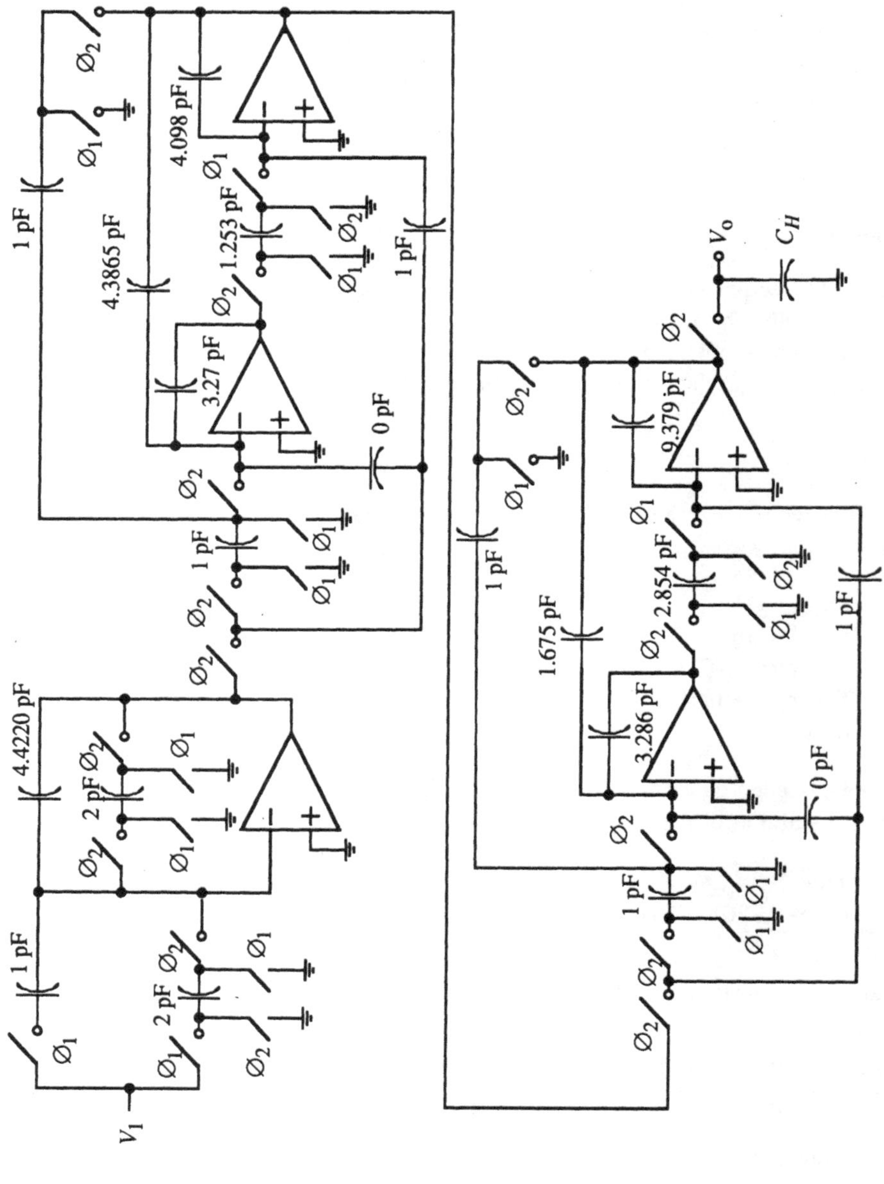

Figure 12.12 SC fifth-order low-pass filter.

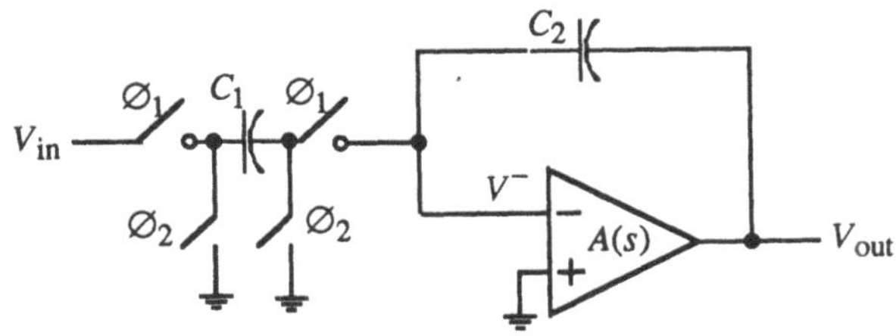

Figure 12.13 SC integrator with finite gain and bandwidth op-amp.

$$A(s) = \frac{A_o}{1 + \dfrac{s}{\omega_1}} = \frac{A_o \omega_1}{s + \omega_1} \tag{12.29}$$

where $A_o\omega_1$ is the gain-bandwidth product. Using Figure 12.13 to write an expression for V_{out} within a single ϕ_1 cycle yields

$$V_{out} = -AV^- \tag{12.30}$$

$$(V_{in} - V^-)sC_1 = I_{in} = (V^- - V_{out})sC_2 \tag{12.31}$$

or

$$\frac{V_{out}}{V_{in}} = \frac{-A(s)C_1}{C_1 + C_2[1 + A(s)]} = \frac{-C_1}{C_1 + C_2} \frac{A(s)}{1 + \left(\dfrac{C_2}{C_1 + C_2}\right)A(s)} = -(1 - \beta)\frac{A(s)}{1 + \beta A(s)} \tag{12.32}$$

If A' and ω' are defined such that

$$A' = \frac{(1 - \beta)A_o}{1 + \beta A_o} \quad \text{and} \quad \omega' = \omega_1(1 + \beta A_o)$$

then the transfer function can be written as

$$\frac{V_{out}}{V_{in}} = \frac{-A'}{1 + \dfrac{s}{\omega'}} \tag{12.33}$$

Applying a step voltage to the input and taking the inverse Laplace transform of equation (12.33) yields

$$V_{out} = \left(\frac{V_{in}}{s}\right)\left(\frac{-A'}{1 + \frac{s}{\omega'}}\right) \tag{12.34}$$

or

$$V_{out}(t) = -V_{in}A'(1 - e^{-\omega' t})u(t) \tag{12.35}$$

From Figure 12.12, the duty cycle, K_d, for the closure of the ϕ_1 and ϕ_2 must be less than half of the sampling period, T; therefore, $K_d \leq 1/2$. The maximum time for the voltages to settle to a steady state condition is given by $T = K_d/f_c$, where f_c is the sampling frequency. The error in the output voltage is given by the exponential term in (12.35) or

$$\text{Error} = e^{-\omega' T} \tag{12.36}$$

If the gain-bandwidth product is assumed to be constant, then at the clock frequency, f_c, where f_c is greater than ω_1, the op-amp open loop gain behaves as shown in Figure 12.14 and can be expressed as

$$A_c = \frac{A_o \omega_1}{2\pi f_c} \tag{12.37}$$

Substituting (12.37) into (12.36) with $\beta A_o \gg 1$ yields

$$\text{Error} = e^{-\omega' K_d \frac{A_c 2\pi}{A_o \omega_1}} = e^{-(1+\beta A_o)\omega_1 K_d \frac{A_c 2\pi}{A_o \omega_1}} \cong e^{-\beta K_d 2\pi A_c} \tag{12.38}$$

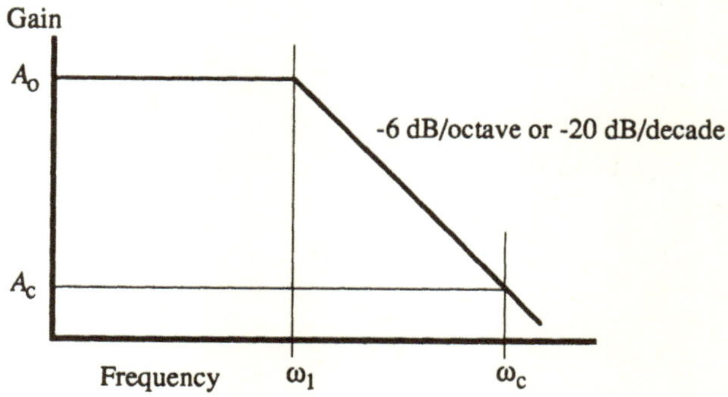

Figure 12.14 Open loop gain of op-amp with the dominant pole at ω_1.

Example 12.3

Given $A_o = 500$ and $\omega_1 = 2\pi(1000)$ with $f_c = 100{,}000$ Hz and $K_d = 0.5$, calculate the ratio of C_1 to C_2 in Figure 12.13 such that the settling error is less than 1%.

Solution

Substituting these values into equation (12.38), the error term is Error = $e^{-\beta(0.5)(5)(2\pi)} \le 0.01$. Solving for β yields $\beta \ge 0.29$. Because $\beta = C_2/(C_1 + C_2)$, therefore $C_2/C_1 \ge 0.415$.

12.5 NONIDEAL CONSIDERATIONS OF SC SWITCHES

The influence of MOS switches shall now be considered. The off-resistance of the MOSFET switches is usually high enough so that they can be considered to be open circuits. Here R_{on} depends on the actual dimensions of the switch. Commercial CMOS analog switches have values for R_{on} between 40 and 70 Ω. A reasonable value of an on-resistance for a MOSFET in an integrated circuit is on the order of 2 to 10 kΩ. The analysis of the effects of the on-resistance, R_{on}, of the switches is directly involved with those of the op-amp. An op-amp with finite gain and limited bandwidth is characterized by (12.29). To evaluate the effect of the op-amp output resistance, R_o, the following pessimistic circuit with values

$$A_o = 500 \qquad \omega_1 = 2\pi(1000) \qquad R_{on} = 4{,}000\ \Omega \qquad R_o = 3{,}000\ \Omega$$

will be considered. The op-amp will be charging one or several capacitors. Using the equivalent circuit in Figure 12.15, the node voltages are

$$V_{in} = \left[\frac{s(R_o + R_{on})\alpha C + 1}{(s\alpha CR_{on} + 1)(1 - A) + s\alpha CR_o}\right]\frac{V_o}{s} \tag{12.39}$$

$$V_{out} = \frac{AV_o/s}{(s\alpha CR_{on} + 1)(1 - A) + s\alpha CR_o} \tag{12.40}$$

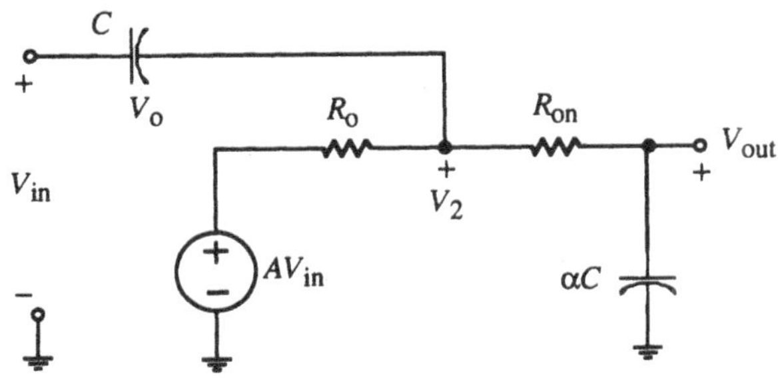

Figure 12.15 Equivalent op-amp circuit for a charging half-period.

Now assuming that $K_d = 1/2$ and substituting (12.29) into (12.40) yields

$$V_{out} = \frac{\dfrac{-A_o\omega_1}{\alpha C(R_{on} + R_o)}(V_o/s)}{s^2 + \dfrac{\alpha R_{on}C(1 + A_o)\omega_1 + 1 + \omega_1}{\alpha C(R_{on} + R_o)}s + \dfrac{\omega_1(1 + A_o)}{\alpha C(R_{on} + R_o)}} \tag{12.41}$$

The important steps are to determine the final value of V_{out} and to measure how fast V_{out} will reach this final value.

The final value is given by

$$V_{out}(\text{as } t \rightarrow \infty) = -\frac{A_o}{1 + A_o}V_o \tag{12.42}$$

The denominator of equation (12.41) determines the poles. Two example cases will be considered.

First, $C = 50$ pF and $\alpha = 1$. The poles are located at

$$s_{1,2} \approx -2.33 \times 10^6 \pm j1.89 \times 10^6$$

Second, $C = 50$ pF and $\alpha = 0.1$. The poles are located at

$$s_1 \approx -3.33 \times 10^6$$
$$s_2 \approx -27.0 \times 10^6$$

The dominant term is the real part of the complex root. Five time constants are necessary to bring the voltage to within 99% of its final value; therefore, the required sampling frequency can be shown to be less than 230 KHz.

Next the analysis of the effect of R_{on} and A while discharging αC through the following amplifier stage will be considered. For the circuit in Figure 12.16, R_{on} is the

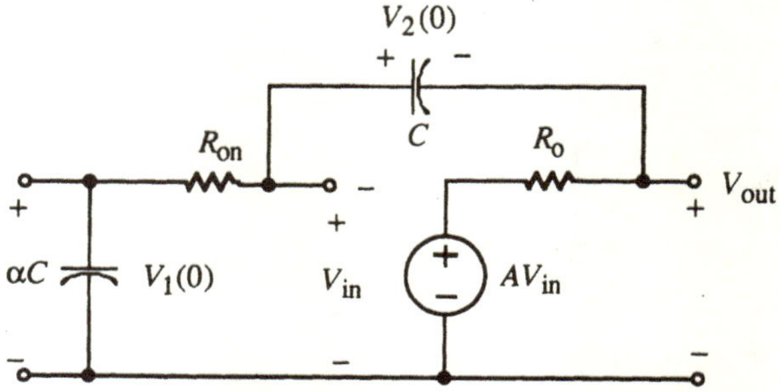

Figure 12.16 Equivalent op-amp circuit for a discharging half-period.

on-resistance of one or possibly two switches and both capacitors αC and C have initial voltages designated by $V_1(0)$ and $V_2(0)$, respectively. The output voltage is given by

$$V_{\text{out}} = \frac{[R_o + A/(sC)]V_1(0) + \{A[R_{\text{on}} + 1/(s\alpha C)] - R_o\}V_2(0)/s}{(1 - A)[R_{\text{on}} + 1/(s\alpha C)] + [R_o + 1/(sC)]} \tag{12.43}$$

The final value of V_{out} is given by

$$V_{\text{out}}\ (\text{as } t \to \infty) = -\frac{\alpha V_1(0) + V_2(0)}{1 + (1 + \alpha)/A_o} \tag{12.44}$$

The second term $(1 + \alpha)/A_o$ in the denominator of (12.44) represents the error. Compensation for this error is possible if a reasonable estimate or measurement of A_o can be obtained. The poles can be found by substituting (12.29) into (12.44) to obtain the characteristic polynomial, $P(s)$, given by

$$P(s) = \frac{\alpha C(R_o + R_{\text{on}})}{\omega_1}s^2 + \left[\frac{1 + \alpha}{\omega_1} + A_o\alpha CR_{\text{on}} + \alpha C(R_{\text{on}} + R_o)\right]s + (1 + \alpha + A_o) \tag{12.45}$$

Considering practical values, equation (12.45) can be simplified to

$$P(s) = \frac{\alpha C(R_o + R_{\text{on}})}{\omega_1}s^2 + \left(\frac{1 + \alpha}{\omega_1} + A_o\alpha CR_{\text{on}}\right)s + A_o \tag{12.46}$$

As in the charging case, if $\alpha = 1$, the dominant (controlling) terms are

$$s_1 = -1.49 \times 10^6$$
$$s_2 = -6.02 \times 10^6$$

Considering that it takes five time constants for the transients to settle, s_1 indicates an upper limit of about 149 KHz for the sampling frequency. For the previous examples, it can be observed that the lower value of the roots of s determines the highest usable sampling frequency. This root is generally proportional to $A_o\omega_1$. Thus high-frequency applications will require op-amps with larger gain-product bandwidths or smaller capacitances and small R_{on} (larger switch geometries).

REFERENCES

[1] Allen, Phillip E., and Sanchez-Sinencio, Edgar, *Switched Capacitor Circuits*, Van Nostrand Reinhold, 1984, p. 183.

[2] Martin, K., and Sedra, A. S., "Exact Design of Switched-Capacitor Bandpass Filters Using Coupled-Biquad Structure," *IEEE Trans. Circuits and Systems,* Vol. CAS-27, No. 6, June 1980, pp. 469–475.

PROBLEMS

12.1 Derive the equivalent resistances shown for Figures 12.1(b), 12.1(c), and 12.1(d).

12.2 Repeat example 12.1 using $C_1 = 240$ pF and $C_2 = 48$ pF as initial values. Show that the equation

$$\frac{\dfrac{-C_{r1}C_{r3}}{C_1C_2}(1 + z'^{-1})^2}{\left(1 + \dfrac{C_{r1} + C_{r2} + C_{r3}}{C_1} + \dfrac{C_{r2}C_{r3}}{C_1C_2}\right)z'^{-2} + \left(\dfrac{2C_{r2}C_{r3}}{C_1C_2} - 2\right)z'^{-1} + \left(1 - \dfrac{C_{r1} + C_{r2} + C_{r3}}{C_1} + \dfrac{C_{r2}C_{r3}}{C_1C_2}\right)}$$

produces a 3 dB attenuation at the passband cutoff frequency.

12.3 Find a solution for equation (12.27) given the biquadratic function

$$\frac{Az^{-2} - Bz^{-1} + C}{z^{-2} + Dz^{-1} + E}$$

12.4 Find the component values in example 12.2 for
(a) the Butterworth SC circuit,
(b) the Tchebyscheff Type I SC circuit,
(c) the elliptic SC circuit.
For the elliptic filter, use the following transfer function:

$$\frac{0.01821z^{-2} - 0.01584z^{-1} + 0.01821}{z^{-2} - 2.122254z^{-1} + 1.1477697}\,\frac{0.277552z^{-2} - 0.4740z^{-1} + 0.277552}{z^{-2} - 1.951709z^{-1} + 1.044326}$$

Appendix A
Mathematical Techniques

The following mathematical techniques are provided as a review.

A.1 POLYNOMIALS FROM COMPLEX ROOTS

If a polynomial $x^n + a_{n-1}x^{n-1} + \ldots + a_2x^2 + a_1x + a_0$ has known real or complex roots, r_1, r_2, r_3, . . . , r_n, then the coefficients, a_i, can be determined by the following method.

$a_0 = (-1)^n$ times the product of all the roots, or $(-1)^n r_1 \cdot r_2 \cdot r_3 \cdot \ldots \cdot r_n$,

$a_1 = (-1)^{n-1}$ times the sum of all possible products of the roots taking $n - 1$ roots at a time,

$a_2 = (-1)^{n-2}$ times the sum of all possible products of the roots taking $n - 2$ roots at a time,

.

.

.

$a_{n-1} = (-1)$ times the sum of all possible products of the roots taking 1 root at a time, or simply (-1) times the sum of all the roots.

Example A.1

Find the fourth-order polynomial that has roots at $x = \{-1, -2, -3, \text{and } -4\}$.

Solution

The polynomial is $x^4 + a_3x^3 + a_2x^2 + a_1x + a_0$, where

$$a_0 = [(-1)(-2)(-3)(-4)] = 24$$
$$a_1 = -1 \cdot [(-1)(-2)(-3) + (-1)(-2)(-4) + (-1)(-3)(-4) + (-2)(-3)(-4)] = 50$$
$$a_2 = [(-1)(-2) + (-1)(-3) + (-1)(-4) + (-2)(-3) + (-2)(-4) + (-3)(-4)] = 35$$
$$a_3 = -1 \cdot [(-1) + (-2) + (-3) + (-4)] = 10$$

or $x^4 + 10x^3 + 35x^2 + 50x + 24$.

A.2 THE TAYLOR SERIES

The Taylor series expansion of $f(x)$ is given by $f(x) = P_n(x) + R_n(x)$, where

$$P_n(x) = f(x_o) + f'(x_o)(x - x_o) + \frac{f''(x_o)}{2!}(x - x_o)^2 + \ldots + \frac{f^{(n)}(x_o)}{n!}(x - x_o)^n \qquad \textbf{(A.1)}$$

$$R_n(x) = \frac{f^{(n+1)}[\xi(x)]}{(n+1)!}(x - x_o)^{n+1} \qquad \text{(A.2)}$$

$P_n(x)$ is called the nth *Taylor polynomial* for f about x_o, $R_n(x)$ is called the *remainder term* (or truncation error) associated with $P_n(x)$, $f'(x_o)$ is the first derivative of $f(x)$ with respect to x evaluated at $x = x_o$, and $f^{(n)}(x_o)$ is the nth derivative of $f(x)$ with respect to x evaluated at $x = x_o$. The infinite series obtained by taking the limit of $P_n(x)$ as $n \rightarrow \infty$ is called the *Taylor series* for f about x_o. For the special case of $x_o = 0$, the Taylor polynomial is often called a *Maclaurin polynomial*, and the Taylor series is called a *Maclaurin series*. The truncation error of a Taylor polynomial can be bounded since $\xi(x)$ will be between x_o and x.

Example A.2

Determine the third Taylor polynomial for $f(x) = \cos(x)$ about $x_o = 0$ and use the Taylor polynomial to approximate cos(0.01).

Solution

$\text{Cos}(x) \cong \cos(0) - [\sin(0)](x - 0) - [\cos(0)](x - 0)^2/2 = 1 - x^2/2$. At $x = 0.01$, then $\cos(x) \cong 1 - 0.0001/2 = 0.99995$. The truncation error term is $(1/24)x^4 \cos[\xi(x)]$, where $0 < \xi(x) < 0.01$, because $f'''(0) = 0$. Therefore the magnitude of the truncation error is bounded by $(1/24)x^4 \cos[\xi(x)] \leq (1/24)(0.01)^4(1) \cong 4.2 \times 10^{-10}$.

A.3 THE NEWTON-RAPHSON METHOD

Newton's method is one of the most powerful numerical methods for solving a root-finding problem $f(x) = 0$ and can be derived from the first Taylor polynomial of $f(x)$ expanded about x_o. Given

$$f(x) = f(x_o) + f'(x_o)(x - x_o) + f''[\xi(x)]\frac{(x - x_o)^2}{2} \tag{A.3}$$

where $\xi(x)$ lies between x and x_0, substitution of $f(x) = 0$ yields

$$0 = f(x_o) + f'(x_o)(x - x_o) + f''[\xi(x)]\frac{(x - x_o)^2}{2} \tag{A.4}$$

Newton's method assumes that $|x - x_o|$ is small, and therefore $|x - x_o|^2$ is negligible. Solving for x gives

$$x \approx x_o - \frac{f(x_o)}{f'(x_o)} \tag{A.5}$$

where x_o is the initial guess and x is the new estimate. Newton's iteration is usually written as

$$\boldsymbol{x_n \approx x_{n-1} - \frac{f(x_{n-1})}{f'(x_{n-1})}} \tag{A.6}$$

and can be used to find both the real and complex roots, p_i, of polynomial equations if the initial guess, x_o, is given as a complex number and the process is carried out using complex arithmetic. Once a root is determined, the polynomial, $f(x)$ can be divided by $x - p_i$ to produce a new polynomial that contains the same roots as the original polynomial minus the root just found. The iterations can then be repeated on the deflated polynomial. This process ensures that the root-finding technique does not repeatedly find the same root.

Bairstow's method is an extension of Newton's method in two variables and is often used in filter routines to determine the poles of elliptic filters. Bairstow's method is used as part of program B3 to find the poles of elliptic filters.

Example A.3

Find the roots of the polynomial $x^3 + x^2 + x + 1 = 0$.

Solution

Step 1 is to find the first root and deflate the polynomial. With $f(x) = x^3 + x^2 + x + 1$ and $f'(x) = 3x^2 + 2x + 1$, using an initial guess of $x_o = 1 + \mathrm{j}$, Newton's method using complex arithmetic produces a root at $x = \mathrm{j}$. Dividing $x^3 + x^2 + x + 1$ by $x - \mathrm{j}$ yields a second-order polynomial $x^2 + (1 + \mathrm{j})\,x + \mathrm{j}$, as shown in the synthetic division that follows:

$$
\begin{array}{r l}
 & x^2 + (1 + \mathrm{j})x + \mathrm{j} \\
x - \mathrm{j} \,\big) & \overline{x^3 + x^2 + x + 1} \\
 & -(x^3 - \mathrm{j}x^2) \\ \hline
 & (1 + \mathrm{j})x^2 + x + 1 \\
 & -[(1 + \mathrm{j})x^2 + (1 - \mathrm{j})x] \\ \hline
 & \mathrm{j}x + 1 \\
 & -(\mathrm{j}x + 1)
\end{array}
$$

Step 2 is to find the second root and deflate the polynomial. Using $f(x) = x^2 + (1 + \mathrm{j})\,x + \mathrm{j}$ and $f'(x) = 2x + (1 + \mathrm{j})$, with an initial guess of $x_o = 1 + \mathrm{j}$, Newton's method using complex arithmetic produces a root at $x = -1$. Dividing $x^2 + (1 + \mathrm{j})\,x + \mathrm{j}$ by $x + 1$ yields a first-order polynomial $x + \mathrm{j}$, which yields the third root at $x = -\mathrm{j}$.

Step 3 is to write the polynomial in factored form. The factored polynomial becomes $(x - \mathrm{j})(x + 1)(x + \mathrm{j}) = (x^2 + 1)(x + 1)$.

A.4 PARTIAL FRACTION EXPANSION

Given a ratio of polynomials in s of the form

$$G(s) = \frac{A_{n-1}s^{n-1} + A_{n-2}s^{n-2} + \ldots + A_1 s + A_0}{B_n s^n + B_{n-1}s^{n-1} + \ldots + B_1 s + B_0} \tag{A.7}$$

where the A_i and B_i are known constants. If the denominator is factored, then

$$G(s) = \frac{A_{n-1}s^{n-1} + A_{n-2}s^{n-2} + \ldots + A_1 s + A_0}{B_n(s - s_1)(s - s_2) \ldots (s - s_n)} \tag{A.8}$$

If the roots, $s_1, s_2, \ldots, s_n$ are all different, then $G(s)$ can be written as

$$G(s) = \frac{K_1}{s - s_1} + \frac{K_2}{s - s_2} + \ldots + \frac{K_n}{s - s_n} \tag{A.9}$$

for any real, or complex, root, $s_i = a_i + jb_i$. Equation (A.9) represents the partial fraction expansion of (A.7). The value of K_1 can be determined as follows. Multiply both sides of (A.9) by $(s - s_1)$. This yields

$$(s - s_1)G(s) = K_1 + (s - s_1)\left(\frac{K_2}{s - s_2} + \ldots + \frac{K_n}{s - s_n}\right) \tag{A.10}$$

Now let $s = s_1$. All the terms on the right-hand side of (A.10) except K_1 become zero because of the factor $(s - s_1)$. The left-hand side becomes the constant $(s - s_1)G(s)|_{s=s_1}$. Note that this factor is not zero, because $G(s)$ has a factor of $s - s_1$ in its denominator, which cancels the $s = s_1$ multiplying term. This cancellation should be performed before $s = s_1$ is substituted. To solve for the other values of K_i, the procedure is the same except that both sides of (A.9) are multiplied by $s - s_i$ and then evaluated at $s = s_i$.

In the case of a multiple root that occurs r times, $G(s)$ becomes

$$G(s) = \frac{A_{n-1}s^{n-1} + A_{n-2}s^{n-2} + \ldots + A_1 s + A_0}{B_n(s - s_1)^r(s - s_2) \ldots (s - s_{n-r+1})} \tag{A.11}$$

or

$$G(s) = \frac{K_{11}}{(s - s_1)^r} + \frac{K_{12}}{(s - s_1)^{r-1}} + \ldots + \frac{K_{1r}}{(s - s_1)} + \frac{K_2}{(s - s_2)} + \ldots + \frac{K_{n-r+1}}{(s - s_{n-r+1})} \tag{A.12}$$

The different K_{1i} are given by

$$K_{1i} = \frac{1}{(i-1)!}\frac{d^{i-1}}{ds^{i-1}}[(s - s_1)^r G(s)]|_{s=s_1} \tag{A.13}$$

Example A.4

Expand $G(s) = (s^4 + s^3 + s^2 + s + 1)/(s^4 + 3s^3 + 4s^2 + 3s + 1)$ as a partial fraction.

Solution

Dividing the denominator into the numerator and using Newton's method to find the roots of the denominator yields

$$G(s) = 1 - \frac{2s^3 + 3s^2 + 2s}{(s+1)^2(s^2+s+1)} = 1 + \frac{K_{11}}{(s+1)^2} + \frac{K_{12}}{(s+1)} + \frac{K_2 s + K_3}{s^2+s+1}$$

$$K_{11} = -\left.\frac{2s^3 + 3s^2 + 2s}{(s^2+s+1)}\right|_{s=-1} = 1$$

from (A.13) K_{12} can be found as

$$K_{12} = -\left.\frac{(s^2+s+1)(6s^2+6s+2) - (2s^3+3s^2+2s)(2s+1)}{(s^2+s+1)^2}\right|_{s=-1} = -1$$

$$K_2 s + K_3 = -\left.\frac{s(2s^2+3s+2)}{(s+1)^2}\right|_{s=-\frac{1}{2}+j\frac{\sqrt{3}}{2}} = -K_2\left[-\frac{1}{2} + j\frac{\sqrt{3}}{2}\right] + 0$$

or $K_2 = -1$ and $K_3 = 0$.

A.5 QUALITY FACTOR

Second-order bandpass transfer functions can be written in the form

$$G(s) = \frac{s}{s^2 + \frac{\omega_o}{Q}s + \omega_o^2} = \frac{s}{s^2 + 2\delta s + \omega_o^2} \tag{A.14}$$

where Q is the quality factor and δ is the damping factor. From (A.14), it can be seen that $G(s)$ has an amplitude maximum near $s^2 + \omega_o^2 = 0$ and is 3 dB down (gain of 0.7071067 from max) for ω in the near vicinity of

$$|\omega_o^2 - \omega^2| = \frac{\omega_o}{Q}\omega \tag{A.15}$$

(Equation (A.15) holds true exactly for low-pass functions, that is, if s were replaced by a constant in the numerator of equation (A.14); however, it is also used as an approximate relation for bandpass functions.) For a given Q and ω_o, denote the two solutions to (A.15) as ω_1 and ω_2. Then

$$Q = \frac{\omega_o\omega_1}{\omega_o^2 - \omega_1^2} = \frac{f_o f_1}{f_o^2 - f_1^2} \tag{A.16}$$

Equation (A.16) is valid for $f_1 < f_o$. Repeating for $f_2 > f_o$ and rearranging the terms yields

$$Q = \left(\frac{f_2}{f_o} - \frac{f_o}{f_2}\right) = 1 \quad \text{and} \quad Q = \left(\frac{f_o}{f_1} - \frac{f_1}{f_o}\right) = 1 \tag{A.17}$$

Eliminating Q and solving shows that

$$f_1 f_2 = f_o^2 \tag{A.18}$$

Substituting into (A.16) to obtain an expression for the bandwidth, B, yields

$$B = f_2 - f_1 = \frac{f_o}{Q} \tag{A.19}$$

Thus the frequencies f_1 and f_2 have geometric symmetry around f_o. The bandwidth varies directly with f_o and inversely with Q. Solving (A.18) for f_1 and substituting into (A.19) produces an expression for f_2 given by

$$f_2 = \frac{f_o}{2Q} + \sqrt{f_o^2\left(1 + \frac{1}{4Q^2}\right)} \tag{A.20}$$

Similarly,

$$f_1 = -\frac{f_o}{2Q} + \sqrt{f_o^2\left(1 + \frac{1}{4Q^2}\right)} \tag{A.21}$$

where the positive solutions are chosen.

A typical response curve showing these parameters is given in Figure A.1.

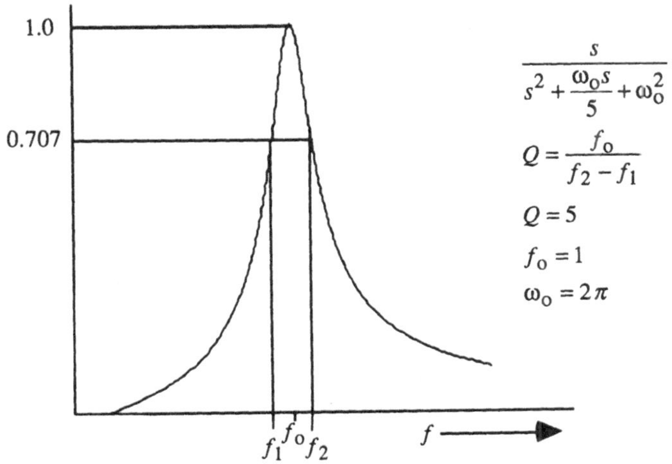

Figure A.1 Response curve for a Q of 5.

Appendix B
Programs

Program B1:	Fast Fourier Transform
Program B2:	Determines the Tchebyscheff rational function given the passband attenuation, stopband attenuation, stopband edge frequency, and the filter order.
Program B3:	Determines the poles of the voltage transfer function, $G(s)$, from the poles of the Tchebyscheff rational function given passband ripple, filter order, and Tchebyscheff rational function pole locations.
Program B4:	Uses the permutation method of Amstutz to find element values for even order passive elliptic filters corresponding to type-a, type-b, and type-c.
Program B5:	Uses the permutation method of Amstutz to find element values for odd-order passive elliptic filters corresponding to type-s.
Program B6:	Kaiser Window Coefficients.
Program B7:	Estimation Order and Coefficient Word length for FIR equiripple filters.

PROGRAM B1
FAST FOURIER TRANSFORM*

```
10 DEFDBL A-C,S,U-Z
20 DIM A(1025,2),U(5,2)
30 DIM W(5,2),T(5,2)
40 REM FFT DEMONSTRATION PROGRAM FOR N POINT TRANSFORM
50 PRINT "THIS PROGRAM EVALUATES THE MAGNITUDE AND PHASE"
60 PRINT "RESPONSE OF A NON-RECURSIVE Z TRANSFORM FILTER"
70 PRINT "USING THE FAST FOURIER TRANSFORM"
80 PRINT
90 PRINT "THE NUMBER OF POINTS IN THE FFT WILL BE 2^M"
100 PRINT
110 INPUT "TYPE M, OR <RET> FOR M = 10; N = 2^M ";M
120 INPUT "TYPE SAMPLING FREQUENCY IN HERTZ ";SFREQ
130 INPUT "TYPE THE NUMBER OF TAPS FOR THE FIR FILTER ";NTAPS
140 REM THE COEFFICIENTS OF THE FIR FILTER ARE IN A(1,1), A(2,1),....
150 REM THE REST OF THE A(I,1) COEFFICIENTS ARE PADDED WITH ZEROES
160 REM AFTER THE FFT, A(I,1) CONTAINS THE REAL PART OF THE
170 REM RESPONSE AT FREQUENCY W = (I-1)*3.14159/512
180 REM AND A(I,2) CONTAINS THE IMAGINARY PART OF THE RESPONSE
190 REM
200 FOR I = 1 TO NTAPS
210 PRINT "TYPE COEFFICIENT OF TAP ";I;TAB(2);
220 INPUT A(I,1)
230 NEXT I
240 REM
250 IF M = 0 THEN M=10
260 PRINT "MAG";TAB(16);"MAGDB";TAB(32);"HZ";TAB(48);"PHASE"
1000 N=2^M
1010 NV2 = N/2
1020 NM1 = N - 1
1030 J = 1
1040 FOR I = 1 TO NM1
1050 IF (I >= J) THEN GOTO 1090
1060 T(1,1) = A(J,1):T(1,2) = A(J,2)
1070 A(J,1)=A(I,1):A(J,2)=A(I,2)
1080 A(I,1)=T(1,1):A(I,2)=T(1,2)
1090 K = NV2
1100 IF ( K >= J) THEN GOTO  1140
1110 J = J - K
1120 K = K/2
1130 GOTO 1100
1140 J = J + K
1150 NEXT I
1160 PI = 3.141592654#
1170 FOR L = 1 TO M
```

*From *Theory and Application of Digital Signal Processing* by Lawrence R. Rabiner and Bernard Gold, © 1975, p. 367. Adapted by permission of Prentice-Hall, Englewood Cliffs, New Jersey.

```
1180 LE=2^L
1190 LE1 = LE/2
1200 U(1,1)=1:U(1,2)=0
1210 W(1,1)=COS(PI/LE1):W(1,2)=SIN(PI/LE1)
1220 FOR J = 1 TO LE1
1230 FOR I = J TO N STEP LE
1240 IP = I + LE1
1250 SAVER = A(IP,1)*U(1,1)-A(IP,2)*U(1,2)
1260 SAVEI = A(IP,2)*U(1,1)+A(IP,1)*U(1,2)
1270 T(1,1)=SAVER:T(1,2)=SAVEI
1280 A(IP,1)=A(I,1)-T(1,1):A(IP,2)=A(I,2)-T(1,2)
1290 A(I,1)=A(I,1)+T(1,1):A(I,2)=A(I,2)+T(1,2)
1300 NEXT I
1310 SAVER = U(1,1)*W(1,1)-U(1,2)*W(1,2)
1320 SAVEI = U(1,2)*W(1,1)+U(1,1)*W(1,2)
1330 U(1,1)=SAVER:U(1,2)=SAVEI
1340 NEXT J
1350 NEXT L
5000 FOR I = 1 TO N/2+1
5010 RE = A(I,1):IM=A(I,2)
5020 MAG = SQR(ABS( (A(I,1)^2+A(I,2)^2)))
5030 IF MAG > 0 THEN MAGDB = 20*LOG(MAG)/LOG(10) ELSE MAGDB = -999
5040 IF RE > 0 THEN PHASE = ATN(IM/RE)*180/3.14159
5050 IF RE = 0 THEN PHASE = 90*SGN(IM)
5060 IF RE < 0 THEN PHASE = ATN(IM/RE)*180/3.14159+180
5070 IF PHASE > 180 THEN PHASE = PHASE -360:GOTO 5070
5080 IF PHASE < -180 THEN PHASE = PHASE + 360:GOTO 5080
5090 PRINT CSNG(MAG);TAB(16);CSNG(MAGDB);TAB(32);CSNG((I-1)*(SFREQ/N));TAB-
(48);CSNG(PHASE)
5100 IF I MOD 16 = 0 THEN GOSUB 6000
5110 NEXT I
5120 A$=INKEY$
5130 IF A$="" THEN GOTO 5120
5140 END
6000 A$=INKEY$
6010 IF A$="" THEN GOTO 6000
6020 RETURN
```

PROGRAM B2*

```
10 DEFDBL A-Z
20 DIM A(100),THETA(100)
30 INPUT "TYPE PASSBAND ATTENUATION IN DB ";RP
40 INPUT "TYPE STOPBAND ATTENUATION IN DB ";RS
50 INPUT "TYPE STOPBAND EDGE FREQUENCY IN RADIANS ( MUST BE >1)";WX
60 PRINT "PASSBAND RESPONSE VARIES BETWEEN 1.00000 AND";10^(-RP/20);"
   VOLTS"
70 E2 = 10^(RP/10)-1:PRINT "E*E =";E2
80 PRINT "RN(W) VARIES FROM 1.0000 TO -1.0000 IN THE PASSBAND."
90 PRINT "RN(W) VARIES FROM ";SQR( (10^(RS/10)-1)/E2);" TO INFINITY IN THE
   STOPBAND"
100 L=SQR( (10^(.1*RS)-1)/(10^(.1*RP)-1))
110 KL = 1/L:KPL = SQR(1-(1/L)^2):KWX = 1/WX:KPWX = SQR(1-(1/WX)^2)
120 K = 1/L:GOSUB 3000:SUM = KK:K = KPL:GOSUB 3000:SUM1 = KK
130 K = KWX:GOSUB 3000:SUM2 = KK:K =  KPWX:GOSUB 3000:SUM3=KK
140 PRINT "N =";SUM1*SUM2/(SUM*SUM3);
200 N = INT((SUM1*SUM2)/(SUM*SUM3))+1:PRINT "         USE N = ";N
210 NI = 0:RN=1:RN1=1
220 IF N = 2*INT(N/2) THEN NI=1
230 FOR I = 1 TO N/2: GOSUB 7000: NEXT I
260 PRINT "B = ";1/RN
270 INPUT "PRESS <RET> TO QUIT ";C$
900 END
3000 REM CALCULATION OF K = COMPLETE ELLIPTIC INTEGRAL
3010 A(0) = ATN(K/SQR(1-K^2))
3020 THETA(0) = 3.14159/2
3030 P = 1:I=0
3040 X = 2/(1+SIN(A(I)))-1
3050 Y = SIN(A(I))*SIN(THETA(I))
3060 A(I+1)=ATN((SQR(1-X^2))/X)
3070 THETA(I+1)=.5*(THETA(I)+ATN(Y/SQR(1-Y^2)))
3080 E = 1-A(I+1)*2/3.14159:I =I+1
3090 IF E > .0000001 THEN GOTO 3040
3100 FOR J = 1 TO I: P = P*(1+COS(A(J))):NEXT J
3110 X = 3.14159/4+THETA(I)/2
3120 KK=LOG(SIN(X)/COS(X))*P/LOG(2.7182818#)
3130 RETURN
4000 REM CALCULATION OF SN(U,K)
4010 SN=0:J=0:Q = EXP(-3.14159*SUM3/SUM2)
4020 V = 3.14159/2*U/SUM2
4030 W = Q^(J+.5):SN=SN+W*SIN((2*J+1)*V)/(1-W*W):J = J + 1
4040 IF W > .0000001 THEN GOTO 4030
4050 SN = SN*2*3.14159/KWX/SUM2
4060 RETURN
7000 REM DETERMINATION OF POLES AND ZEROS
```

*From *Approximation Methods for Electronic Filter Design* by Richard Daniels, McGraw-Hill, Inc., © 1974, by permission of the publisher.

```
7010 U = (2*I-NI)*SUM2/N
7020 GOSUB 4000
7030 XO(I)=(SN):XZ(I)=(WX/SN)
7040 PRINT "WO(";I;")=";XO(I);"         WZ(";I;")=";XZ(I)
7050 RN=(XO(I)^2-1)/(XZ(I)^2 -1)*RN
7060 RETURN
```

Example B2.1

With regard to Figure 5.1, the following pole-zero locations were produced given the initial specifications that follow.

Type passband attenuation in dB ? 3.0103
Type stopband attenuation in dB ? 60
Type stopband edge frequency in radians ? 1.133
Passband response varies between 1.00000 and 0.7071067840850258
E*E = 0.9999999836037312
$R_N(w)$ varies from 1.00000 to −1.00000 in the passband
$R_N(w)$ varies from 999.9995081980054 to infinity in the stopband
N = 6.989079739548641 Use N = 7
$w_o(1)$= 0.5663460991116648 $w_z(1)$= 2.000543515686992
$w_o(2)$= 0.8789775435076472 $w_z(2)$= 1.288997682114965
$w_o(3)$= 0.9887024751136662 $w_z(3)$= 1.145946373889889
B= −179.2322414868699

Example B2.2

With regard to equation (7.27), the following pole-zero locations were produced given these initial specifications.

Type passband attenuation in dB ? 3.0103
Type stopband attenuation in dB ? 40
Type stopband edge frequency in radians ? 1.0589145
Passband response varies between 1.00000 and 0.7071067840850258
E*E = 0.9999999836037312
$R_N(w)$ varies from 1.00000 to −1.00000 in the passband
$R_N(w)$ varies from 99.9950006947662 to infinity in the stopband
N = 5.997256856797012 Use N = 6
$w_o(1)$= 0.4010918254705045 $w_z(1)$= 2.64007998357441
$w_o(2)$= 0.8674691561889573 $w_z(2)$= 1.220694122027482
$w_o(3)$= 0.9897913078420552 $w_z(3)$= 1.069836127687004
B= −100.2528435022799

PROGRAM B3*

```
10 DEFDBL A-Z
20 DIM C1(100),B1(100),A1(100),D1(100),Z1(100),F1(100)
30 INPUT "Passband ripple in db ";AX
40 INPUT "Filter order ";FILORD
50 FOR I = 1 TO FILORD/2
60 INPUT F(I)
70 NEXT I
80 RP=AX:FB=1/(2*3.141592654#):NIN=FILORD MOD 2:N=INT(FILORD/2)
90 FOR I = 1 TO N
100 F1(I)=F(N+1-I)/(2*3.141592654#)
110 NEXT I
120 EE=10^(.1*RP)-1:E=SQR(EE):FBB=FB*FB
130 M=NIN+2*N:EM=2*INT(M/2):TP=2*3.14159:K=0:T=M:R=0
140 FOR I = 1 TO N:Z1(I)=SQR(1-FBB/F1(I)/F1(I)):NEXT I
150 FOR I = 1 TO N:W1(I) = TP*F1(I):NEXT I
160 GOSUB 370: GOSUB 690
170 PRINT "                POLES                  ZEROES"
180 IF M>EM THEN C1(2*M)=0
190 FOR I = 0 TO 2*M STEP 2:A1(M-I/2)=C1(I)+D1(I):NEXT I
200 GOSUB 790
210 R=R+1:P(R)=P(R)/10:Q1(R)=Q1(R)/100:D=1+P(R)+Q1(R)
220 B1(R)=(1+P(R)/2)*FBB/D
230 ZF1(R)=FB/D^.25
240 ZQ1(R)=1/SQR(ABS(2*(1-B1(R)/(ZF1(R)*ZF1(R))   ))):ZW1(R)=TP*ZF1(R)
250 GAIN2=GAIN2/ZW1(R)^2
260 ROOT(R,1)=-.5*( ZW1(R))/ZQ1(R):ROOT(R+EM/2,1)=ROOT(R,1)
270 ROOT(R,2)=.5*SQR(ABS(ZW1(R)^2/ZQ1(R)^2 - 4*ZW1(R)^2)):ROOT(R+EM/2,2)=-
ROOT(R,2)
280 PRINT ROOT(R,1),ROOT(R,2)
290 IF R<EM/2 THEN GOTO 210
300 IF A=0 THEN GOTO 340
310 SIGMA=SQR(FBB/(.1*A-1))*TP
320 GAIN2=GAIN2/SIGMA
330 ROOT(R+1+EM/2,1)=-SIGMA:PRINT -SIGMA
340 REM
350 REM
360 STOP
370 REM determination of f(z)
380 FOR I = 1 TO NIN:S1(I)=1:NEXT I
390 FOR I = NIN+1 TO NIN+N:S1(I)=Z1(I-NIN):NEXT I
400 FOR I = NIN+N+1 TO NIN+2*N: S1(I)=Z1(I-NIN-N):NEXT I
410 SN=I-1 :GOSUB 540
420 FOR I = 0 TO EM STEP 2:A1(I)=E*B1(I):NEXT I
430 FOR I = 0 TO 2*EM STEP 2
440 GOSUB 630
```

*From *Approximation Methods for Electronic Filter Design by Richard Daniels,* ©1974, McGraw-Hill, Inc., by permission of the publisher.

```
450 NEXT I
460 RETURN
540 REM prod(x+s(i)
550 B1(0)=S1(1):B1(1)=1:J=1
560 J=J+1
570 A1(0)=S1(J)*B1(0)
580 FOR I = 1 TO J-1:A1(I)=B1(I-1)+S1(J)*B1(I):NEXT I
590 FOR I = 0 TO J-1:B1(I)=A1(I):NEXT I
600 B1(J)=1
610 IF J<SN THEN GOTO 560
620 RETURN
630 REM determination of f(z)^2
640 IF I<(EM+2) THEN JI=0:JF=I
650 IF I>EM THEN JI=I-EM:JF=EM
660 C1(I)=0
670 FOR J=JI TO JF STEP 2:C1(I)=C1(I)+A1(J)*(A1(I-J)*10^(M-I/2) ):NEXT J
680 RETURN
690 REM determination of q(z)
700 FOR I = 1 TO NIN
710 S1(I)=-10
720 NEXT I
730 FOR I = NIN+1 TO NIN+N:S1(I)=-10*Z1(I-NIN)^2:NEXT I
740 FOR I = NIN+N+1 TO NIN+2*N:S1(I)=S1(I-N):NEXT I
750 SN=M:GOSUB 540
760 DD=(-1)^NIN
770 FOR I = 0 TO 2*M STEP 2:D1(I)=DD*B1(I/2):NEXT I
780 RETURN
790 REM factor findre
800 FOR I = 1 TO T:A1(I)=A1(I)/A1(0):NEXT I
810 A1(0)=1:B1(0)=1:C1(0)=1:I1=0
820 P=0:Q=0:I1=I1+1
830 B1(1)=A1(1)-P:C1(1)=B1(1)-P
840 FOR I = 2 TO T:B1(I)=A1(I)-P*B1(I-1)-Q*B1(I-2):NEXT I
850 FOR I = 2 TO T-1:C1(I)=B1(I)-P*C1(I-1)-Q*C1(I-2):NEXT I
860 X1=T-1:X2=T-2:X3=T-3:X4=C1(X2)^2+C1(X3)*(B1(X1)-C1(X1))
870 IF X4=0 THEN X4=.001
880 DDP=(B1(X1)*C1(X2)-B1(T)*C1(X3))/X4:P=P+DDP
890 DQ=(B1(T)*C1(X2)-B1(X1)*(C1(X1)-B1(X1)))/X4:Q=Q+DQ
900 IF ABS(DDP+DQ) >.000001 THEN GOTO 830
910 P(I1)=P:Q1(I1)=Q:A1(1)=A1(1)-P:T=T-2
920 FOR I = 2 TO T:A1(I)=A1(I)-P*A1(I-1)-Q*A1(I-2):NEXT I
930 IF T>2 THEN GOTO 820
940 IF T=2 THEN I1=I1+1:P(I1)=A1(1):Q1(I1)=A1(2)
950 IF T=1 THEN A=-A1(1)
960 RETURN
```

Example B3.1

The parameter, NIN, is the number of zeroes at infinity and has been programmed so that NIN = 1 for odd order elliptic filters, and NIN = 0 for even-order type-a elliptic filters. The program would require modification to accommodate type-b and type-c elliptic filters. The input data for program B3 can be taken directly from the output of program B2.

Using the values given in equation (7.27),

Passband ripple in dB ? 3.0103
Filter order ? 6
? 1.0698388
? 1.2206966
? 2.64008339

Poles	Zeros
−1.166030792904782D−02	±0.9937640520822193
−6.642337324484707D−02	±0.8808912341937725
−0.20341647439428	±0.4197740742320909

From the pole locations given by program B3, the transfer function corresponding to example 7.5 is

$$G(s) = \frac{0.01(s^2 + 1.0698388^2)(s^2 + 1.2206966^2)(s^2 + 2.64008339^2)}{(s^2 + 0.407s + 0.2178)(s^2 + 0.133s + 0.7804)(s^2 + 0.234s + 0.9877)}$$

where

$$(s^2 + 0.407s + 0.2178) = [s - (-0.2034 + j\ 0.41977)][s - (-0.2034 - j\ 0.41977)],$$
$$(s^2 + 0.133s + 0.7804) = [s - (-0.0664 + j\ 0.88089)][s - (-0.0664 - j\ 0.88089)],$$
$$(s^2 + 0.234s + 0.9877) = [s - (-0.01166 + j\ 0.9938)][s - (-0.01166 - j\ 0.9938)].$$

PROGRAM B4*

```
10 DEFDBL A-Z
20 PRINT "ANTIMETRIC ELLIPIC FILTERS, CCT12/78, 1008"
30 FILTY=0:FILCHAR=0:SLR=1:DUAL=1:DENORM2=1
1030 DIM B(16),C(16),D(16),E(30),F(16),R(15),S(15),DB(16),TB(16)
1031 DIM  ZER(60,2),POLE(60,2)
1032 IZERO=0:IPOLE=0
1040 DN=LOG(10)/10:PI=3.1415926#
1050 PRINT
1060 PRINT "REJECTION, RIPPLE(DB),1/2-DEG(2-15), TYPE (A,B, OR C):"
1065 INPUT AS2,AP,M,T$
1066 NUMX = 20
1070 IF AS2 <= AP THEN STOP
1090 N=2*M
2010 ES=EXP(DN*AS2)-1
2020 EP=EXP(DN*AP)-1
2030 V=SQR(ES/EP)+SQR(ES/EP-1)
2040 U=PI*PI/(2*LOG(V+V))
2050 V=V/(SQR(ES)+SQR(ES+1))
2060 W=U*LOG(V+SQR(V*V+1))/PI
2070 W=SIN(W)/COS(W):AO=W:W=W*W
2080 Y=EXP(-U):Z=Y:K=M-1
2090 FOR J = 1 TO N
2100 E(J)=1:NEXT J
2110 FOR J = 1 TO 1024
2120 IF K<>M THEN GOTO 2150
2130 X=((1-Z)/(1+Z))^2
2140 AO=AO*(W+X)/(1+W*X)
2150 E(K)=E(K)*(1-Z)/(1+Z)
2160 Z=Z*Y: IF Z<2.5E-19 THEN GOTO 2180
2165 K=K-1
2170 IF K = 0 THEN K=N
2175 NEXT J
2180 E(M)=0:E(N)=E(N)*E(N)
2190 PRINT "U=";U;"AO=";AO;"EP=";E(N)
2200 FOR J = 1 TO M-1
2210 E(J)=-E(J)*E(N-J)
2220 PRINT "E=";-E(J)
2230 E(N-J)=-E(J):NEXT J
2250 X= SQR(AO*AO+1/(AO*AO)+E(N)*E(N)+1/(E(N)*E(N)))
2260 FOR J = 1 TO M-1 STEP 2: K=(J+1)/2
2270 Y=AO*E(J):Y=Y+1/Y
2280 Z=E(N)*E(J)
2290 R(K)=E(M-J)*(1/Z-Z)/Y:S(K)=-X/Y
2300 PRINT "RE=";R(K);"SE=";S(K)
2310 R(M-K+1)=R(K)
```

*From *Circuit Design Using Personal Computers* by Thomas R. Cuthbert, Jr., © 1983, John Wiley & Sons, Inc., by permission of the publisher.

```
2320 S(M-K+1)=-S(K):NEXT J
2330 IF K+K=M THEN GOTO 3010
2340 R(K+1)=-AO:S(K+1)=0
2350 PRINT "RE=";-AO
3010 IT=2:IF T$="A" THEN IT=1
3020 EB=-E(1):IF T$="A" THEN EB=E(N)
3030 EO=E(N): IF T$="C" THEN EO=-E(1)
3040 FP=SQR((E(N)+EO)/(1+E(N)*EB))
3050 FS=SQR((1+E(N)*EO)/(E(N)+EB))
3060 D(1)=0
3065 FOR J = IT TO M
3070 D(J)=(E(2*J-1)+EB)/(1+E(2*J-1)*EO)
3080 F(J)=SQR(1/D(J)):NEXT J
3100 SR=0:TQ=0:T0=0:B(1)=0:I=1
3110 FOR J = 1 TO M
3120 W=(AO^2+E(2*J-1)^2)/(1+(AO*E(2*J-1))^2)
3130 X=(1+EO*EB)*S(J)+EO+EB*W
3140 Y=EO^2+2*EO*S(J)+W
3150 Z=1+2*EB*S(J)+EB^2*W
3160 U=SQR(Y/Z):V=X/Z
3170 R(J)=SQR((U-V)/2):S(J)=SQR((U+V)/2)
3175 IPOLE=IPOLE+1
3177 POLE(IPOLE,1)=-R(J)/FP:POLE(IPOLE,2)=S(J)/FP
3179 IPOLE=IPOLE+1
3180 PRINT "RF=";-R(J)/FP;"SF=";S(J)/FP
3181 POLE(IPOLE,1)=-R(J)/FP:POLE(IPOLE,2)=-S(J)/FP
3200 SR=SR+R(J)/U
3210 I=-I:W=I*R(J)/S(J)
3220 TQ=(TQ+W)/(1-TQ*W)
3230 IF T$<>"A" GOTO 3270
3240 U=(F(2)-S(J))/R(J):V=(F(2)+S(J))/R(J)
3250 W=I*(V-U)/(1+U*V)
3260 T0=(T0+W)/(1-T0*W)
3270 B(1)=B(1)+R(J):NEXT J
4010 IF T$="A" THEN T0=T0/(1+SQR(1+T0*T0))
4020 FOR K = IT TO M
4030 DB(K)=0:TB(K)=T0:I=1
4040 FOR J = 1 TO M
4050 DB(K)=DB(K)+1/(R(J)+(F(K)-S(J))^2/R(J))+1/(R(J)+(F(K)+S(J))^2/R(J))
4070 I = -I:W=(F(K)-I*S(J))/R(J)
4080 TB(K)=(TB(K)+W)/(1-TB(K)*W):NEXT J:NEXT K
5010 D(M+1)=D(M):F(M+1)=F(M):DB(M+1)=DB(M):TB(M+1)=TB(M):C(1)=0
5020 FOR J = 1 TO M+1-IT STEP 2
5030 TB(M+1-J)=-1/TB(M+1-J):NEXT J
5040 FOR J = IT TO M+1
5050 B(J)=(1+TB(J)^2)*DB(J)/(4*D(J))-TB(J)*F(J)/2
5060 C(J)=TB(J)/F(J):NEXT J
6010 FOR L = 1 TO 2
6020 FOR K = L+2 TO M+1 STEP 2
6030 FOR J = L TO K-2 STEP 2
```

```
6040 U=C(J)-C(K)
6050 V=1/(U/(B(J)*(D(K)-D(J)))-1)
6060 B(K)=(B(K)-B(J))*V*V-B(J)*(V+V+1)
6070 C(K)=U*V:NEXT J:NEXT K:NEXT L
7010 W=1:IF T$<>"C" THEN W=((1-TQ*T0)/(TQ+T0))^2
7020 FOR J = 1 TO M+1 STEP 2:B(J)=B(J)*W
7030 C(J)=C(J)*W:NEXT J
7040 PRINT "LD RESIS =";W;"    (";1/W;"    )"
7050 PRINT TAB(10);"L(C)";TAB(30);"C(L)";TAB(50);"PEAK"
7052 IF T$<>"A" THEN PRINT "  1          ";FP/B(1)
7060 REM
7062 REM FOR TYPE C ELLIPTIC FILTERS, STOPBAND EDGE = FS/FP AND NOT
     PZSCALE2
7064 PZSCALE2 = FS/FP
7066 REM
7085 V=0
7090 FOR J = IT TO M
7095 V=V+C(J)
7096 PRINT J;FP*C(J);FP/B(J);F(J)/FP
7097 NEXT J
7098 PRINT M+1;FP*C(M+1);"STPDB EDGE = ";FS/FP
7099 PRINT "TESTS ";B(M)/B(M+1)-1;(W+1)*SR-V-C(M+1)
8000 STOP
```

The following examples correspond to Figure 7.6, Table 7.2, equation (7.27), (7.29) and (7.31), and example 7.5 for elliptic type-a, type-b, and type-c implementations. The values in parentheses would be used for implementation of a network dual (reference Chapter 6, Duality and Reciprocity). The values given for RF and SF are the poles of the voltage transfer function, $G(s)$.

For Type a

```
REJECTION, RIPPLE(DB), 1/2-DEG(2-15), TYPE(A,B, OR C)
? 40,3,3,A
U= 0.82332693553922           AO=0.2296353108218732      EP=0.7106330752644
E=0.6263291057810176
E=0.374908159728896
RE=-9.517945787731751D-02     SE=0.6534081024395546
RE=-0.2296353108218732
RF=-1.168111217932183D-02     SF=0.9937797945078053
RF=-6.654472809240142D-02     SF=0.8809429907526448
RF=-0.2038118378411714        SF=0.4198476596368662
LD RESIS = 5.808899444425789  (0.1721496489252536)

                L(C)                    C(L)                     PEAK
1         -0.2486535799918705     0.5442446714777546     2.640083397927738
2          2.41298146685094       0.4121297384782688     1.220696624744963
3          2.755554805462586      0.2542370375954297     1.069838852537977
4          2.112437472774575      stpdb edge = 1.058914522870861
```

For Type b

```
REJECTION, RIPPLE(DB), 1/2-DEG(2-15), TYPE(A,B, OR C)
? 40,3,3,B
U= 0.82332693553922          AO=0.2296353108218732      EP=0.7106330752644
E=0.6263291057810176
E=0.374908159728896
RE=-9.517945787731751D-02    SE=0.6534081024395546
RE=-0.2296353108218732
RF=-1.359506620399677D-02    SF=0.9927096281135206
RF=-7.347347652631826D-02    SF=0.8638289465452303
RF=-0.195340838186894        SF=0.3898992196756554
LD RESIS = 5.808899444425789  (0.1721496489252536)

              L(C)                C(L)                  PEAK
1                            0.6095748249089779
2       2.536746048131171    0.4508445435835205    1.274114538352594
3       2.937673034828391    0.2742942351119816    1.083031150550583
4       2.278798027486919    stpdb edge = 1.069838852537977
```

For Type c

```
REJECTION, RIPPLE(DB), 1/2-DEG(2-15), TYPE(A,B, OR C)
? 40,3,3,C
U= 0.82332693553922          AO=0.2296353108218732      EP=0.7106330752644
E=0.6263291057810176
E=0.374908159728896
RE=-9.517945787731751D-02    SE=0.6534081024395546
RE=-0.2296353108218732
RF=-1.584642399560751D-02    SF=0.9915254673231737
RF=-.0879419143137565        SF=0.8402218817147896
RF=-0.325416647904805        SF=0.2724816311468957
LD RESIS = 1(1)

              L(C)                C(L)                  PEAK
1                            2.329889028482919
2       1.836813453833249    0.8234170159417645    1.313671011294058
3       0.9116017598193474   0.7382454897654718    1.096064676798804
4       1.14313632053807     stpdb edge = 1.080875883444048
Tests -2.518985020572018D-12  -2.939870569207415D-13
```

PROGRAM B5*

```
10 PRINT "SYMMETRICAL ELLIPTIC FILTER, C&S12/78, 1009"
20 DEFDBL A-H,O-Z
1010 DIM B(16),C(16),D(16),E(15),F(30)
1020 DN=LOG(10)/10:PI=3.1415926#
1030 PRINT "STOPBAND EDGE = ";:INPUT FS
1040 FS = FS/1000
1050 PRINT "PASSBAND EDGE = ";:INPUT FP
1060 FP=FP/1000
1070 IF ABS(FS-FP)<=0 THEN GOTO 1030
2040 PRINT "NUMBER OF PEAKS (1-15) = ";:INPUT N
2050 IF N<=0 THEN GOTO 2040
2060 M=2*N+1
2080 FC=SQR(FS*FP)
2090 R=FC+FC
2100 FOR K = 1 TO 2
2110 S=FS+FP
2120 FOR J = 1 TO 6
2130 P=SQR(S*R)
2140 S=(S+R)/2
2150 IF 1E+08*(S-P)<S THEN GOTO 2170
2160 R=P:NEXT J
2170 IF K >=2 THEN GOTO 2200
2180 Q=M/S
2190 R=ABS(FS-FP):NEXT K
2200 Q=Q*S
2210 S=EXP(-PI/Q)
2220 Y=S
2230 PRINT "CRITICAL Q =";CSNG(Q/(4*(1-S)*S^N))
2250 PRINT "STOPBAND REJECTION IN DB = ";:INPUT S
2260 IF S <=0 THEN GOTO 1030
2270 S=EXP(S*DN/2)
2280 R=EXP(PI*Q)
2290 P=(LOG(1+(S*S-1)/(R/4+1/R)^2))/DN
2310 R=R/(2*(S+SQR(S*S-1)))
2320 R=LOG(R+SQR(R*R+1))/(2*Q)
2330 R=SIN(R)/COS(R)
2340 W=R
2360 PRINT "3 DB (KHZ) ABOUT = ";FP+(FS-FP)/(1+FC/(FP*R*R))
2365 PRINT "NOMINAL OHMS RESISTANCE = ";:INPUT R
2370 IF R <=0 THEN GOTO 2365
2390 Z=Y:E(N)=W:W=W*W
2400 FOR J = 1 TO M-1
2410 F(J) =1:NEXT J
2420 K=1
2430 FOR J = 1 TO 1024
```

```
2440 F(K)=F(K)*(1-Z)/(1+Z)
2450 IF K < M-1 THEN GOTO 2500
2460 Z=Z*Y
2470 X = (( 1-Z )/(1+Z))^2
2480 E(N)=E(N)*(W+X)/(1+W*X)
2490 K=0
2500 Z=Z*Y
2510 IF Z < 2.5E-19 THEN GOTO 2530
2520 K = K + 1: NEXT J
2530 FOR J = 1 TO N
2540 F(J)=F(J)*F(M-J)
2550 F(M-J)=F(J):NEXT J
3010 FOR J = 1 TO N
3020 D(J)=F(2*J)*(1-F(J)^4)/F(J)
3030 B(J) = E(N)*F(J): NEXT J
3040 C(1) = 1/B(N)
3050 FOR J = 1 TO N-1
3060 C(J+1) = (C(J) - B(N-J))/(1+C(J)*B(N-J))
3070 E(N-J) = E(N+1-J) + E(N)*D(J)/       (1+B(J)*B(J)):NEXT J
4010 FOR J = 1 TO N
4020 B(J) = ((1+C(J)*C(J))*E(J)/D(J)-C(J)/F(J))/2
4030 C(J)=C(J)*F(J)
4040 D(J)=F(J)*F(J):NEXT J
4050 B(N+1) = B(N):C(N+1)=C(N):D(N+1) = D(N)
5010 IF N = 1 THEN GOTO 6020
5020 L = 1
5030 FOR K = L + 2 TO N+1 STEP 2
5040 FOR J = L TO K-2 STEP 2
5050 Y=C(J)-C(K)
5060 Z=1/(Y/(B(J)*(D(K)-D(J)))-1)
5070 B(K) = (B(K)-B(J))*Z*Z - B(J)*(1+Z+Z)
5080 C(K) = Y*Z: NEXT J
5081 NEXT K
5082 IF L = 2 THEN GOTO 6010
5083 L=2: GOTO 5030
6010 S=B(N)/B(N+1)-1
6020 Q = .0005/(PI*FC)
6030 P=Q*R:Q=Q/R
6040 IF FS<FP THEN GOTO 6150
6060 PRINT "                    LOW PASS FILTER"
6070 FOR J = 1 TO N
6080 C(J)=Q*C(J)
6090 D(J)=Q*B(J)*D(J)
6100 B(J)=P/B(J)
6110 F(J)= FC/F(J):NEXT J
6120 C(N+1) = Q*C(N+1)
6130 GOTO 6220
6150 PRINT "                    HIGH PASS FILTER"
6160 FOR J = 1 TO N
6170 C(J)=Q/C(J)
```

```
6180 D(J)=Q/(B(J)*D(J))
6190 B(J)=P*B(J)
6200 F(J)=FC*F(J):NEXT J
6210 C(N+1)=Q*C(N+1)
6220 PRINT "      KHZ          FARAD        HENRY"
6230 FOR J = 1 TO N STEP 2
6240 PRINT TAB(27);CSNG(C(J))
6260 PRINT CSNG(F(J));J;CSNG(B(J));CSNG(D(J)):NEXT J
6270 PRINT TAB(27);CSNG(C(N+1))
6280 IF N = 1 THEN STOP
6290 L=(INT((N+1)/2))*2:K=M-1-L
6300 FOR J = L + 2 TO M-1 STEP 2
6310 PRINT CSNG(F(K));K;CSNG(B(K));CSNG(D(K))
6320 PRINT TAB(27);CSNG(C(K))
6330 K=K-2:NEXT J
6340 PRINT "PRECISION TEST: ";S
7020 STOP
```

The following example corresponds to Figure 7.6, Table 7.2, , and example 7.5 for an elliptic type-s implementation. Passband edge and stopband edge frequencies are given in Hertz corresponding to a normalized passband edge of 1 rad/sec and a stopband edge of 1.0254313 radians/sec.

For Type s

Stopband Edge = ? 0.1632072
Passband Edge = ? 0.159155
Number of Peaks (1-15) = ? 3
CRITICAL Q = 82.5433
Stopband rejection in dB = ? 40
3 dB (KHz) about = 1.593657924521486D-04
Nominal ohms resistance = ? 1

KHZ		FARAD	HENRY
			2.883957577357127
2.38148145D-04	1	0.56107323689744	0.7960243620193951
			2.21833237844857
1.63944286D-04	3	0.1731455577117	5.442975665387524
			1.484482601462442
1.73695577D-04	2	0.2686676708329	3.124981064153786
			1.931047238798716

PROGRAM B6
KAISER COEFFICIENTS

```
10 INPUT "STOPBAND ATTENUATION,N";ATT,N
20 GOSUB 1000              'COMPUTE BETA FROM ATTENUATION
30 X = BETA
40 GOSUB 3000               'CALCULATE IO(B)
50 IOB=X
60 IF N MOD 2 = 0 THEN GOSUB 100
70 IF N MOD 2 = 1 THEN GOSUB 200
80 END
100 FOR N1 = -N/2 TO N/2-1
110 X = BETA*SQR(1-(4*(N1+1/2)^2)/(N-1)^2)
120 GOSUB 3000
130 PRINT X/IOB
140 NEXT N1
150 RETURN
200 FOR N1 = -(N-1)/2 TO (N-1)/2
210 X = BETA*SQR(1-(4*(N1)^2)/(N-1)^2)
220 GOSUB 3000
230 PRINT X/IOB
240 NEXT N1
250 RETURN
1000 IF (ATT > 50) THEN BETA = .1102*(ATT-8.7)
1010 IF (ATT >= 20.96) AND (ATT <= 50) THEN GOSUB 2000
1020 IF (ATT < 20.96) THEN BETA = 0
1030 RETURN
2000 BETA = .58417*(ATT-20.96)^.4+.07886*(ATT-20.96)
2010 RETURN
3000 Y = X/2
3010 T = 1E-08
3020 E = 1
3030 DE = 1
3040 FOR I = 1 TO 25
3050 DE =DE*Y/I
3060 SDE = DE^2
3070 E = E + SDE
3080 IF E*T-SDE > 0 THEN GOTO 3100
3090 NEXT I
3100 X = E
3110 RETURN
```

PROGRAM B7
ESTIMATED ORDER AND COEFFICIENT WORD LENGTH FOR FIR EQUIRIPPLE FILTERS

```
10 REM PROGRAM FOR EMPIRICAL ORDER ESTIMATION OF FIR FILTERS.
20 DEFDBL A-Z
30 INPUT "LOWPASS(0), BANDPASS(1), BANDSTOP(2), OR HIGHPASS(3) ? ";ITYPE
40 INPUT "TYPE MAXIMUM PASSBAND ATTENUATION IN DB ";AF1
50 INPUT "TYPE MINIMUM STOPBAND ATTENUATION IN DB ";AF2
60 INPUT "TYPE SAMPLING FREQUENCY IN HERTZ ";SFREQ
70 IF (ITYPE=1) OR (ITYPE=2) THEN PRINT " THE FILTER RESPONSE IS ASSUMED TC
BY SYMMETRICAL ABOUT";SFREQ*.25;"HZ"
80 INPUT "TYPE FIRST PASSBAND EDGE FREQUENCY IN HERTZ ";F1
90 INPUT "TYPE FIRST STOPBAND EDGE FREQUENCY IN HERTZ ";F2
100 F1 = F1/SFREQ
110 F2=F2/SFREQ
120 IF ITYPE = 1 THEN F1=.5-2*F1:F2=.5-2*F2
130 IF ITYPE = 2 THEN F1=2*F1:F2=2*F2
140 IF ITYPE = 3 THEN F1=.5-F1:F2=.5-F2
150 D1 =  .5*(1-10^(-AF1/20))
160 D2 = 10^(-AF2/20)
170 DELOMG = -(F1-F2)
180 REM FILTER LENGTH ESTIMATION
190 DL1 = LOG(D1)/LOG(10)
200 DL2 = LOG(D2)/LOG(10)
210 DQ1 = DL1*DL1
220 D = .005309*DQ1+.07114*DL1-.4761
230 D = D*DL2
240 D = D-.00266*DQ1-.5941*DL1-.4278
250 D = 2+D/DELOMG
260 NF = D
270 NF = INT(NF)
280 IF (ITYPE=1) OR (ITYPE=2) THEN NF=2*D
290 IF ((ITYPE=2) OR (ITYPE=3)) AND (NF MOD 2 = 0) THEN NF = NF + 1
300 PRINT "THE FILTER LENGTH IS ";NF
310 ALD=LOG(D1)/LOG(10)
320 ALS=LOG(D2)/LOG(10)
330 ALO=LOG(DELOMG*360)/LOG(10)
340 W=7.22116-.589268*ALD+(.112354*ALD-2.328539)*ALO-3.559051*ALS
350 PRINT "THE WORD LENGTH IS ";INT(W+1)
```

Answers to Selected Questions

CHAPTER 1

1. $L_{a1} = 1.06666$ H, $C_a = 0.6$F, $L_{a2} + L_b = 1.6$ H, $C_b = 2$ F
   $C_c = 1.2$F, $L_{c2} = 0.533333$ H, $L_{c1} + L_{a2} = 1.2$ H, $L_b + L_{c1} = 1.6$ H

2. $$\frac{d\left[\tan^{-1}\left(\frac{\omega - \beta}{\alpha}\right)\right]}{d\omega} = \frac{\alpha}{\alpha^2 + (\omega - \beta)^2}$$

6. $b_1 = 1.2732406$, $b_2 = 0$, $b_3 = 0.4244135$, $b_4 = 0$, $b_5 = 0.2546481$,
   $b_6 = 0$, $b_7 = 0.1818915$, $b_8 = 0$, $b_9 = 0.1414711$, $b_{10} = 0$

CHAPTER 2

1. $N = 13.29$, use $N = 14$

2. $N = 14.26$, use $N = 15$

3. $-0.9876887 \pm j0.1564345$ $\quad -0.8910068 \pm j0.4539906$
   $-0.707107 \pm j0.707107$ $\quad -0.4539906 \pm j0.8910068$
   $-0.1564345 \pm j0.9876887$

4. $$G(s) = \frac{1}{s^4 + 2.613126s^3 + 3.414213s^2 + 2.613126s + 1}$$

5. $g_1 = g_{12} = 0.2610524$ $\quad g_2 = g_{11} = 0.7653669$
   $g_3 = g_{10} = 1.217523$ $\quad g_4 = g_9 = 1.586707$
   $g_5 = g_8 = 1.847759$ $\quad g_6 = g_7 = 1.98289$

CHAPTER 3

1. $N = 8.03$, use $N = 9$

2. $\arccos(10) = \pm j2.993$, $\log_{10}(-3) = 0.47712 \pm j1.364$

4. $Q = 0.748$ $Q = 1.819$ $Q = 3.28$ $Q = 5.52$ $Q = 10.20$ $Q = 32.07$
$\omega_o = 0.1768$ $\omega_o = 0.4007$ $\omega_o = 0.6203$ $\omega_o = 0.8022$ $\omega_o = 0.9315$ $\omega_o = 0.999$

5. $\omega = 1.42$

6. $N = 5.77$

CHAPTER 4

1. $-1.556272 \pm j0.3844607$ $-1.132861 \pm j0.8915767$
$-0.67189 \pm j1.047978$ $-0.346397 \pm j1.060382$
$-0.106405 \pm j1.047861$

2(e). $\omega_o = 1.603$ $Q = 0.5150312$
$\omega_o = 1.433$ $Q = 0.6386697$
$\omega_o = 1.245$ $Q = 0.9263918$
$\omega_o = 1.1155$ $Q = 1.610185$
$\omega_o = 1.053$ $Q = 4.949245$

4. $\omega = 1.42$

5.
$$G(s) = \frac{0.5222329s^2 + 2.0889317}{s^3 + 2.4342s^2 + 2.8263s + 2.0889317}$$

6. (a). $\omega = 0$
(b). $\omega = 1.431577, 1.536271, 1.789025, 2.331502, 3.708879, 10.87381$

CHAPTER 5

1. For $N = 5.00$, $\omega_x = 1.507$, for $N = 6$, $\omega_x = 1.254$, for $N = 7$, $\omega_x = 1.133$

2. $N = 5.609$, use $N = 6$

4.
$$G(s) = \frac{0.000099069s^6 + 0.0049344s^4 + 0.0392713s^2 + 0.080425}{s^6 + 0.9246231s^5 + 1.9972s^4 + 1.263747s^3 + 1.0402s^2 + 0.3598873s + 0.0902388}$$

CHAPTER 6

2. Capacitors are 3,398 pF, 729.3 pF, 4776 pF.
   Inductors are 3.727 uH, 17.36 uH, 2.65 uH.

5. Capacitors are 849.4 pF, 182.3 pF, 1194 pF.
   Inductors are 3.727 uH, 17.36 uH, 2.65 uH.

6. (a) One possible network: Capacitors are 1.35 pF, 899 pF; inductors are 12.18 uH, 19.54 uH
   (b) Capacitors are 3108 pF, 340.9 pF, 5352 pF.
   Inductors are 4.08 uH, 37.16 uH, 2.37 uH.

7. $C_1 = 2,\ 2\ L_1 = 1,\ \dfrac{2C_1C_2}{C_1 - C_2} = \dfrac{4}{3},\ \dfrac{L_2}{2} = 1$

CHAPTER 7

5. $L_1 = 1,\ C_1 = 1.5,\ L_2 = 2,\ R_2 = 2$

6. $R_2 = 1.100747\ \Omega$, or $R_2 = 0.9084739\ \Omega$

7. (a) Inductors are 15.7375 H, 15.78556 H, 7.705338 H.
   Capacitors are 0.1787592 F, 0.1221632 F, 0.02631489 F.
   (b) Inductors are 13.05441 H, 16.11181 H, 14.45075 H.
   Capacitors are 0.2120224 F, 0.2109683 F, 0.1174199 F.

8. Inductors are 0.2588189 H, 1.201627 H, 1.759305 H.
   Capacitors are 0.7578743 F, 1.552914 F, 1.552918 F.

9. (c) Inductors are 1.342029 H, 1.689034 H, 1.741812 H. Capacitors are 1.643242 F, 0.103897 F, 1.784059 F, 0.1543088 F, 1.345573 F.
   (d) Inductors are 1.04624 H, 0.8553013 H, 0.8920336 H.
   Capacitors are 2.058496 F, 0.1084611 F, 2.565264 F, 0.5318165 F, 2.380019 F, 0.3621172 F, 1.842229 F.

CHAPTER 8

1. (a) Gain = −10 (b) Gain = 11

2. (a) $\dfrac{s}{s+1}\ \dfrac{s,}{s^3 + 3s^2 + 3s + 1}$

3. (b) VCVS: $R_1 = 30\text{K}$, $R_2 = 1.034K$, $R_3 = 1K$, $R_4 = 10K$, $R_5 = 20K$;
$C_1 = 1$ uF, $C_2 = 1$ uF [reference Figure 8.7(b)].

4. $A = 1$, $C = (2\ \pi\ 1000)^2$, $B = \pi\ 1000$

8. Stage 1: $R_1 = 9243$, $R_2 = 2688$, $R_3 =$ infinity, $R_4 = 0$, $R_5 = 1{,}004{,}579$
$C_1 = 0.033$ uF, $C_2 = 0.01$ uF
Stage 2: $R_1 = 13{,}734$, $R_2 = 6315$, $R_3 =$ infinity, $R_4 = 0$, $R_5 = 273{,}089$
$C_1 = 0.022$ uF, $C_2 = 0.001$ uF
Stage 3: $R_1 = 3115$, $R_2 = 1053$, $R_3 =$ infinity, $R_4 = 0$, $R_5 = 53{,}200$
$C_1 = 0.33$ uF, $C_2 = 0.001$ uF

9. $R_1 = 3388.75$, $R_2 = 6777.502$, $R_3 = 135887$, $R_{4a} = 95399$,
$R_{4b} = 516095$, $C_1 = 0.03309$ uF, $C_2 = 0.007354$ uF, $C_3 = 366.795$ pF
$C_4 = 183.398$ pF

13. $Q^2 > 0.125$

18. $\omega_o = 10{,}000 \qquad Q = 1/30$

19. $R_1 = R_2 = 21{,}200$

20. $K_1 = -0.8$

CHAPTER 9

1. (a). 906.04 Hz, 1008.3 Hz, 2068.5 Hz, 2291.97 Hz
(b). 903.86 Hz, 1005.30 Hz, 2043.2 Hz, 2257.82 Hz

2. 2.200781 V

3. Section 1: 0.0859751 (B01H), 0.0859751 (B01H)
−0.8280498 (9603H), 1 (7FFFH)
Section 2: 0.0237369 (30AH), 0.0474738 (614H), 0.0237369 (30AH)
0.4677633 (3BDFH), −1 (8001H), 0.6271843 (5047H)
Section 3: 0.058247 (775H), 0.1164941 (EE9H), 0.058247 (775H)
0.5840412 (4AC1H), −1 (8001H), 0.6489469 (5310H)

5. (a). $$\frac{0.078643\ z^{-1} + 0.078643}{-0.842714\ z^{-1} + 1}\ \frac{0.008106\ z^{-2} + 0.016211\ z^{-1} + 0.008106}{0.44604\ z^{-2} - z^{-1} + 0.58638}$$

$$\frac{0.01545\, z^{-2} + 0.0309\, z^{-1} + 0.01545}{0.50371\, z^{-2} - z^{-1} + 0.55809}$$

(b). $$\frac{0.267824\, z^{-2} - 0.0855\, z^{-1} + 0.267824}{0.13602\, z^{-2} - 0.68588\, z^{-1} + 1}$$

$$\frac{0.2047\, z^{-2} + 0.0613\, z^{-1} + 0.2047}{0.3386\, z^{-2} - 0.8679\, z^{-1} + 1} \frac{0.127\, z^{-2} + 0.2082\, z^{-1} + 0.127}{0.6122\, z^{-2} - z^{-1} + 0.8499}$$

(c). $$\frac{0.05055 z^{-2} - 0.0848\, z^{-1} + 0.05055}{0.50528\, z^{-2} - z^{-1} + 0.5441}$$

$$\frac{0.05505\, z^{-2} - 0.0982\, z^{-1} + 0.05505}{0.5062\, z^{-2} - z^{-1} + 0.5302} \frac{0.5103\, z^{-2} - z^{-1} + 0.5103}{0.3417\, z^{-2} - 0.6608\, z^{-1} + 0.3634}$$

$$\frac{0.1362\, z^{-2} - 0.2689\, z^{-1} + 0.1362}{0.5121\, z^{-2} - z^{-1} + 0.5193} \frac{-0.08722938\, z^{-2} + 0.08722938}{0.53325\, z^{-2} - z^{-1} + 0.5446799}$$

CHAPTER 10

2. (b). Using a rectangular window

$h(0)$	$= -3.784133E - 02$	= (FB28H)	$= h(32)$
$h(1)$	$= 6.357829E - 08$	= (0H)	$= h(31)$
$h(2)$	$= 4.324725E - 02$	= (589H)	$= h(30)$
$h(3)$	$= 0$	= (0H)	$= h(29)$
$h(4)$	$= -3.118298E - 02$	= (FC02H)	$= h(28)$
$h(5)$	$= 0$	= (0H)	$= h(27)$
$h(6)$	$= 2.384186E - 08$	= (0H)	$= h(26)$
$h(7)$	$= 0$	= (0H)	$= h(25)$
$h(8)$	$= 4.677445E - 02$	= (5FDH)	$= h(24)$
$h(9)$	$= 0$	= (0H)	$= h(23)$
$h(10)$	$= -0.1009102$	= (F315H)	$= h(22)$
$h(11)$	$= 0$	= (0H)	$= h(21)$
$h(12)$	$= 0.1513653$	= (1360H)	$= h(20)$
$h(13)$	$= 0$	= (0H)	$= h(19)$
$h(14)$	$= -0.1870979$	= (E80DH)	$= h(18)$
$h(15)$	$= 0$	= (0H)	$= h(17)$
$h(16)$	$= 0.2$	= (1999H)	$= h(16)$

CHAPTER 11

1.

$\bar{b}(0) = 0.484169,$ $\bar{b}(1) = 0.201558,$ $\bar{b}(2) = 0.639549,$ $\bar{b}(3) = -0.182396$

$b(1) = 0.584948,$ $b(2) = 0.420554,$ $b(3) = 0.228577,$ $b(4) = -0.091198$

$h(0) = h(7) = -0.0455987,$ $h(1) = h(6) = 0.1142884$

$h(2) = h(5) = 0.2102768,$ $h(3) = h(4) = 0.2924741$

4. Estimated filter length = 10, estimated word length = 8.
Actual response with infinite precision is ±0.73 dB passband ripple, and minimum stopband attenuation of 21.89 dB.

5. (b). $h(0) = -4.705222E - 03$ = (FF66H) = $h(15)$

$h(1) = -2.554202E - 02$ = (FCBDH) = $h(14)$

$h(2) = -4.194388E - 02$ = (FAA5H) = $h(13)$

$h(3) = -3.900376E - 02$ = (FB06H) = $h(12)$

$h(4) = 7.281327E - 03$ = (EEH) = $h(11)$

$h(5) = 9.646545E - 02$ = (C50H) = $h(10)$

$h(6) = 0.199046$ = (1968H) = $h(9)$

$h(7) = 0.268014$ = (2236H) = $h(8)$

CHAPTER 12

2.

(a). $C_{r1} = 3.85$ pF, $C_{r2} = 3.85$ pF, $C_{r3} = 2.96$ pF, $C_1 = 240$ pF, $C_2 = 48$ pF
(b). $C_{r1} = 25.34$ pF, $C_{r2} = 25.34$ pF, $C_{r3} = 19.49$ pF, $C_1 = 240$ pF,
$C_2 = 48$ pF

3.

$$K_5 = \sqrt{\frac{D}{E} + 1 + \frac{1}{E}}, \qquad K_4 = K_5, \qquad K_6 = \frac{1 - \frac{1}{E}}{K_5},$$

$$K_3 = \frac{C}{E}, \qquad K_2 = \frac{\frac{-A}{E} + K_3}{K_5}, \qquad K_1 = \frac{\frac{B}{E} - 2K_3 + K_2K_5}{-K_5}$$

Bibliography

The books listed below provide additional information on many of the topics treated in this text. These volumes cover the subjects of filter theory, filter tables, analog-sampled data concepts, MOS technology, analog MOS design, and many other related subjects.

Budak, Aram, *Passive and Active Network Analysis and Synthesis,* Boston: Houghton Mifflin Company, 1974.

Burden, Richard L., and Faires, J. Douglas, *Numerical Analysis,* Boston: PWS-KENT, 1989.

Chirlian, Paul, M., *Basic Network Theory,* New York: McGraw-Hill Book Company, 1969.

Cuthbert, Thomas R., Jr., *Optimization Using Personal Computers,* New York: John Wiley & Sons, Inc., 1987.

Encyclopedia Britannica, ''Elliptic Functions,'' 1938.

Geffe, Philip R., *Simplified Modern Filter Design,* New York: John F. Rider Publisher, 1963.

Ghausi, M. S., and Laker, K. R., *Modern Filter Design-Active RC and Switched Capacitor,* Englewood Cliffs, NJ: Prentice-Hall, 1981.

Graeme, Jerald G., *Application of Op Amps-Third Generation Techniques,* New York: McGraw-Hill Book Company, 1973.

Gray, Paul R., and Meyer, Robert G., *Analysis and Design of Analog Integrated Circuits,* New York: John Wiley & Sons, 1977.

Hamming, R. W., *Numerical Methods for Scientists and Engineers,* New York: McGraw-Hill, 1973.

Hnatek, E. R., *A User's Handbook of D/A and A/D Converters,* New York: John Wiley & Sons, 1976.

Huelsman, Lawrence P., and Allen, Phillip E., *Introduction to the Design and Applications of Active Filters,* New York: McGraw-Hill, 1980.

Jong, M. T., *Methods of Discrete Signal and System Analysis,* New York: McGraw-Hill, 1982.

Kuo, F. K., and Kaiser, J. F. (eds.). *System Analysis by Digital Computer,* New York: John Wiley & Sons, 1967.

Mitra, Sanjit K., *Active Inductorless Filters,* New York: IEEE Press, 1971.

Moschytz, George S., *MOS Switched-Capacitor Filters: Analysis and Design,* New York: IEEE Press, 1984.

Rhea, Randall W., *RF & Microwave Filters,* Atlanta: Eagleware Corporation, 1991.

Roberge, James K., *Operational Amplifiers,* New York: John Wiley & Sons, 1975.

Schmid, H., *Electronic Analog/Digital Conversions,* New York: Van Nostrand Reinhold, 1975.

Szentirmai, George, *Computer-Aided Filter Design,* New York: IEEE Press, 1973.

Temes, Gabor C., *Integrated Analog Filters,* New York: IEEE Press, 1987.

Temes, Gabor C., and Mitra, Sanjit K., *Modern Filter Theory and Design,* New York: John Wiley & Sons, 1973.

Vlach, Jiri, and Singhal, Kishore, *Computer-Aided Design of Analog Networks,* New York IEEE Press. 1987.

Vlach, Jiri, and Singhal, Kishore, *Computer Methods for Circuit Analysis and Design,* New York: Van Nostrand Reinhold, 1983.

About the Author

Michael G. Ellis is a research electrical engineer for the U.S. Army Engineers in Vicksburg, Mississippi. Dr. Ellis received his Ph.D. from Mississippi State University in 1994 and previously studied at Georgia Institute of Technology and Vanderbilt University. Dr. Ellis has been an engineer at the Georgia Tech Research Institute, Scientific-Atlanta, Inc., and Hayes Microcomputer, Inc. He is a registered professional engineer and a member of the IEEE.

Index

The Artech House Microwave Library

Acoustic Charge Transport: Device Technology and Applications, R. Miller, C. Nothnick, and D. Bailey

Algorithms for Computer-Aided Design of Linear Microwave Circuits, Stanislaw Rosloniec

Analysis, Design, and Applications of Fin Lines, Bharathi Bhat and Shiban K. Koul

Analysis Methods for Electromagnetic Wave Problems, Eikichi Yamashita, editor

Automated Smith Chart Software and User's Manual, Leonard M. Schwab

C/NL2 for Windows: Linear and Nonlinear Microwave Circuit Analysis and Optimization, Software and User's Manual, Stephen A. Maas and Arthur Nichols

Capacitance, Inductance, and Crosstalk Analysis, Charles S. Walker

Design of Impedance-Matching Networks for RF and Microwave Amplifiers, Pieter L. D. Abrie

Digital Microwave Receivers, James B. Tsui

Electric Filters, Martin Hasler and Jacques Neirynck

Electronic Filter Analysis and Synthesis, Michael G. Ellis, Sr.

E-Plane Integrated Circuits, P. Bhartia and P. Pramanick, editors

Feedback Maximization, Boris J. Lurie

Filters with Helical and Folded Helical Resonators, Peter Vizmuller

Fundamentals of Distributed Amplification, Thomas T. Y. Wong

GaAs FET Principles and Technology, J. V. DiLorenzo and D. D. Khandelwal, editors

GaAs MESFET Circuit Design, Robert A. Soares, editor

GASMAP: Gallium Arsenide Model Analysis Program, J. Michael Golic *et al.*

Handbook of Microwave Integrated Circuits, Reinmut K. Hoffmann

Handbook for the Mechanical Tolerancing of Waveguide Components, W.B.W. Alison

HEMTs and HBTs: Devices, Fabrication, and Circuits, Fazal Ali, Aditya Gupta, and Inder Bahl, editors

High-Power Microwave Sources, Victor Granatstein and Igor Alexeff, edtiors

High-Power GaAs FET Amplifiers, John Walker, editor

High-Power Microwaves, James Benford and John Swegle

High-Temperature Superconducting Microwave Circuits, Zhi-Yuan Shen

Introduction to Microwaves, Fred E. Gardiol

Introduction to Computer Methods for Microwave Circuit Analysis and Design, Janusz A. Dobrowolski

Introduction to the Uniform Geometrical Theory of Diffraction, D. A. McNamara, C. W. I. Pistorius, and J. A. G. Malherbe

LOSLIN: Lossy Line Calculation Software and User's Manual, Fred E. Gardiol

Lossy Transmission Lines, Fred E. Gardiol

Low-Angle Microwave Propagation: Physics and Modeling, Adolf Giger

Low Phase Noise Microwave Oscillator Design, Robert G. Rogers

MATCHNET: Microwave Matching Networks Synthesis, Stephen V. Sussman-Fort

Matrix Parameters for Multiconductor Transmission Lines: Software and User's Manual, A. R. Djordjevic *et al.*

MIC and MMIC Amplifier and Oscillator Circuit Design, Allen Sweet

Microelectronic Reliability, Volume I: Reliability, Test, and Diagnostics, Edward B. Hakim, editor

Microelectronic Reliability, Volume II: Integrity Assessment and Assurance, Emiliano Pollino, editor

Microwave and RF Circuits: Analysis, Synthesis, and Design, Max Medley

Microwave and RF Component and Subsystem Manufacturing Technology, Heriot-Watt University

Microwave Circulator Design, Douglas K. Linkhart

Microwave Engineers' Handbook, 2 Volumes, Theodore Saad, editor

Microwave Materials and Fabrication Techniques, Second Edition, Thomas S. Laverghetta

Microwave MESFETs and HEMTs, J. Michael Golio *et al.*

Microwave and Millimeter Wave Heterostructure Transistors and Applicatons, F. Ali, editor

Microwave and Millimeter Wave Phase Shifters, Volume I: Dielectric and Ferrite Phase Shifters, S. Koul and B. Bhat

Microwave and Millimeter Wave Phase Shifters, Volume II: Semiconductor and Delay Line Phase Shifters, S. Koul and B. Bhat

Microwave Mixers, Second Edition, Stephen Maas

Microwave Transmission Design Data, Theodore Moreno

Microwave Transition Design, Jamal S. Izadian and Shahin M. Izadian

Microwave Transmission Line Couplers, J. A. G. Malherbe

Microwave Tubes, A. S. Gilmour, Jr.

Microwaves: Industrial, Scientific, and Medical Applications, J. Thuery

Microwaves Made Simple: Principles and Applicatons, Stephen W. Cheung, Frederick H. Levien *et al.*

MMIC Design: GaAs FETs and HEMTs, Peter H. Ladbrooke

Modern GaAs Processing Techniques, Ralph Williams

Modern Microwave Measurements and Techniques, Thomas S. Laverghetta

Monolithic Microwave Integrated Circuits: Technology and Design, Ravender Goyal *et al.*

Nonlinear Microwave Circuits, Stephen A. Maas

Nonuniform Line Microstrip Directional Couplers, Sener Uysal

Optical Control of Microwave Devices, Rainee N. Simons

PLL: Linear Phase-Locked Loop Control Systems Analysis Software and User's Manual, Eric L. Unruh

Scattering Parameters of Microwave Networks with Multiconductor Transmission Lines: Software and User's Manual, A.R. Djordjevic *et al.*

Solid-State Microwave Power Oscillator Design, Eric Holzman and Ralston Robertson

Terrestrial Digital Microwave Communications, Ferdo Ivanek *et al.*

Time-Domain Response of Multiconductor Transmission Lines: Software and User's Manual, A.R. Djordjevic *et al.*

Transmission Line Design Handbook, Brian C. Waddell

Yield and Reliability in Microwave Circuit and System Design, Michael Meehan and John Purviance

For further information on these and other Artech House titles, contact:

Artech House
685 Canton Street
Norwood, MA 01602
617-769-9750
Fax: 617-762-9230
Telex: 951-659
email: artech@world.std.com

Artech House
Portland House, Stag Place
London SW1E 5XA England
+44 (0) 71-973-8077
Fax: +44 (0) 71-630-0166
Telex: 951-659

www.ingramcontent.com/pod-product-compliance
Lightning Source LLC
LaVergne TN
LVHW101500200926
848483LV00001B/173